Voyage to
MARS

LAURENCE BERGREEN

RIVERHEAD BOOKS/NEW YORK 2000

Voyage to
MARS

NASA'S SEARCH FOR LIFE BEYOND EARTH

James Garvin's e-mails to Laurence Bergreen are reprinted with his permission.
Permission to reprint excerpts from the JPL online field journals of Jennifer Harris, Pieter Kallemeyn, Bridget Landry, Rob Manning, David Mittman, and Donna Shirley is provided through the courtesy of the Jet Propulsion Laboratory, California Institute of Technology, Pasadena, California.
Excerpts from *The Search for Life on Other Planets* by Bruce Jakosky are reprinted with the permission of Cambridge University Press.
Excerpts from Carl Sagan's writing are reprinted from:
Sagan, Carl. *Billions and Billions: Thoughts on Life and Death at the Brink of the Millennium.* New York: Random House, 1997.
Sagan, Carl. *Cosmos.* New York: Random House, 1980.
Sagan, Carl. *Pale Blue Dot: A Vision of the Human Future in Space.* New York: Random House, 1994
Sagan, Carl, and John Norton Leonard. *Planets.* New York: Time Inc., 1966.

RIVERHEAD BOOKS
a member of
Penguin Putnam Inc.
375 Hudson Street
New York, NY 10014

Library of Congress Cataloging-in-Publication Data

Bergreen, Laurence.
    Voyage to Mars: NASA's search for life beyond
Earth/Laurence Bergreen.
        p.   cm.
    Includes index.
    ISBN 1-57322-166-X
    1. Space flight to Mars.   2. United States. National
Aeronautics and Space Administration   3. Manned
space flight.   I. Title.
TL799.M3 B47   2000   00-042502
629.45'53'00973—dc21

Printed in the United States of America

10 9 8 7 6 5 4 3 2 1

This book is printed on acid-free paper. ♾
Book design by Gretchen Achilles

To Betsy, Nick, and Sara

"Ever since I was a small child, I've believed there was life out there. When I look at the magnitude of the universe, with its billions of stars, I believe that if life developed here on Earth, it must have developed elsewhere. We simply can't be unique. I really don't think we're the most intelligent life-form in the universe, but that's just my gut feeling."

— DR. CLAIRE PARKINSON, *NASA scientist*

CONTENTS

PART ONE

The Mission

1	Mars on Earth	3
2	Message in a Bottle	38
3	Ground Truth	58
4	From Outer Space to Cyberspace	76

PART TWO

Code S

5	Shootout at Caltech	101
6	The Honor of the Team	123
7	Goddard	137
8	The Genesis Question	181

PART THREE

Discovering Mars

9	Rocket Science	219
10	Ghosts and Ghouls	256

11   Human Error   279

12   Mars or Bust   309

Epilogue: Stargazing   337

Acknowledgments   *340*

Select Bibliography   *343*

Part One

THE MISSION

MARS ON EARTH

Subject: ICELAND
Date: Thu, 16 Jul 1998 00:48
From: Laurence Bergreen <bergreen@NYCnet.net>
To: Jim Garvin <jgarvin@nasa.gov>

Hi Jim,

It's late Wednesday night, and I am back home from Houston. With time growing short, what can you tell me about Iceland? Last I heard, there was a strong chance of postponement till October. Looking forward to hearing from you as soon as possible. Thanks.

Larry

Subject: Re: ICELAND
Date: Thu, 16 Jul 1998 09:25:53
From: Jim Garvin <jgarvin@nasa.gov>
To: Laurence Bergreen <bergreen@NYCnet.net>

Larry,

We're GO for Iceland. As of now, we are booked to arrive in Iceland early on the 20th, and quickly pick up a helicopter ride to Surtsey for a 6-hour working visit.

I am trying to be sure we can catch the Iceland Coast Guard helicopter, given that we land between 6 and 6:30 AM and must get thru customs and get the rental Jeep.

Get set for Mars on Earth.

Jim

It's 6:15 in the morning when Jim Garvin, a planetary geologist who works for the National Aeronautics and Space Administration, meets me at Iceland's Keflavik Airport. As arranged, he's flown in from Baltimore, and I've come from New York. Jim is forty-one, talks in torrents, and is plainly Type A, endowed with the passion and restlessness of an old-fashioned genius. Although he has two small children, he puts in eighty-hour work weeks. He is intense. There is no such thing as a short conversation with Garvin. His replies to simple questions have a way of digressing into hour-long ruminations on the nature and origins of the universe, but he gets away with it mostly because he is unfailingly polite. Once he launches into a monologue, he gestures emphatically, as if visualizing and touching everything he describes. He is fit and compact, with black hair, handsome Irish features, and a perpetually worried voice. He looks clean-cut, at least compared to other scientists, and his skin is slightly irritated in patches, as though he's been vigorously applying aftershave lotion. A friend once told me it is often hard to get Jim Garvin's attention, but once you do, it can be overwhelming. Now I have his attention.

After we retrieve our bags, Jim sets out to find Oscar, the pilot of the plane we've hired to take us from Keflavik to the island of Heimaey, off the southern coast of Iceland, where we are to rendezvous with the Iceland Coast Guard, weather permitting. Oscar, when we catch up with him, looks too young to drive a car, let alone pilot a plane. We cram ourselves into his single-engine Aerospatiale, a lightweight aircraft of French design. The co-pilot's seat I occupy is so cramped that my knees interfere with the controls. We are battling fatigue, Jim and I. We have been up all night, and the inside of my mouth tastes like kerosene from the Aerospatiale's tank.

We have come all this way because geologists studying Mars have designated Iceland a Mars analogue. In 1976, when the Viking Lander returned color images of the Red Planet, scientists realized that Mars bears a striking resemblance to the landscape sliding below Oscar's little airplane. Iceland is, in many places, an arctic desert devoid of vegetation and untouched by humanity. These days, NASA-supported scientists regularly visit to study this volcano-ridden island to compare it to its distant relative, Mars. The theory is that by studying Iceland, scientists can better understand the workings of the Red Planet. Iceland is only twenty million years old, a geological babe, and thus relatively unweathered, a primeval landscape. The absence of trees on the Icelandic landscape is a blessing, revealing the island's geological makeup. Mars is similarly bare. Iceland festers with active and dormant volcanoes—just as Mars does. The resemblance makes it possible to work out significant aspects of the geologic history of both places by comparing the two.

Mars is so reminiscent of Earth that it is considered "semi-habitable." The atmosphere is only one percent as dense as ours, but breathable air could be extracted from it. The Martian day, or "sol," lasts about as long as a day on Earth; a Martian year consists of 687 Earth days. Like Earth, Mars has its seasons, but they last twice as long. And Martian weather is anything but monotonous or predictable. In 1997, when Pathfinder landed on Mars, its tiny weather station gathered data on the local Martian weather, which NASA posted on the Internet. The reports showed that temperatures range from 60° F at noon to −100° F at night. Travelers' advisory: because of the much lower atmospheric pressure on Mars, surface temperatures differ drastically from air temperatures. If you were standing on the surface in midday, your feet would be warm and snug, but the fluids in your head would freeze. Mars' atmosphere has fog, wind, and red dust, lending pink tints to a sky accented by two small, misshapen moons, Phobos ("fear") and Deimos ("terror").

Mars resembles Earth in other ways. Its polar ice caps wax and wane seasonally. There are clouds. There is ample geologic evidence that rivers once flowed freely on its surface. The stage has long been set for life to appear there. Yet the Earth teems with life, while Mars appears barren, at least on the surface. Why? No one really knows, yet the answers may lurk in the perplexing differences between the two planets.

The Earth's surface consists of overlapping, often ill-fitting plates covering its molten interior. They form a crust similar to an eggshell, thin and brittle. They bump and grind against each other; occasionally they pull apart, as they are doing now in Iceland, giving rise to earthquakes and volcanoes and mountain ridges lurking beneath the oceans. Iceland sits right on the spine of the Mid-Atlantic Ridge, a segment of the Mid-Ocean Ridge, which is the longest mountain range on Earth, extending 40,000 miles, or one-and-a-half times around the planet. Iceland's unique placement means that half of it belongs, in a geological sense, to the European continent, and half to the American. And the two halves are pulling apart at the rate of one centimeter a year. That doesn't sound like a lot, but when this movement occurs over the course of ten or twelve million years, it eventually becomes a very big deal. Iceland could break apart and be absorbed by other, larger land masses. Or if it surges in volcanic activity, it could enlarge itself, adding enough real estate to accommodate many more hardy souls. For now, a seam runs right through Iceland, clearly marked in some places by a narrow chasm and in others by small streams and little cracks. If you jump across one of the cracks, you jump from one continent to another.

At this moment, no one knows for certain if Mars has or had plates similar to Earth's or, if the Red Planet did have them, how they operated. If Mars never had crustal plates, their absence poses interesting questions about how it developed without them. And if it did, we see no direct evidence of them—not yet, at any rate. The geologic processes associated with crustal plates would have affected the way life did, or did not, develop on Mars.

"Nothing you see here is more than ten thousand years old," Jim shouts over the whine of the engine, as we pass over the Reykjanes Peninsula region of southwest Iceland, "and some of it is only five thousand years old, or less." Jim lives by the geological clock, which extends billions of years, all the way back to the formation of the universe. The universe is an old, old place, perhaps 15 billion years old, possibly more, and the planets of our solar system are old, too, something on the order of 4.7 billion years. When you measure time in billions of years, you dismiss a million years as a hiccup. A span of five or ten thousand years is insignificant. The concept of a year, the time it takes for the Earth to complete a revolution around the Sun, scarcely seems an adequate yardstick for measuring the development of the universe and the planets. Iceland's arriviste status in the geological scheme of things is rare and intriguing; the place teems with clues about the formation of Earth, of Mars, and of the entire solar system. To understand the Red Planet, even partially, is to understand something about the nature of the universe, to catch glimpses of our distant past and our future, to extend perception to a scale much larger than ordinary human comprehension, to harness the imagination to the intellect, and the intellect to the stars.

These days, planetary scientists like Jim regard the geology of Mars as crucial for understanding Earth and the other rocky planets in the solar system—Venus and Mercury (and the moon as well). Jim reminded me that the geologic prizes on Mars are rich. Although it is forty percent smaller than Earth, Mars has peaks and valleys that are far more extreme. The continental United States could fit nicely into one of its canyons. Its volcanoes are awesome. The largest, Olympus Mons, is almost 90,000 feet high. It would tower over Mt. Everest, and it's large enough to occupy the state of Arizona. It is one hundred times larger than the biggest volcano on Earth; in fact, Olympus Mons is the largest mountain in the entire solar system. Mars is a planet of geological superlatives.

Oscar levels off the Aerospatiale at 2,000 feet. Beneath us, the primeval landscape—gray and brown and black, rocky and dusty and nearly tree-less—extends toward the horizon. Is this what it would be like to fly over the scarred surface of Mars? Eventually, we cross a beach, and the island of Heimaey, our stopover point, lies ahead, gradually gathering substance in the blue mist. It is a remarkably tranquil day, so calm that a limp windsock on the ground barely swivels as we veer toward the island's tiny runway, a strip of asphalt running uphill between two volcanic peaks. Ever since leaving New York, I've been placing my life in the hands of complete strangers, and now, sitting beside Oscar as he casually maneuvers his small aircraft, I won-der if I've finally gone too far.

"Move your legs! Please!"

Oscar orders me to contract so he can freely guide us to a safe landing. The plane taxis to a standstill. We are almost there.

Jim hasn't managed to coax NASA into funding this leg of the jour-ney—which comes to about $300. As we slap down our plastic to pay the bill, Jim cites NASA's "faster-better-cheaper" way of doing business to explain why we must pay the airfare to conduct scientific research. Dan Goldin, NASA's mercurial administrator, instituted the policy after he took over the agency in 1992. NASA, like any federal bureaucracy, has indulged in its share of waste and redundancy, and Goldin, coming out of private in-dustry, wanted to trim the bureaucratic flab and refocus NASA. Essentially, he wanted to do more with less. He increased the number of planetary missions under the "faster-cheaper-better" regimen; instead of one expen-sive mission, the agency would send two, or even four, cheap ones, and the returns would be correspondingly greater. And they were! But planetary exploration at any price is an exceedingly risky business, and more missions has also meant more failures. In the grip of "faster-better-cheaper," NASA didn't realize that the American public would fasten onto the fail-ures of its recent missions to the Red Planet—Mars Climate Orbiter and Mars Polar Lander—and forget the successful ones. The notion that NASA was exploring the planets on the cheap and occasionally bungled the job alarmed the media, and it alarmed Congress—how could this have *happened?*—yet it was Congress who, year by year, imposed the budget cuts on NASA that led the agency to adopt "faster-cheaper-better." The result is NASA Lite.

The cuts have been playing havoc with Jim's work life. For weeks, the Iceland expedition has been in doubt because of the fragile health of the reconnaissance plane, a modified P-3. This is a large four-engine turboprop originally meant to fly low over the ocean to detect submarines lurking below the surface. NASA adapted this aircraft for remote sensing: measuring geological, oceanographic, and atmospheric features with instruments used in conjunction with satellites. But NASA's P-3 is a thirty-year-old rust bucket, and it has seen hard use. Jim has reminisced about the crew's Technicolor yawns as the plane followed the rolling terrain at a low altitude, like an airborne roller coaster. He has described the spiderweb cracks that developed in the windshield during an Iceland mission in May 1996. The windshield threatened to crack wide open, jeopardizing the mission. One pilot gave an order to don emergency gear, but the other pilot disagreed, and besides, they had no emergency gear or crash helmets or parachutes. To make matters worse, they were carrying too much fuel to land, and the Icelandic government prohibits dumping fuel into the Atlantic. They had to fly for hours at slow speed, burning fuel, until they could land safely and legally. More recently, the plane developed a chronic fuel leak and lost an engine in flight over Greenland. The accumulated weight of these stories worried me. Even Jim, who does this kind of thing for a living, was anxious. I checked out the P-3 with my friend Peter, a commercial pilot who has flown all over the world in dicey equipment. Peter explained that, worst-case scenario, if an engine or two quit, the plane could coast more or less gently to the ground, unlike a helicopter, which would drop from the sky. I was not completely reassured.

NASA keeps the rust bucket aloft, despite everything, "to facilitate cost-effective essential remote sensing that has inexorably been rewriting textbooks associated with atmospheric science, climate change, and the lay of the land," as Jim puts it. In other words, this rust bucket is changing the way scientists think about how our planet works.

Despite the significance of its science missions and the public dismay when they go wrong, relentless budget cutting continues to afflict NASA. The agency now receives less than 14 billion dollars a year, less than one percent of the overall federal budget, and each year its budget shrinks a little more. The unkindest cuts of all affect people, not hardware. Administrator Dan Goldin earns about $150,000 a year, and scientists like Jim Garvin, who hold one or more advanced degrees and are often among the leading figures in their fields, earn less, something equivalent to a college professor's salary. Unlike academics, they work six or seven days a week, year-round, without sabbaticals. And NASA has stringent rules governing outside income from consulting or lecturing, so moonlighting is out of the question, even if the NASA scientists had time for such activities, which they don't. Willingly or not, Jim and his colleagues must emulate the example of Louis Agassiz, the famous naturalist, who stated, "I cannot afford to waste my time making money."

Why do they do it? Why do these driven scientists, who could be earning several times more than their current salaries in private industry, stick with stingy old NASA? Why do they remain oblivious to imploring spouses and former colleagues who have gone to seek their fortunes in private industry? The most these NASA scientists can reasonably hope for is recognition from their peers, if they make a major discovery. They'll have an easier time getting grants, lots of impressive plaques to hang on the wall, and that's about it. Despite the influence of their ideas on the course of science and exploration, obscurity is often their lot. Who can name the members of the team that in 1996 announced possible evidence of fossilized life in a Martian meteorite—a discovery that, if correct, will stand as one of the most significant breakthroughs of all time? Who can name *any* NASA-supported scientist, for that matter, with the possible exception of Carl Sagan? And who, outside of the scientific community, is aware of Sagan's actual role in NASA's exploration of space?

Sagan's success as a popularizer of the cosmos obscures his real achievements as a scientist, thinker, and writer. A productive scientist and winner of the Pulitzer Prize, he frequently appeared on *The Tonight Show;* he didn't fit into neat categories. He was cursed with charisma. An astronomer by training, he gave a convincing impression of being at home with a number of disciplines ranging from mathematics to history. His fascination with space offered reassurance rather than terror of the unknown. He developed a benign, Jeffersonian vision of the universe as the last frontier, the ultimate, infinite West, where humanity would be able to seek refuge after fouling this planet and possibly destroying itself in the process. Sagan's outer space, like Thomas Jefferson's West, offered sufficient scope to alleviate humanity's ills. He was pessimistic about the future of mankind if we were confined to Earth for too long. It seemed to him a near certainty that, sooner or later, we would blow ourselves up. The only escape from his Malthusian nihilism was the vastness of space and the promise of distant planets, where humankind could start anew. This vision of space as the new frontier influenced NASA from its inception, imparting a sense of purpose, and it inspired younger scientists by giving them a larger context for their research. In the midst of bureaucratic setbacks and budget battles, Sagan knew what was at stake in the exploration of space: over the short term, enlightenment; over the long term, the survival of humanity.

Throughout his career, he cultivated a special fascination with Mars. For him, it was a touchstone of all heavenly bodies and possibly the salvation of humanity. He wrote about it for scientists and for general readers, artfully mixing speculation and scientific fact. He prodded NASA to explore. And he held out hope for life on Mars. As early as 1966, when the conventional wisdom in the scientific community, chastened by the barren photographs resulting from the Mariner missions, held the chance of life on Mars to be zero, Sagan, almost alone among prominent scientists, speculated that such a phenomenon might still be possible.

Sagan influenced a generation of younger scientists, who have their hands on the levers of the future and who fervently believe that now is their time to change scientific thinking about the nature of the universe and our place in it. They stick with their work for many reasons: because they can't do without it; because NASA gives them the means to do what they've yearned for since they were children growing up in the heyday of the space race, watching John Glenn go into orbit; because NASA will let them send something of their own design—a part of them—into space; because NASA has the rockets and the launch facilities and the infrastructure to get it done; because NASA will validate their work in the eyes of the scientific community and the world. Because, when it comes to planetary exploration, NASA is the only game in town.

The little airport on Heimaey is deserted; the Iceland Coast Guard helicopter has yet to arrive. Oscar sits at a table in an empty café, smoking a cigarette. Jim, wired, munches on a Mars bar ("It's my planet, Larry. I may as well") and reminds me that twenty-five years ago, this quaint little island ("Heimaey" means "Home Island") had to be evacuated when that volcano—over *there*—erupted, and lava poured down its slopes into the village. When the eruption ended, Eldfell, as the volcano is known, had transformed the island. It was fifteen percent larger and contained thirty million additional tons of lava, which the local populace later used for roads and buildings. Nor was that eruption unusual for Iceland. Every five years, Iceland witnesses a major volcanic eruption, some capable of sending enough ash into the atmosphere to darken the hemisphere's skies and lower global temperatures. In 1783, the largest volcanic eruption observed in modern times occurred in Iceland. It lasted for months and disgorged more than two hundred square miles of lava. That explosion hurled sulfur dioxide particles into the lower atmosphere; they in turn caused acid rain that polluted the ground, poisoned cattle, starved a quarter of Iceland's population, and darkened the skies over Europe. (A natural catastrophe of that magnitude has likely occurred on Mars.) Iceland is overdue for another eruption, Jim remarks casually, and an active volcano dominates the island where we will spend the day. The island's name is Surtsey.

The newest place on Earth, Surtsey is even more Mars-like than the rest of Iceland. It was formed in a mammoth undersea volcanic eruption that lasted from 1963 until 1967 and, during its early phases, lit up the night skies for miles around. It was named for Surtur, the fire-bearing giant of Norse mythology. The island is younger than Jim, who was seven years old when it erupted into being. At the time, his family was living in Beirut, where his father worked for IBM. Jim's maternal grandmother, who was living with them, became fascinated by the eruption. She collected all the newspaper clippings about it she could find and showed them to her grandson, giving an unexpected direction to his life. Jim has remained in Surtsey's thrall ever since. The vanity plates on his ten-year-old Jeep announce, "SURTSEY."

Only a couple of hundred people have ever set foot on the island. Access is extremely difficult. The North Atlantic currents surrounding it are too rough for most boats to negotiate; swells around the island are frequently twenty feet; waves have been said to reach eighty feet. The island is off-limits to everyone but a few heavily credentialed scientists who have obtained permission from the Icelandic government to conduct research there on a "non-biologically interfering basis." In practice, this restriction has made Surtsey into one of the largest and most carefully studied natural laboratories on the face of the Earth, of interest to geologists because of its recent, well-documented formation; to botanists and biologists, who track the development of life; and to Marsists like Jim Garvin who regard a visit to Surtsey as the closest they'll ever come to the Red Planet.

The only practical way to reach the island is by helicopter, weather permitting. "There are no guarantees, as helicopters only go out that far once a month," Jim alerted me several weeks earlier, when we were starting to get serious about the field trip. "Please note there are NO insurance provisions. ANYONE going to Surtsey does so at his or her own risk, and there is some, as the island is still HOT and there are hydrothermal systems with 120-degree centigrade water just beneath the ground. Also, the weather can change, and people have been stranded. I was urged to remind you of this. Also, what I must do out there will require vigorous hiking over lava and volcanic ash. Anyway, there is a chance we will get there for a day that I do believe you will thoroughly enjoy."

Jim once mentioned to me that his colleagues considered him a bit eccentric. After pondering his disclaimers and warnings concerning Surtsey, I recalled a strange story I'd heard about him. In the heyday of the Apollo program, NASA was thinking seriously about sending people to Mars, yet the agency hesitated. Guiding a robotic spacecraft to the Red Planet is an intricate, ambitious, and unpredictable undertaking; a human mission would be far more risky and complex. NASA was stymied by the problem of getting its astronauts home. Jim came up with a unique solution: he offered to go to Mars on a one-way basis.

When I asked Jim about this story, he was mildly abashed. "I'm not proud of this now, but when I was younger, before I married and had kids, I volunteered to go."

"One way?"

"Well, yes. The only way you could get a man there was one way. It would be too costly to get him back to Earth. So I would go there, have enough life support to explore and survive for two years, and then . . ."

"And then?"

"That would be all. And it would have been worth it, the scientific returns would have been spectacular, but that was before I had kids. Now I have other responsibilities I didn't have then."

The thin tin walls of the Heimaey airport terminal start to vibrate. There's a sound almost below the threshold of hearing. We feel it in our guts as it gets louder and more intense. *Thwacka-thwacka-thwacka . . .* Rotors whirring, the Iceland Coast Guard helicopter, an impressively large and sturdy Bell Jet Ranger, descends into view. It is a noisy piece of equipment, manned by a crew outfitted with brilliant orange flight suits. These are the men of the Iceland Coast Guard, and they swarm around us like giant luminescent insects. A small group of Icelandic botanists has joined us, and our little group approaches the helicopter, deafened by the whine of the motor and the thumping of the rotor. A crew member, smiling crazily, hands each of us a life vest and a helmet equipped with a microphone. I place the helmet snugly over my head, and the unbearable *thwacka-thwacka-thwacka* subsides to a distant drumbeat. We strap ourselves into the seats, and the helicopter slowly rises from the tarmac to a height of about six feet. We delicately revolve until the nose suddenly pitches forward, and we take off like a shot. This is flying. We swoop over the ocean at an altitude of about 500 feet, until we reach the island of Surtsey, fifty miles from nowhere. From the air, the place looks so barren and primitive and devoid of anything recognizable that even Heimaey, by comparison, seems civilized. The jagged gray lava formations of Surtsey rise to greet us. The helicopter sets down lightly on a concrete landing pad considerably smaller than my living room.

Less than twelve hours before, I was sitting in the back of a taxicab in New York City, and Jim was fighting traffic on the way to Baltimore-Washington International Airport. It is now 10:30 A.M. local time on a rare, beautiful morning on sub-arctic Surtsey. Our coordinates are 63° 13′ North, 20° 31′ West.

The rotors slow almost to a stop. We emerge from the helicopter and wave merrily to the crew. The helicopter begins to whine, the rotors fling gritty volcanic ash into the air, and the machine lifts off. It tilts toward the mainland and disappears, leaving a therapeutic silence.

We are alone on the newest place on Earth.

The temperature is in the high fifties, about as warm as it gets on the surface of Mars. True, the atmosphere here is saturated with oxygen and the gravity is approximately two-and-a-half times greater than the Red Planet's, but as far as Jim is concerned, he's *on* Mars.

Jim believes this is a particularly auspicious day to begin his mission, since July 20 is "Space Exploration Day," also known as "Moon and Mars Day." On this date in 1969, the Apollo 11 astronauts stepped onto the lunar surface. Seven years later to the day, the Viking 1 lander transmitted the first images of the plains of Chryse to Earth. Quite a date, when it comes to planetary exploration. Someday, he believes, July 20 will be designated a national holiday for space exploration.

The botanists disperse ahead of us. Although he is exhausted from a sleepless night, Jim heads out with a bundle of topographic maps under one arm and a pair of bright orange Swarovski laser binoculars around his neck. "They are good to three-tenths of a centimeter. I have about fifteen different programs I can set it on, and if I want, I can connect it to my laptop," he says. He wears hiking shoes designed for walking across lava, an olive drab NASA flight jacket, and a white baseball cap bearing the legend "MARS OB-SERVER," a reference to the billion-dollar NASA spacecraft that vanished in 1993. No trace of it has ever been found nor has a wholly satisfactory explanation been offered for its disappearance. Jim refers to the incident as an "act of God" and wears this cap as a casual memorial to the lost mission.

As he sets out on this glorious Surtsey morning, he experiences an overwhelming sense of déjà vu. The last time he was here was 1991, and now he suddenly feels as though he's home again, and he wipes his eyes. There's a spiritual dimension to his scientific exploration that he can't quite put into words. It's an epiphany—a scientific epiphany, if there can be such a thing. "I feel as if I were back at a good place for learning and experiencing how life gains a foothold on a previously unborn, sterile world," he later told me, when he was better able to verbalize, but now, at the moment he sets out, it's the adrenaline rush he's feeling, the intoxicating sense that the world, or this little volcanic part of it, anyway, belongs to him for the time being. There's so much going on his mind—calculations, memories, nascent hypotheses—that I think of Garvin as a highly emotional computer.

He allows the nostalgia to wash over him and goes to work. Although Surtsey looks stonily barren from the air, close inspection yields different results. "Observing change is the central theme of Earth and planetary science," Jim tells me. "What is changing? What *has* changed? Has life on the Red Planet—if it's there, that is—changed or been changed by Martian conditions over time? Have environments on Mars eradicated the footholds of life that may have been established at one or more times in Martian history? With Surtsey, I am struck by the incredible pace of biological change in only seven years. What appeared as a nearly sterile, Mars-like vista in 1991 is now an alien landscape complete with kitchen-table-sized mounds of flourishing higher-order plants, dunes covered by grasses." Jim mentions that ecological succession has changed the face of Surtsey at a rate ten, or even a hundred, times faster than typical terrestrial landscapes, and perhaps a million times faster than change on Mars. Thanks to the action of wind and water, Surtsey is destined to vanish beneath the surface of the ocean almost as quickly as it arose. "I'm going to have an equation by October that will contain the predictive lifetime for Surtsey," he says. "We'll have the landscape volume erosion rate for the first time for an isolated volcanic island." If the present erosion rates hold, Jim predicts Surtsey will vanish beneath the waters of the North Atlantic by 2045, and his subsequent calculations confirm this date. Think of this process as geological time-lapse photography. A world arises and vanishes before your eyes.

Ahead lies an impossibly steep curving wall of solid rock, 460 feet high; to the left, a craggy, broken crater—a tephra ring about fifty yards across. A tephra ring is a partial crater formed as exploding volcanic ash falls to the ground in a semicircle, usually molded by the prevailing winds, which on Surtsey can be fierce. Geology-speak is a babel of languages. *Tephra* is Greek for ash. *Lava* is Hawaiian. Many other terms are Icelandic, which is among the oldest continuously spoken languages in the West, the language of the Sagas, and, at times, the language of geology.

Lava, in Icelandic, is *hraun.* "It's the oldest word for lava there is," Jim says. There are several subsets of *hraun: apalhraun,* which is rough lava, and *helluhraun,* which is smooth. A small volcano in Icelandic is a *dyngja. Hlaup* means "flood," and *jökull* means "glacier." If you put those two words to-gether—*jökulhlaup*—you get something for which there is no exact equiv-alent in English: a catastrophic outburst flood caused by water trapped under a glacier, which cracks open the ice and violently disgorges.

This catastrophic event occurred in 1993 on an Icelandic flood plain called the Skeidararsandur. Blocks of ice as large as houses tumbled for miles across the flooded black primeval landscape in an orgy of geologic violence. A similar geological disaster also occurred on Mars in the distant past. The scale was immense. It is estimated that the Martian *jökulhlaup* released as much as 100,000 cubic meters of water per second, more than the entire flow of the Amazon river.

At the moment, we are standing on *hraun,* or, more precisely, *helluhraun,* with a little *apalhraun* scattered here and there. Looking into the tephra ring, Jim says he's stunned "to observe the development of erosional canyons massive enough to drive a Hummer through." He didn't see anything like this on his last trip to Surtsey. The erosional scars remind him of features shown in the latest images from Mars Orbiter Camera, now circling the Red Planet.

"These mini-canyons, technically erosional gullies, expose the under-belly of Surtsey, the volcano. They give clues about its future and the processes that formed it. The sheer beauty of these signs of geologic aging and their abundance are remarkable!" He takes a closer look at the black windblown tephra. "See how it's sorted? See how the small rocks have risen to the top? That sorting is common. Some of them are rounded." Those smooth contours, he tells me, are diagnostic of wind and water, and he looks for similar shapes on Mars. "So far, we haven't found a lot of really rounded ones on Mars," he admits. But he keeps looking because evidence of water is essential to the detection of life beyond Earth. In fact, water has assumed such importance that the question of extraterrestrial life has been reframed; where scientists once inquired, "Is there life elsewhere in the universe?" they now ask, "Is there liquid water elsewhere in the universe?"

Many planetary geologists, Garvin included, now see convincing evi-dence that Mars once had lots of water, and it may still have a tremendous amount of water even now. Their goal is to follow the water because they hope it will lead them to life. So they seek distinctive water signatures. They look for evidence of dried-up rivers and oceans and shorelines; they theorize about subsurface water, and they measure glaciers—anything associated with water.

Stepping lightly on tephra, Jim makes his way across the eastern side of the island, squinting and kneeling, taking measurements, orienting and re-orienting himself, studying the landscape, observing the reverse sorting of the soil, in which "coarser fragments the size of popcorn nubs rise to the top of the soil horizon, leaving the finer, claylike fraction below." He notes the fragmentation of large blocks of volcanic rocks. On the right, Jim confronts a landscape studded with pitted blocks ranging in size from softballs to bas-ketballs. Those pits grab his attention. He spent a good deal of his graduate career at Brown in the mid-1970s studying patterns of pits and surface tex-tures on terrestrial rocks and on Martian boulders photographed by the two Viking landers, trying, as he put it, "to unravel the geologic secrets of Mars." Here is a banquet of strikingly similar boulders on which he is ready to feast. He notes unpitted gray rocks with angular shapes—so-called "country rocks"—as well as pitted rocks, whose morphology speaks to him, telling of displacement from a lava flow.

Jim explains how this local landscape came to be: "Once the seawater was kept out of where the lava was bubbling up, a carapace of lava formed. And that lava is very important, because it protects the vent. The vent is where the hot rock comes up. That is the reason this island survives today." He displays what looks to me like an ordinary rock, but to Jim, it's a geo-logic sonnet. "This is tephra that tumbled downhill. See how it's made up of bits of other stuff? It's actually a breccia. A breccia is stuff made of other stuff, little welded bits as strong as concrete." As I hold the raw geological ma-terial in my hand, Jim reminds me that this is what the rocks on Mars look like; the main difference is that they're coated with a brown dust. He feels around the edges. "It's a smooth little rock," he says. "That means it's been worn by erosive agents, so we look at the rounding of the corners to get an indication of what's going on." I carefully replace the rock so as not to dis-turb the course of Surtsey's geological evolution.

The *hraun* we traverse feels like soft beach sand. Jim tells me that on Mars, the soil is ten times finer than what we're walking on now. "It would be more like walking on talcum powder."

We press on, and the terrain subtly shifts. "Now we have coarse stuff lying on the surface," Jim remarks, as he tries to read the landscape. "Here's a little piece of basaltic pumice. That's a good one," he says, slipping it into his pocket, which is, perhaps, not quite kosher. "That's one for the spectrometer," he explains. There's an honor system in force here. You're not supposed to disturb anything. You try not to leave footprints in this haven for scientists if you can possibly avoid it. "Now, this looks like—aha! This—" he announces, "is a little lava bomb."

"What?"

"A lava bomb is something that flew through the air and went *splat!* And then it started to break. Already, it's weathering away. See how it's crumbling. Again, this is what we looked for at the Pathfinder landing site." He calls my attention to smooth rocks inside smooth rocks, and he begins to interpret. "You can piece together the history of this rock," he says. "These rocks were *always* smooth; they got pasted together at the time of the eruption."

He sees some similarities between the geology underfoot and the Pathfinder landing site on Mars. NASA sent Pathfinder to a location on Mars where it was believed that a great outpouring of water once occurred. "Some people think the rocks in Pathfinder's vicinity came to rest there as the result of one big flood, but that's ludicrous. It's a mixed population of rocks around Pathfinder," which suggests, to him at any rate, that the geological history of the area has been fairly complex. Water might have come and gone around the Pathfinder site more than once over the eons. I look around; if you photographed a replica of Pathfinder here on Surtsey, you could persuade a fair number of people that the spacecraft was actually on Mars. The more Jim talks, the more I feel a geological kinship between the two planets; Mars seems so Earth-like, or is it more accurate to say that Earth is so Mars-like?

Garvin kneels to inspect a delicate lava formation. "See the thin carapace of lava? This black stuff?"

"It's very soft."

"Right, very soft underneath."

"It's falling apart."

"Not all of it. And that's important, because that's the action of a process that tears down rocks and makes clays. We take clays for granted. On Mars, there's likely to be a lot of clay."

"And water is necessary for clay."

"Yes. You have to break rocks. Look at this." He points to where the hillside is collapsing. "What you see is little mudflows. And look at this. Here is a beautiful little lava rock! Very angular. This is a classic, coated with fine-grain stuff. It's almost a pentagon."

Jim points to the volcano's peak looming overhead and recollects the last time he climbed it. "The wind was blowing at forty-five miles per hour the whole time, and it was very hard even to talk." That windspeed was moderate, by Surtsey standards; the island endures 200 days of gale force winds a year. "When we were here in ninety-one, this area was a desert, but plants are taking over now." Now, the main plant in evidence is the lowly sandwort, a simple succulent that has proliferated on Surtsey with astonishing speed; small, dense, and tenacious, it can boldly go where other vegetation can't. Even mosses can't get a grip on Surtsey; the wind rips them out of the ground and flings them away.

"Look! There is a gorgeous breccia. Notice it's in a little hollow, okay? That's called an apron. We look for those kinds of things on Mars. Outside, you can see there's a layering to it that's caving in. See the carapace of lava up there? It's starting to break off. In a big storm, that could fall." It looks like the burned crust of a pie at the edge of a pan. "Now, see how these rocks are perched? Notice the pits. That's where Mars comes in." You see something similar in images from Pathfinder, Garvin says—pits left by primary gas bubbles in the lava. He snaps a picture of the pitted rocks on Surtsey as he continues. "Look at these pitting textures! All different. It's exquisite."

He zeroes in on a block of lava that speaks to him in a private language. Crouching, he declares, "Now, this is not primary lava. It's softer, and it's been coated with a bright alteration stain caused by chemical weathering. It's allochthonous. That means it's out of place, been moved away." I step back to take in the scene, and I realize the site looks like the Grand Canyon in miniature. "This could be the beginnings of a little Martian canyon system," Garvin exults. "It's gorgeous. Oh, God, wouldn't I love to measure that with a laser!"

We've been picking our way around the base of the volcano, and now we turn away from it and face the ocean. Before long, Jim again shouts. "Look at that." He points to a slight discoloration on a mound of stones, in which he sees vast implications. "That's the high-water mark from a wave, where the fine dust coated the rocks. Now *that* is the kind of shoreline we are looking for on Mars." The subject of ancient shorelines on Mars carries the charge of controversy and borderline heresy. Several scientists have tried mapping the shorelines of ancient Martian oceans that vanished a billion or more years ago, but their work has yet to gain widespread acceptance. I try to imagine Mars as a wet place, covered with oceans, teeming with possibilities, but this is like trying to visualize oceans in the Sahara, for Mars is red and dry and cold.

A large empty plastic bottle catches my eye, disturbing my reverie. The object seems as incongruous here as it would on the surface of Mars. We notice pieces of plastic, and buoys, and rope, and blocks of wood studded with rusty nails. "The garbage of humanity," as Jim calls it, has drifted out here, fifty miles from nowhere, a mocking reminder of home. All day long, he has been scrupulous about not disturbing plants or lava or rocks, to the point of walking in old footprints. Avoiding the detritus, we cross a hard, crusty portion of the beach. "Hard pan clays," he remarks. "See how they crack? They're desiccated. We look for things like that on Mars. More indirect evidence of water. Here on Surtsey, we have a microcosm. We have a scale where it's easier to see things. One of the things about Mars to remember is that it's a *big* planet, about forty percent as large as Earth. If we land in three or four places on Mars, we'll learn about them, but we won't get the big picture of Mars that way, so we study sites on Earth that we believe operate in a similar way."

We approach the water's edge, but a formidable barrier repels us: a giant collection of round, basketball-sized rocks. "If we ever saw a field of dense, interconnecting rocks like this on Mars, we'd know the action of water was responsible. But we haven't seen this, yet." As a geologist, Jim looks for patterns, distributions, colors, textures, and shapes. He is the detective, and they are the fingerprints. If he successfully unravels the geological mysteries of Surtsey with them, he will also know more about the development of the Red Planet.

Turning away from the beach, Jim and I finally begin the ascent to the volcano's summit. I've been trying to put off this chore, but here it is, the thing we must do. Jim reminds me that we are climbing an active volcano, and there's always a chance that it could blow without warning. I recall Iceland's uninterrupted pattern of volcanic outbursts every five years for the last 1,100 years, and I remind myself that it's due for another eruption. I feel as though we're crawling up the side of a giant, overstressed pressure cooker.

Jim tells me that a series of sensitive seismometers has been placed on the volcano; in fact, all the volcanoes in Iceland are similarly equipped, and the seismometers are so sensitive that they can detect microseizures involving magma, or molten rock. I'm somewhat relieved to hear about this detection system, but in the event of a warning, I wonder how anyone would be able to convey the news to us. Six months after our visit, a big volcano finally did erupt beneath Iceland's largest glacier, Vatnajökull, located on the southeast coast, home to most of the country's population.

The gray lava and rounded rocks give way to a smooth, steep incline. Jim estimates it's twenty degrees, but it feels more like thirty to me, very steep, indeed. We zigzag our way across and look down on the larger of Surtsey's craters, a craggy rusty red configuration filled with volcanic ash that from this height resembles a soft, inviting mattress. The wind picks up, and we crouch to avoid being flung down the slippery side of the mountain. Wind, incidentally, figures prominently in the Martian environment. On the surface, dust devils are everywhere. In the upper atmosphere, winds can reach 350 miles per hour, and windstorms occasionally engulf the entire planet, obscuring the surface for days.

Jim reaches a seep, a place where the ground comes apart, as if it were fabric that has been rent. A faint plume of steam rises from the wound, and the smell of sulfur permeates the air. Kneeling beside this smoldering, malodorous seep, I begin to think of Hell as a realistic notion, based on observable geology. Jim asks me to place my hand on the soil near the edge, and it feels like hot clay. A fine white crust along the rim contains bacteria that thrive in the heat and sulfur. This is the most primitive type of life on Earth, Jim reminds me. Life may have begun in volcanic seeps similar to the one at our feet, and it might have started the same way on Mars, on other planets, and on countless moons and asteroids—if it ever did.

These bacteria are examples of extremophile life, primitive life forms that have recently been discovered in places where biologists once assumed life could not survive because the conditions were too hostile—too hot, too cold, too dark, too salty, too deep. In recent years, many of the assumptions about the requirements for life on Earth—and, by implication, the possibility of life on Mars and other celestial bodies—have been overturned.

"We are finding out about the tenacity of life," Jim said before the trip, "and it's startling. We're finding creatures that live at five times atmospheric pressure two miles deep in the ocean in places where the water would boil if there weren't tons of pressure on top of it. We're finding giant simple worms that look like garden hoses that live under those conditions. They don't need any light; they scavenge the sulfur produced in volcanic eruptions deep in the ocean. They live off sulfur; they eat bacteria that grow in the sulfur, and that sustains them. Is there sulfur on Mars? Likely." Life flourishes just about *anywhere*, it turns out, no matter how extreme the conditions. "Can you stick life a mile down in rocks and have it survive and bloom? Yes. Can you put it two miles deep in the ocean where there is no light of day, ever? Yes. Stick it on the coldest place on the planet and it will at least remain dormant there? Yes! Now, if you can form niches of life on Earth in such horrid environments, with pressure that would crush a human being to pulp and temperatures that would boil our skin—if you have life forms under those conditions, then it gets quite interesting. In fact, the question now in biology is: can you even produce a sterile environment?"

The question got me thinking about the famous Miller-Urey experiment designed to illuminate the origins of life. In 1953, two scientists at the University of Chicago, Stanley L. Miller and Harold Urey, put gaseous methane, ammonia, water vapor, and liquid water—ingredients thought to simulate a primitive Earth atmosphere—into a closed system, and sent an electrical discharge spark through the mixture. The gases interacted, and a gummy residue formed; analysis showed it contained organic molecules, including many amino acids, which are the building blocks of life. It had been previously thought that the prerequisites for life were rather complex and rare, but the experiment suggested that all you needed to produce life were a few simple, readily available chemicals and an energy source. These things could be found on other planets, on some asteroids, and most probably on Mars. You don't need oxygen for life to develop, and you don't even need the Sun; the heat source could be volcanic or subterranean. I asked Jim, "If you put together all the necessary ingredients, does life *inevitably* develop?" Because if it did, it could be developing on Mars and throughout the universe, wherever those things are found.

"Larry," he said, "you've just asked the Genesis Question. We don't know the answer. Some people believe it could; some believe it couldn't. A few billion years ago in the history of this planet, and in the history of Mars, and possibly in the history of other places, there may have been very sporadic conditions that might have been able to sustain life. But that was at the time when the planets were being constantly bombarded by junk left over from when the solar system formed. There was a lot of leftover crap, and it eventually smashed into the planets. We think all the planets formed about four point seven billion years ago in a relatively commonplace little spinning nebula of dust that collapsed to produce them and also spun off stuff that didn't quite make it, like the materials in the asteroid belt that occasionally crash into us." And then he said: "There was even an idea that life sprang forth on those objects, and there was a great so-called 'panspermia' wherein life spread from one place to another from some unknown source. Not us. We weren't the source, according to panspermia theory. We were just one of the places where it landed and survived."

I casually remarked that panspermia sounded like the answer to the question of life in the universe.

"Be careful," Jim said. "The idea is very controversial, and often misunderstood. A lot depends on whom you talk to." Although there is no consensus about life on Mars now, he told me, many scientists have come to believe that it's very hard to imagine that Mars didn't have a failed attempt at life forms at some point in its history. "The question is: where did it go? Seeing the existence of life on Mars would be like finding the Rosetta stone. We may be alone *now* but not in the past." Jim thinks of Mars as the mother of all control experiments. "The theory goes like this: the Earth is a very messy, complicated, intersecting set of systems, but we also need a sandbox to play in, and the best sandbox we have is Mars. It's a natural control experiment for things we want to understand about our own planet, if we were able to strip away and isolate some of the variables. For instance, Mars is colder and drier. Water exists there as ice or as a gas in the atmosphere. When it did exist as a liquid, it probably did so only briefly. There is no biosphere altering the planet, as we have on Earth. If it ever started, it failed."

It's possible that we could end up like Mars, as the Sun fades. Jim tells me that if all the water on Earth froze and then evaporated, we could very well have conditions that would suck the oxygen out of our atmosphere without renewing it.

I begin to think of Mars as Earth reduced to the essentials. For purposes of scientific research, it's more promising than the moon, even though it is much harder to reach. "Back in the days of Apollo, we could use military-class technology to zip up to the moon and fly around and be very clever because we had unlimited funds and a national commitment from our president to put human beings there. We don't have that commitment for Mars," Jim reminds me, making the idea of regular transits to Mars suddenly sound sensible. "People argue that NASA will never have carte blanche like that again. Nowadays, you have to keep the price way down. It means that when you go to Mars you can't carry enough fuel to go into the orbit you want. You have to use the gravity and atmosphere of Mars itself to get you there."

Jim takes heart from historical precedents for these difficulties. "Think of them in terms of the exploration of our own planet," he says. "Think of the early sailors willing to risk their lives sailing from Greece to Crete, an island about a day away, if the wind blows right. They might be willing to do that because, what the heck? That's analogous to going to the moon, which we can reach in a matter of a few days. Now imagine sailing not from Greece to Crete but from Greece to North America. That's the scale of difference we're talking about when we send spacecraft out to Mars." At that scale, the celestial sailors will have to learn to improvise in order to survive, just as their maritime forbears did.

While we linger at the seep, Jim reminds me that only thirty-five years ago there was nothing here but the Atlantic Ocean and fresh air. And now we are standing on rock containing copious evidence of bacteria. Could life have spread as quickly on a Martian volcano? Well, why not? No one knows. Questions like these form the basis for "astrobiology," the search for extraterrestrial life—generally in the form of primitive bacteria invisible to the naked eye. Although the questions posed by astrobiology—or, as it is sometimes called, exobiology—have concerned NASA scientists for more than twenty years, the field has suddenly entered a period of rapid expansion, as it moves from the realm of the purely speculative to the potentially demonstrable.

Biologists are coming around to the idea that Earth, while complex and idiosyncratic, is hardly unique. Our planet does not necessarily contain a divine, magical, or fluke recipe for life. On the contrary, life emerged here when our planet was less than a billion years old, as the outcome of geologic and chemical processes. It might have been the inevitable outcome; if so, it could easily appear throughout the solar system and the universe.

In that case, why has extraterrestrial life been so hard to find? One thing is now clear to many scientists. As the song goes, they've been looking for life in all the wrong places—mainly in moderate, sunlit, moist environments. As biologists develop a greater understanding of all the unlikely, remote places where life exists on Earth, it has become apparent that there is much greater latitude. Life forms can be so hardy and unpredictable that they will find a way to exist just about anywhere. And at the microbiological level, it can be so simple it seems barely alive at all. Still, to qualify as life, the stuff has to satisfy at least two widely accepted conditions. It must be able to replicate, and it must be able to mutate and evolve. Darwin's principles of natural selection apply at all levels of life, and if life is discovered on Mars, or anywhere else in the universe, natural selection will apply there, as well.

We make our way along shallow erosional gullies, which provide a foothold on the volcano's sheer upper reaches, until we arrive at the summit of Surtsey, a precarious location high above the surface of the North Atlantic. Jim, who's lighter and more agile, is a lot better adapted to climbing than I. The jet lag and lack of sleep are taking their toll; my heart thumps wildly, and the wind pushes me off balance. I look up, trying to orient myself. Heimaey, so solid and inviting by comparison, floats in the distance, and beyond, Iceland itself. After a brief rest, we head down the steep slope.

By midafternoon, we reach a small research hut at the base of the volcano, where the Icelandic botanists who flew in with us have gathered. A pot of water comes to a boil on a little propane stove, a welcome sight, a bit of Earth on Mars. Over a mug of instant coffee, I converse with a botanist, Sturla Friðricksson, who, Jim explains, is considered the grand old man of Surtsey research. Sturla's face is seamed and cured to a leathery condition by the Northern sun. He looks as though he's served time on the *Kon-Tiki*. As he launches into a complete geological history of Surtsey, a saga in itself, the Icelandic Coast Guard returns to rescue us. Their helicopter touches down with a great throbbing racket; the rotors feel like they're sucking the air right out of my nostrils. Silent and overwhelmed with impressions from our day's exploration, Jim and I begin the journey back to the mainland, as though returning to Earth.

When he's not climbing active volcanoes, Jim Garvin often roams the hallways at his place of work, NASA's Goddard Space Flight Center in Greenbelt, Maryland. That was where we met, exactly one year earlier, when I was visiting a friend who also works there. Jim was standing in a busy corridor, holding forth on the subject of Mars, and within minutes, the sound of his voice attracted a crowd of curious scientists, who drifted away from whatever they were doing to listen. Somebody ought to be getting this down, I thought, and I started to take notes as fast as I could. When we began to talk, he identified himself as a co-investigator for the Mars Global Surveyor (MGS), a state-of-the-art spacecraft designed to orbit Mars and conduct a number of pioneering experiments, including mapping the surface of the Red Planet in more detail than is available for Earth.

His special area of interest, he explained, is an instrument on MGS known as a laser altimeter—a laser designed to fire impulses at the surface of Mars. Minute fluctuations in the time it takes for the impulses to return create a three-dimensional picture of the surface, accurate to within a few meters. This is an incredibly intricate engineering feat—akin to extending a tape measure all the way from New York City to Washington, D.C., to determine the surface variations on the dome of the Capitol, while recording the results in a moving car back in New York.

At that first encounter, Jim invited me—as he does everyone he meets—to share his obsession with Mars. He is a rigorous scientist, but underneath the rigor lurks a romantic explorer. Mars is not just a planet to him; it holds, potentially, the answers to the riddles of the universe. At the time of this meeting, in July 1997, the Pathfinder spacecraft had just landed on the Red Planet, and its tiny rover, Sojourner Truth, had captured the imagination of the scientific community and people around the world, who were able to follow the extraterrestrial proceedings closely on the Internet. As I talked with Jim about the development of Mars exploration, it occurred to me that Pathfinder belonged to a much larger story—mankind's exploration of Mars—and that the exploration was itself part of an even larger story: the search for the origins of life on Earth and throughout the universe.

Despite the sophistication of the new missions to Mars, Jim waxes nostalgic about the Viking program of the mid-1970s—"the Cadillac of missions," he says. "They actually had better equipment then." Of course, it cost the American taxpayer about ten times as much as the current hardware does. He became involved with the Viking missions when he was still an undergraduate at Brown; a geology major, he helped to analyze images from the Viking 2 lander spacecraft, and he got hooked on the study of Mars. (Planetary spacecraft come in three basic varieties—flybys, landers, and orbiters. The flybys whiz past a planet on their way to somewhere else. An orbiter circles a planet. And a lander touches down on the surface.)

Just when he thought he'd found his vocation, the Viking missions ended, and NASA closed the book on Mars exploration. The missions, Jim often says, were the victims of their own success. They sent back thousands of stunning color images, and provided enough data to keep scientists occupied for two decades. They accomplished so much it seemed there was nothing left to do except send people to Mars, and there wasn't enough money in the budget for that.

After graduation, Jim went to Stanford for an advanced degree in computer science. The life of a geek was not his style. So what if he could debug his colleagues' programs and make them run faster? The work was too routine, too solitary, too stationary. He returned to Brown for his Ph.D. in geology, where he studied under Tim Mutch and Jim Head, who also taught a popular undergraduate course known as "Rocks for Jocks." One day, Mutch said to Head, "You know, there are no fundamental problems left on Earth." Mutch turned his attention to the planets and published an important—one is tempted to call it groundbreaking—book, *The Geology of Mars,* in 1976. This was a revolutionary idea, to study the geology of the Red Planet in a scientific manner. Geology claimed a gigantic new turf: the solar system and, by extension, planets and asteroids everywhere. All at once, geology became an integral part of the exploration of space, and Mutch was leading the way, training a new generation of planetary geologists, including Jim Garvin.

"At first glance," Jim says, "Tim Mutch might have been perceived as a Jimmy Stewart type of character: tall, thin, amiable, and always aboveboard, almost self-deprecating. Deeper inside the man was his passion and resolve." Occasionally, he'd remark to Jim, in an offhand way, "You're a Mars person.

Did you know that?" And at a party, he buttonholed his fast-talking young graduate student and said, "Jim, you and a few others are the future of Mars exploration, so it is yours to make it happen." That was, he says, "heavy stuff" for a twenty-one-year-old grad student to hear.

In defiance of conventional geological practice, Mutch concentrated on the enigmatic landforms of Mars. "This was revolutionary thinking to me, as most geologists argue that studying typical landforms is the best way to learn how a surface was formed," Jim says. "But Tim argued that finding those enigmatic landscapes might be more pivotal in the workings of Mars than background normal landscapes."

In 1980, Tim Mutch led an expedition to the Himalayas. He made a successful ascent accompanied by two graduate students, but the weather turned foul during their descent. One of his crampons broke, and it was impossible for him to continue. The students wanted to carry him down, but he told them, "No way. Strap me in here. Go back to base camp and get help and come back for me." By then, he might have been delirious from lack of oxygen. The students went down to base camp, and he probably thought they'd return in an hour to rescue him, but they had a rough time getting through the storm, and by the time they made it back, eight hours had passed, and there was no sign of Tim. His body was never found. The best guess is that the storm blew him off the mountain.

About a year later, Tim's widow, Madeline, held a memorial gathering to which Jim was invited. She showed slides taken during her husband's fatal descent. It was unbearably moving, especially for Jim, who had been Tim Mutch's last graduate student. In an obscure but deeply felt way, Jim believed that as Mutch's disciple, he was supposed to carry the torch—but where? He didn't know, and even today, he still doesn't know where, exactly, but he always hears Mutch's voice in his ear, pointing the way to the Red Planet. And NASA was the only way to get there.

During Jim's early career at the agency, an unofficial Mars Underground developed within NASA's bureaucracy. This was a loosely knit affiliation of scientists and engineers who maintained a keen interest in Mars, despite the agency's lack of Mars programs, and who also maintained a fervent desire to return to the Red Planet, first with robotic spacecraft, and later, with people, if the money and the motivation could somehow be found. The Mars

Underground published papers, held symposia, and tended the flame through difficult times.

These were not easy years for Jim. An instrument he'd proposed, a radar altimeter, was initially selected for a Mars mission, but later deselected, or dropped. Soon after, in January 1986, the Space Shuttle Challenger disaster threw the agency into crisis. A period of soul-searching ensued within NASA. He worked for Sally Ride, the astronaut, on a project designed to renew the agency. Out of copious discussions, the Ride committee produced a grand new vision for NASA: the United States must return to the moon and, beyond that, establish a permanent lunar base. Their recommendations were never acted on. After the group disbanded, Jim's laser altimeter was selected for the Mars Observer mission, which ended in catastrophe in September 1993.

Finally, in 1996, Mars's time came round again. First, there was NASA's announcement of the discovery of nanofossils in a meteorite from Mars. Suddenly, as one scientist put it, NASA was bitten by the life-on-Mars bug. The discovery, by a team of NASA scientists, gave the agency a focus it had been lacking since the Challenger disaster a decade earlier. The following year, the Pathfinder spacecraft settled on Mars on the Fourth of July, and its miniature rover rolled down a ramp and inched across the surface of the Red Planet, acting as a robotic geologist. "We can now get to the Red Planet for the price of a big-budget Hollywood movie," NASA claimed. Jim puts it even more simply: *"Mars is back."* It's his mantra.

The Keflavik Naval Air Station, where Jim and I are billeted in Iceland, is a sprawling NATO base that once served as an essential Cold War outpost. These days, it's mostly a stopover for young European pilots who bring their planes in from France or Italy; they drink a lot, sleep a little, and depart at first light. Although Jim is a civil servant, his quasi-military status becomes evident the moment he enters the base. He salutes *everybody,* and they salute back—at least, some do. "My civil service status grade is equivalent to a colonel's," he says, "but no one here is aware of that."

We're assigned to the Bachelor Officers Quarters, cement barracks strongly reminiscent of college dormitories. The penetrating odor of burned pizza crust wafts through the halls; the walls reverberate with blasts of heavy metal music. Occasionally, you hear squeals and shouts from girls who may or may not belong here. When you look out the window, you see a landscape so flat and featureless it could be Nebraska. There are schools, playgrounds, pickup trucks, a movie theater, a bowling alley, and a Wendy's where they play "God Bless America," country-style, over the PA system. The unofficial motto of the base might be: "Keflavik, a Nice Place to Raise a Family."

In July, it's light all the time, and the only way you can tell it's late in Keflavik is that it gets very quiet. For a few hours, there are no cars zipping around the roads, no fighter jets streaking overhead. Around midnight, there's a sort of dusk, a suggestion of darkness like a shadow across the sky, but it soon passes, and brightness returns by 2 A.M. or so.

A few days after our Surtsey expedition, Jim goes forth in search of glaciers to measure. We head out in a Land Rover Discovery across the treeless, craggy, doom-laden landscape, in which people or, for that matter, all life-forms, even grass, seem out of place. Mars on Earth. "You have to remember, Iceland, except in the highlands, looks like the ocean floor," Jim says. "Now, what if I were Spock in 'Star Trek,' looking at the Earth from the Starship Enterprise? Captain Kirk says, 'Spock, what do you see? Put the scanners on.' I'd say, 'I see a watery planet. It's a planet dominated by oceans.' The land is an insignificant fraction of what makes up this planet. If we could peel away the water and look at the Earth from space, planetary scientists would say, 'I see what the Earth does. It has a large system of very thin crustal blocks that are moving and being eaten up in some places and being regenerated in others.' "

Jim catches his breath and swerves to avoid a small herd of scrawny Icelandic sheep. "Now we are starting to add a tapestry of new measurements from Mars Global Surveyor, as we try to understand all these different surface units on Mars. Scientists want to find hot pits, if there are any, just like the ones you saw on Surtsey. Now, how big were they on Surtsey?"

Just a few inches wide, I remind him, and he points out that it would be very difficult to see such tiny formations from space, even at high resolution. "You would need an extremely sensitive thermal scanner in orbit." Such a device actually exists, but it would not, on its own, be able to detect alien life. Scientists also look for biomarkers, that is, distinctive signatures of life. And they seek signs of an energy or nutrient system capable of sustaining life. "On Mars, we want to find playas, dried-up seabeds, where there might have been standing bodies of water. We see playas on Earth, in the dried lake beds of the western United States, the dry lakes of Australia. On Mars, these playas may be even bigger. The topography measured by the laser going around Mars can find those areas for us." So playas may hold clues to life on Mars, and volcanoes may also lead scientists to Martian gardens of Eden. It may just be Garvin's bias, because he is a crater expert, but he thinks volcanoes are an important component in the design for living—another reason that Iceland appeals to him. "Iceland has volcanoes that are active, with ice, certainly something that happened on Mars. We have volcanoes interacting with groundwater, very important, because there may be groundwater on Mars. We don't know. And we have volcanoes here producing new lava at great rates. Some of the volcanoes on Mars have sustained high eruption rates for hundreds or even thousands of years. That's what it takes to make an Olympus Mons"—and Olympus Mons is so big that it couldn't exist on Earth. "There's too much gravity here, and anything aspiring to Olympus-Mons-like grandeur would collapse under its own weight." He likens its shape to the much smaller lava shield volcanoes of Iceland. The term is meant to suggest a Viking shield turned on its side; a lava shield volcano slopes very gently. "It's the most common landform made by volcanism in the solar system. Mother Nature does not know how to do it any simpler."

Later, we coast past an immense, dry lake bed studded with pebbles. We get out and walk across its dusty surface. It would not be surprising to see a pterodactyl winging overhead or a spacecraft descend from the skies. This is Nature's rough draft, a land of possibilities. It's not as polished as later versions, but the crude landscape yields its secrets and intentions to geologists. "When the water dries up, it leaves behind a lag deposit of rocks," Jim remarks. The rocks range in size from small cobbles up to large boulders. "And anything bigger," he announces, "is called a real big boulder! The bright stuff you see here is a layer of desiccated, cemented dust made of clay. That is what comes out of suspension when water evaporates. We expect to see signatures of that kind of stuff on Mars." He points to a fissure in the soil. "See this desiccation crack? This is what we hope to see on Mars."

We head north until we reach an enormous glacier: Langjökull. On the other side, its summit obscured by cloud cover, is the great volcano known as Ok. The stony, dusty ground, reddish brown, contrasts with the huge wall of ice. I slowly become aware of the landscape's resemblance to images of the Martian ice caps, those vast dull white fields rising out of the reddish Martian desert. The more we look, the more striking the resemblance to the northern latitudes of Mars. Our isolation feels complete. No birds or cars disturb the pure silence. No airplanes streak by overhead; the atmosphere is untarnished by plumes of smoke. The spectral glacier rises impressively from the dark red rock, its facade reaching into the clouds and mist, massive, gloomy, impersonal, hypnotic. Nearly everything looks alien and supremely indifferent to the two tiny human figures in the midst of this vast, primeval sanctuary. Take all the measurements you want of Mars, but walking through this strange and unnerving place suggests, as nothing else can, what it would be like to traverse the surface of the Red Planet.

Suddenly a pair of bicyclists disturbs us, a man and a woman, en route to a distant town or campsite. It's a relief to share the oppressively majestic Martian landscape with others, even briefly. And then they're gone, gliding into the distance on their bicycles, and we're alone again. For once, Jim is speechless. We return to the car in silence.

When we reach Keflavik, Jim drives us over to the tarmac, saluting smartly whenever he passes a military guard. At last, the afflicted P-3 aircraft is here. It looks all right from a distance, but a closer inspection reveals oil leaking from the nose, creating an embarrassing, 125-foot-long stain on the ground, beginning directly beneath the aircraft. There's talk that the Keflavik Naval Air Station may insist the P-3 leave immediately so that it does not foul the runway.

Jim trots to the base's weather station, where the latest satellite data are available. The weather station glitters with state-of-the-art equipment; the place is so big and solid it looks like the bridge on an aircraft carrier. Although it's sunny here in Keflavik today, the instruments reveal there is a weather front moving in, and steel gray clouds are visible on the horizon. It's now about 3:30 in the afternoon, and sometime after 5:00, the plane is supposed to be in the air, on a six-hour mapping mission. Right away Jim sees it will likely be too cloudy to take data over Surtsey, so instead they'll survey a floodplain known as the Sandur, located in the eastern portion of Iceland. Given the weather and mechanical constraints, this will likely be the only day they will be able to take measurements.

Jim sits at a computer terminal in the weather station and begins composing a report to the base commander about his activities here in Iceland; at the same time, he chatters away with me and an affable young naval attaché. He types rapidly, never making a mistake—". . . As part of NASA's continuing research interests in Iceland as a microcosm for global Earth environmental change and as a natural analogy for landscapes on Mars, an aircraft remote sensing campaign was conducted during the period from 20 July to 26 July, 1998. A NASA P-3 aircraft, outfitted with two scanning airborne laser altimeters, an ice-penetrating radar, a nadir-viewing digital video imaging system, and multiple GPS receivers, was deployed to Iceland . . ."— and when he's done, he rips his report from the typewriter, drops it off at the base commander's office, and trots back to the P-3.

He bounds up the ladder to the cabin, which looks like the inside of an Eyewitness News van, crammed with television monitors and wires, strewn with Styrofoam coffee cups, and devoid of creature comforts. Within this funky high-tech cave, he confers with the navigator, Jon Sonntag, and the pilots. They plot coordinates. They discuss backup plans. They propose flight paths. "Here's the game plan," Jim says, tracing the route on the map with his finger. "Take off, come around . . . *here* . . . and then straight to the Sandur. Surtsey looks really good. Now come over here and do this middle line. That's the number-one priority. If that looks good, see if we can do the north line. At that point, we call the option for doing the south line. If it looks like there are no clouds over this ice cap, we might be able to sneak up and come around. In the past, we always went way up here and came down. I'm afraid that, unless we can throw a real sharp turn, we can't do it. We fly at eighteen hundred feet."

All they need is a working airplane. Pedro (as I will call him) is the mechanic responsible for maintaining the leaky P-3. He stands about five foot three or maybe four, stocky, with a scruffy, uncertain beard, and a good-natured grin. Pedro, who's American, likes his wine, and he likes his beer. In the evenings, he's the first to hit the strip bars of Reykjavik, such as they are. The fate of the P-3 now rests in his hands. Even as they tell me stories of his wild doings, everyone about to fly on this plane expresses confidence in him. (Frankly, I wouldn't let him near my car.) He works slowly and methodically on the plane, and when he's done, declares it good to go. The engines roar to life.

The white P-3, with Jim inside, taxis far out onto the tarmac and ascends to the skies over Iceland. While they're in the air, the pilots do most of the work, for they have to maintain alignment not only with the surface of the planet below but with the Global Positioning System satellite above. This means the plane can't wander more than six feet off course. They map the Sandur, and, weather predictions to the contrary, they do Surtsey as well. They do their mapping with reference not to the Earth's surface, which is always shifting, but to the Earth's center, which is about as close to an absolute, fixed point as you can get.

Mission accomplished, the P-3 returns to Keflavik after midnight. To everyone's surprise, the aircraft has performed flawlessly that evening, laying down precise tracks over Surtsey and a glacier to the north. "We laid down fourteen lines!" Jim announces. "It was fantastic. Staggered just the way we wanted them. And the weather was great. It was sunny and clear on the island. I've got digital video, nose cone video. We're going to have the best map of Surtsey ever made, no question. The flight was as tranquil as bathwater. Even the leak stopped by the time we landed."

To celebrate, Jim and Jon Sonntag and I go out for a beer. After a long day's work in the field, no self-respecting NASA scientist thinks of anything but a beer. The drinking etiquette is to avoid brand names and even recognizable microbreweries in favor of obscure local output. Eventually, we find a little place overlooking the large bay, where they serve frosted glass tankards of Viking, an Icelandic beer. We sit in front of a picture window overlooking the steel gray expanse of Keflavik Bay. A few lights flicker across the water, but not many.

Having just spent the last five hours in geological nirvana, Jim talks on and on about what a great mission it's been, while Jon brings up a slightly different subject: the kind of woman he'd like to meet and settle down with. He's from Houston, but he wants to meet a different kind of girl from the ones he's known in Texas. Maybe here, in Iceland. Maybe even in New York City. He asks me about the women in New York, where I live, and I tell him the best thing to do is visit and see for himself. He pauses and smiles shyly, contemplating the prospect. He just might do that.

This type of talk makes Jim uncomfortable. Throughout our time in Iceland, a lot of stray remarks have escaped his lips about power tools, about the power washer he was using just the other day with his son, Zack. ("That thing was so powerful," he said with genuine conviction, "that it could take the paint right off a car"), about the Ford F-150 pickup truck he'd like to own, and about a Hummer ("How much do those things cost?"), but nothing about women. Which doesn't mean that women don't look at *him*. They do, indeed. His handsome dark Irish looks, his snappy NASA flight jacket, and his politeness combined with an occasional air of confusion tend to attract women.

He met his wife, Cindy, by accident in 1990 when she was working at NASA for a contractor. It seems there was another J. B. Garvin with whom she was doing business, and Jim kept getting e-mails intended for the other one. So he got in touch with her to clear up the confusion. He found himself talking on the phone with her, and she coaxed him gently into asking her on a date. "I often get too focused on my work. I wished I didn't, but that's the way it is." Cindy began to challenge all that.

When they came face-to-face, Cindy already knew what he looked like; to this day, Jim is not sure how she knew. They went to a hockey game and, not long after that, moved in together. Cindy recognized that his interests were a little unusual. Here's a Phi Beta Kappa from Brown, a Ph.D., who says his most valuable possession in the world is a complete set of Jim Bunning baseball cards. He owns practically every Bunning card ever issued, tracing his pitching career from 1954 through the 1980s. Cindy liked shopping, dining out, and other normal activities. That was fine with Jim; he wanted to be with someone normal, someone who would keep him in touch with daily life. They married in 1992. We sit drinking and talking until the sky begins to brighten almost to daytime intensity, and we return to base a little after two in the morning.

The next day, while packing to leave, Jim ponders what to do with the data he's collected during the week in Iceland. He must get it out—in the new NASA, nothing is secret—so he will post it on various Internet sites, for starters. He will write multiple papers, some of which will appear in scientific journals. He will give lectures. He will share the information with Icelandic scientists. "It's my job as a research scientist at NASA to publish the results in *Science, Nature,* and other journals. That's my job, to disseminate." He will have a great deal to discuss, for a crowded schedule of Mars exploration lies ahead. "We're launching again in December and sending a small probe to the south polar ice cap on Mars, which we think is all frozen carbon dioxide. It's so cold, one hundred to two hundred and fifty degrees centigrade below zero. We're also asking other, very fundamental questions: Did life start on Mars? If it did, is it dormant, frozen, fossilized? Is it still there? Is it all microbial? What can it tell us about extremophile life on Earth?" Jim asks, savoring each question.

"I think the potential for Mars is totally untapped, and that's something of a surprise," he continues. "When we first got there in the sixties with the Mariner spacecraft, we thought, 'Oh, my God, there are going to be Martians, canals; it's going to be great.' But when we got there, it looked like the moon. Mars puzzled us. We returned with Viking in the mid-seventies, looking for life, and instead we found the great arctic desert of Mars. We saw frost form in the winter, and we saw snow. We saw rocks and pits that reminded us of gas bubbles in the volcanic rocks you see here in Iceland, but we didn't see the obvious signatures of life. We've got to go back. We've got to understand this place. We'll have a series of robotic voyages to set the stage for bringing back samples of Mars to Earth to investigate the chemistry and—maybe—signs of life. And then someday we'll put human beings there, God and the great American economy willing."

The taxicab heading home from JFK Airport feels as cramped as Oscar's co-pilot seat. I'm probably in more danger barreling along the Grand Central Parkway than I was aloft in Oscar's little Aerospatiale. Night falls for the first time in a week; how strange the darkness seems. After experiencing Iceland's white nights and thinking intensely about what it's like to walk across the surface of Mars, I find that nothing on Earth looks quite the same. The initiation is over, and back home, I reflect on a whimsical passage from Ray Bradbury's sci-fi novel, *The Martian Chronicles,* published in 1950:

The ship came down from space. It came from the stars and the black velocities, and the shining movements, and the silent gulfs of space. It was a new ship; it had fire in its body and men in its metal cells, and it moved with a clean silence, fiery and warm. In it were seventeen men, including a captain. . . . Now it was decelerating with metal efficiency in the upper Martian atmospheres. It was still a thing of beauty and strength. It had moved in the midnight waters of space like a pale sea leviathan; it had passed the ancient moon and thrown itself onward into one nothingness following another. The men within had been battered, thrown about, sickened, made well again, each in his turn. One man had died, but now the remaining sixteen, with their eyes clear in their heads and their faces pressed to the thick glass ports, watched Mars swing up under them.

"Mars!" cried Navigator Lustig.

"Good old Mars!" said Samuel Hinkston, archeologist.

According to Bradbury, this landing, the third human expedition to Mars, was supposed to occur in April 2000. NASA is running a little behind schedule but sends spacecraft to Mars as often as budgets and orbits allow. For now, they are robotic missions; in time, they will bring people to the Red Planet.

Welcome to the new Martian chronicles.

Chapter Two

MESSAGE IN A BOTTLE

Jim Garvin's collection of Jim Bunning baseball cards got me to thinking about what is perhaps the most famous baseball card of all—the 1909 portrait of Honus Wagner in a Pittsburgh Pirates uniform. When I gaze at the face of this young man, who seems to be staring into space, I find myself asking, "What was it like to be alive in 1909?" I have little idea, although it was just the other day, in geological terms. All I have to go on are artifacts, such as this famous baseball card. I can't watch Wagner play baseball, and I can't hear his voice; all I can see is a fuzzy image of the athlete in a uniform, a trick of light and shadow, an impression of life as it once was. To fill in the gaps, I would have to look beyond the card, but if I'm relying on the card, and only the card, I have precious little data.

The card, and its limitations, call to mind the tantalizing images of fossilized bacteria in a 3.9-billion-year-old meteorite from Mars—images that may be the first scientific evidence of life beyond Earth. Fossilization occurs when minerals replace organic elements in once-living things. The morphology remains, although the chemistry is different. Still, scientists can learn a lot from fossils. They can detect the approximate age, which is crucial, and, by studying fossils in their natural setting—*in situ*—they can extrapolate a great deal about the geological, chemical, and biological circumstances surrounding them. "Fossils are the autographs of time," wrote the American astronomer Maria Mitchell. For these reasons, fossilized bacteria from Mars—if that's what they are—have great appeal; they are our best indication of life beyond Earth. Like the antique baseball card, they offer only a very narrow glimpse into the past. The year 1909, that of the Honus Wagner card, wasn't very long ago, but it's long enough past to seem quite mysterious. How much more difficult it is, then, to construct a scenario for the existence of life on Mars several billion years ago from the evidence contained in a meteorite.

If the fossilized bacteria are genuine artifacts of Martian life, they raise more questions than they answer. If life started on Mars, how did it begin? Is it still there? If not, when did it end, and why? If it's on Mars, where else in the universe might it be, and what form does it take? Did it originate on Mars, on Earth, or somewhere else? All these Genesis Questions point up how much scientists have yet to learn about how life began. In the course of asking questions of various scientists who study these problems for a living, almost every reply I received began this way: "No one knows, but . . ." That's followed closely by, "There are several possible scenarios," and "Well, current speculation has it that . . ." The answers are all variations on the theme that no one knows, yet. But scientists have hypotheses. They have scenarios. The meteorite from Mars has inspired a widely accepted scenario—I'll call it the Best Guess Scenario—concerning the origins of life on the Red Planet and how life came to be transported to Earth.

Four and a half billion years ago, the solar system was in its infancy, and the planets were new. In its first billion years of existence, Mars was a warmer and wetter place than it is now. Water flowed freely over its surface and pooled underground, in reservoirs. The flood channels carved into the surface of Mars, some of them many miles in width, left an eloquent record of catastrophic outpourings of water. In all likelihood, the water on Mars was quite salty. Eventually, the floods subsided, and the water drained into Mars' vast northern plains, where it might have frozen. In the process, it reshaped the Martian terrain until it resembled a desert that had once been flooded, but became bone dry. Nevertheless, its contours preserved geological memories of rivers and oceans and lakes.

The large Martian pools of standing water were subject to peculiar tides caused by the planet's two small moons, Phobos and Deimos. And they were subject to the Martian winds; when they blew, reaching speeds of hundreds of miles an hour, they generated waves with peculiar shapes, higher and steeper, with more pronounced peaks than exist on Earth. The lower Martian gravity, less than half of Earth's, allowed the slender waves to tower until they resembled the watery shapes in a drawing by Dr. Seuss; they would flop over and spatter, as if in slow motion. The marine scene on Mars was all oddly familiar, and strangely different.

The Martian sky was blue a few billion years ago, and there were a few clouds, just as Mars has now. It was mostly cold, and extremely cold at the poles, except for the equator, where it was warm. Martian volcanoes erupted with regularity, and in the Red Planet's low gravity they assumed formations that couldn't exist on Earth; they were larger and higher. In these ancient

Martian conditions of two or three billion years ago, life could have formed and evolved, just as life appeared on Earth within a billion years of this planet's existence. The volcanoes, especially the ones close to reservoirs of water, or polar ice, created hot spots where life would most likely have formed on Mars. No one knows how far it developed, or if it ever got underway. It might have remained dormant most of the time, for tens of millions of years at a stretch. Or it might have progressed beyond simple bacteria; there might have been Martian insects crawling around, adapted to the Red Planet's lower gravity, lower density atmosphere, and cooler temperatures. These variations suggested life-forms that were spindly, similar to insects. The skeletons might have been external, with many legs to take advantage of the lower gravity. As for the cooler temperatures, life on Earth has shown remarkable adaptive creativity. "Some insects winter-proof themselves with glycerol, a common antifreeze used in automobile radiators," Carl Sagan theorized. "There is no conclusive reason why Martian organisms should not extend this principle, adding so much antifreeze to their tissues that they can live and reproduce in the extremely cold temperatures occurring on Mars." The ancient Martian atmosphere would have required similar creativity in the creatures' breathing apparatus. If they had evolved to the point of multicellular differentiation, they might have developed enormous gills or lungs, relative to their size. Even if life never reached this advanced stage of evolution on Mars, it is still possible that tiny organisms formed in the water-drenched Martian rock, and then, for some reason, died off, leaving fossilized remains hidden beneath the surface. It was as though Nature initiated an experiment but abandoned it in the early stages.

Ancient Mars was more turbulent than Mars is now. In the young and volatile solar system, it was constantly bombarded by chunks of asteroids. It is possible that at some point in Martian history, an asteroid struck Mars and created cataclysmic changes in the planet's climate and geography, extinguishing whatever life-forms had managed to take hold. Or perhaps the death of Martian organisms came about slowly, as the planet lost its atmosphere a little at a time to space, and its water eventually disappeared below the surface, or vanished with the atmosphere, leaving behind a desiccated, celestial sandbox.

If we had been able to observe the first few billion years in the life of Earth and Mars from a vantage point in distant outer space, we might have noticed several common trends. We would have seen watery places on both planets. We would have seen volcanoes on both planets, their plumes of smoke, their pollution of the atmosphere. We would have seen clouds on both planets, and we would have detected seasonal waxing and waning of the polar caps. As the eons passed, subtle differences between the two planets would have become apparent. If we had been looking closely, we might have noticed the atmospheric changes. We might have seen the dramatic increase in oxygen in Earth's atmosphere, and a corresponding spread of vegetation on its surface; if we were very perceptive, we might have noticed the spread of plant life in its oceans, in the form of algae.

At roughly the same time, we would have seen that Mars was losing its nitrogen-rich atmosphere. It was thinning out, disappearing into the frigid vacuum of space. More obviously, we would have seen the great Martian standing bodies of water recede, exposing a complex erosional system of gullies and playas and rearranged boulders, many of them acting as signposts to the water's former whereabouts and actions. During the last few hundred million years, if we were sufficiently attentive, we might have noticed the spread and intensification of vegetation on Earth, as the biomass increased and diversified, and various life forms competed for natural resources or evolved ways to cooperate or both. At about the same time, we might have watched Mars continue to regress to its early state, with some important differences. It contained geological traces of water and perhaps traces of biology—clues, ultimately, to its origins and to ours.

Where did the elusive Martian water and its life-giving properties go? Come to think of it, where did the water come from? Where did water on *Earth* come from, for that matter? At the Lunar and Planetary Institute in Houston, Steve Clifford has spent years studying water on Mars, and he told me there are several schools of thought concerning the origins of water on this planet and on Mars. "One is that after the Earth was formed, comets bombarded the planet, adding volatiles over perhaps the first billion or half billion years." A comet is basically a celestial snowball, bearing ice from somewhere—God knows where—to here. "The other school of thought is that much of the water we have on the Earth was contained in the early material that formed the planet. As the Earth started to accrete asteroidal material and dust in the early solar nebula, it gradually reached a size where the quantity of radioactive material was sufficient to heat up the planet and cause it to differentiate. The heavier stuff sank toward the middle, which is how we got an iron core, while the lighter stuff, which may have contained water, was released during the formation of the crust and atmosphere." A similar process may have occurred on ancient Mars at about the same time it happened on Earth.

Steve surprised me by suggesting that water on Mars may still linger beneath the surface. He wasn't talking about a little water, but about "sizable reservoirs of ground ice and groundwater." The evidence he has for large volumes of water on Mars is mostly indirect. He calculates the amount of "pore space" to be found in Martian rocks and soil; water could be stored there. If it is, some of the water could be in liquid form, especially well below the surface. "Like the Earth, Mars is thought to be radiating internal heat due to the decay of radioactive elements, which means the deeper you go below the surface, the warmer the temperature gets," he says. "And if you go down several kilometers, you could easily get temperatures that are consistently above freezing," which means liquid water might exist on Mars today. In fact, he thinks there might be two types of water reservoirs on Mars, a region of permafrost near the surface as well as larger and warmer reservoirs at greater depths. There you might find liquid water in great quantities, and water, of course, leads to life. This subsurface system would act as a powerful preservationist of life, no matter how harsh the conditions on the surface are. "If life ever evolved on Mars, and adapted to a subterranean existence, then its survival would be assured for the indefinite future."

Subterranean life on Mars could survive the loss of the atmosphere, it could survive intense cold on the surface, and it could even survive the largest life extinguisher we know about, the impact of a large asteroid. "Imagine what would happen if an asteroid the size of Manhattan collided with Earth," Steve says. "It would certainly be calamitous and likely sterilize the surface of the planet down to a depth of several meters or so, but not below that. It would kill life on the surface, but it wouldn't eliminate *all* life on Earth, because the impact's thermal and chemical effects would be limited to the uppermost part of the surface, with the exception of the area where the impact occurred." The same holds true on Mars, Steve says: "Any microorganisms that might have evolved four billion years ago, when the planet might have been warmer and wetter at the surface, could readily survive to the present day at depths of several kilometers." If he's correct, or even partly correct, life could very well exist within the Red Planet's ancient reservoirs, awaiting discovery.

The Best Guess Scenario assumes that some kind of simple life did exist on Mars a few billion years ago, as Steve described. And it assumes that asteroids have bombarded Mars ever since, pockmarking its surface with craters. Sixteen million years ago, according to the scenario, a refugee from a disorganized asteroid belt struck the surface of Mars with tremendous force. Since Mars has only thirty-eight percent of Earth's gravity, the impact was sufficient to drive pieces of rock buried beneath the surface high into the Martian sky and beyond. The pieces shot into space five times faster than a bullet—fast enough to escape the Red Planet's gravity. Until a few years ago, it was thought the shock of impact would vaporize or severely deform the ejected material—the ejecta—along with everything in it, including any signs of life, but recent computer modeling has shown that the physics involved would allow the ejecta to remain intact. In the model, the asteroid comes in at an angle, strikes, and creates a vapor cloud that sweeps across the surface at an extremely high speed and carries material off Mars into solar orbit. The whole process might take five or ten seconds, long enough for some fragments of Mars to remain intact. "That's a fairly gentle way to get stuff off the surface," says Mark Cintala, who studies craters at NASA's Johnson Space Center—gentle enough to launch a fossil-bearing meteor on a trajectory to Earth.

There's a variation in the Best Guess Scenario that incorporates another method of ejecting material safely from the surface of Mars: spallation. Much of the work on the spallation model was done by Jay Melosh at the University of Arizona, and it's extremely simple, in theory. You put a quarter on a tabletop, rap the underside of the table, and bump the quarter into the air; that's spallation. "You're sending a stress wave through the table, and that stress wave is transmitted to the quarter. Imagine that happening on Mars," Mark Cintala says. In that case, the asteroid's impact would create a compressional wave, as if it were a depth charge below the surface. The resulting shock wave would bounce a rock fast enough to escape the relatively weak Martian gravity and send it on its way.

In the Best Guess Scenario, one of these dislodged pieces of Martian rock sped off in the direction of Earth, a cosmic message in a bottle floating through an ocean of outer space. The journey lasted millions of years, and the ancient chunk of Mars crashed to the surface of Earth a mere 13,000 years ago. The size and shape of a potato, it buried itself in the Allan Hills of Antarctica. The meteor had become a meteorite.

Meteorites have held a special fascination as relics from the heavens, mute messengers from parts unknown. In the Middle Ages, meteors falling to Earth generated superstition and concern. Where did they come from? What did they mean? The faithful brought them to the authorities, and in time, the Catholic Church acquired a large repository of these curiosities. In 1969, the study of meteorites underwent a quiet revolution when Japanese researchers found high concentrations of them preserved in arctic ice. Since 1977, NASA, a technological Vatican, has been collecting meteorites from Antarctica and housing them at the Johnson Space Center in Houston. Each year, there are hundreds of new arrivals, and when there's a promising delivery, scientists clamor to get a piece to study. There are now nearly 10,000 rocks under lock and key in Building 31 at Johnson, many of them preserved in nitrogen. By measuring the radiation absorbed by the meteor during its space travels, scientists can determine approximately when the rock arrived on the Earth and even how long it spent in space before it arrived on our planet.

In December 1984, Roberta Score was hunting for meteorites in Antarctica. At the time, she was employed by Lockheed Martin and working at the Johnson Space Center. Around Johnson, a meteorite-collecting mission is not exactly choice duty; join one, and you were said to have become part of the "Houston weight loss program." Walking across an apparently endless sheet of ice, Robbie came across a greenish stone about the size of a potato. Once she removed her sunglasses, she saw that the meteorite was not greenish after all; it was gray and brown, but she knew it looked different from the ordinary meteorites she found in the field. Along with other samples, it was kept in a freezer aboard the ship that brought it from Antarctica to Point Magu, California, where it was packed in dry ice and sent to Johnson, where it was stored in cabinets that once held moon rocks. The meteor curators, including Robbie Score, designated it ALH 84001—their way of saying that this was the most interesting meteorite collected in 1984 in the Allan Hills of Antarctica. But after being delivered to Johnson, ALH 84001 was misidentified as an asteroid fragment, a diogenite, rather than a piece of Mars, and stored in Building 31. It was not ignored, however; small sections were allocated to the scientific community for further study over the years. In all, almost a hundred "investigators" examined it, and everyone continued to misclassify it as a diogenite—with one exception.

In late 1993, David Mittlefehldt, a veteran Lockheed Martin scientist also working at Johnson, reexamined ALH 84001. Mittlefehldt was an expert on diogenites, and this particular rock didn't look like one to him. It seemed to have more oxidized minerals in it than your normal diogenite, for one thing. Using new technology in the form of high-resolution laser spectrometry, two other scientists, Donald Bogard and Pratt Johnson, extracted gases trapped inside the strange meteorite and discovered that their very idiosyncratic characteristics exactly matched gases on Mars as measured by the Viking spacecraft in 1976. Mittlefehldt published his findings in 1994 in a scientific journal and attracted the notice of the science community. Although this wasn't the first meteorite from Mars to have been discovered, the reclassification created a stir. Of the thousands of meteorites that have been cataloged, only fourteen are believed to have come from Mars; the overwhelming majority come from asteroids, and a few from the moon. Meteorites are named for the places where they have fallen to Earth, so the Martian meteorites have some fairly exotic names—Shergotty (India), Nakhla (Egypt), and Chassigny (France), among them—and are known collectively as SNC or "snick" meteorites. "SNC meteorites" is an elaborate way of saying "meteorites from Mars."

Carefully considering his find, Mittlefehldt noticed minuscule reddish-brownish areas deep within ALH 84001; they looked a lot like carbonates, and on Earth, carbonates such as limestone tend to form close to water. What made this all so curious was that no one had detected carbonates—and their suggestion of water—in the other Martian meteorites. They were billions of years newer; they probably came from a more recent era in the geologic history of Mars, after the water that once flowed freely across its surface had disappeared. This one, however, apparently harkened back to that warm and wet golden age on Mars. Dating confirmed that the meteorite was indeed very old: 4.5 billion years old, much older than other known Martian meteorites, and it contained carbonates that were 3.9 billion years old. Mittlefehldt wanted to get some idea of the temperature range in which the carbonates had formed billions of years ago, so he went to yet another NASA scientist, Everett Gibson, who examined the very curious meteorite with Chris Romanek. Gibson and Romanek published a paper in the December issue of *Nature* in which they said the carbonates had formed at temperatures below 100° C, in other words, at moderate, Earth-like temperatures—"well in the range for life processes to operate," as Gibson put it.

By now a line of reasoning was beginning to take shape. The team had their meteorite; it was from Mars. Almost no one disputed that singular fact. And it was very old, when water was thought to exist on the Red Planet. And it had carbonates, suggestive of water, formed at moderate, Earth-like temperatures. With each new discovery, the stakes became exponentially higher. ALH 84001 had gone from being a curiosity to an interesting and instructive case study to a potential harbinger of a scientific revolution. Each new link had been more difficult to fashion than those that had preceded it, and the final link—to life on Mars—would be the most difficult of all to fashion.

Other scientists soon began angling for a piece of the curious, potato-shaped Martian meteorite, among them David McKay. Over the years at Johnson, he'd become known as a solid and reliable scientist, not the type to go out on a limb. Carl Sagan he was not. If you asked around about McKay, you often heard words like "cautious" and "self-effacing," yet he had a distinct air of authority; he'd published hundreds of scientific papers, and he knew his way around Johnson and around NASA. Over the decades, he'd learned about science and about maneuvering in the world of scientists. He knew about the pitfalls, how quick others were to leap on "discoveries" and tear them to bits. Yet with all his experience, he seemed destined to retire in honorable obscurity, until ALH 84001 came to his attention.

"I'm going to get a piece of that meteorite and look for signs of life in it," he told his wife.

"Sure you are," she said.

McKay had a vast storehouse of information and impressions about rocks on which to draw. In his long career, he had looked at perhaps 50,000 of them, and he spent many hours studying the most intriguing he'd ever seen, ALH 84001, with a scanning electron microscope capable of magnifying objects 30,000 times. With this instrument, McKay identified a bunch of—well, they looked a little like miniature subterranean carrots or worms or tubes. Whatever they were, they didn't look like something you'd expect to find in a meteorite.

Again, he turned to another scientist for assistance. Kathie Thomas-Keprta was a biologist who had spent almost a decade studying extraterrestrial particles—space dust—before she focused her attention on the meteorite from Mars. She was accustomed to making do with very little. A specially modified B-57, flying at high altitude for an hour, might collect just one extraterrestrial particle from an asteroid, a particle too small to see but big enough for her to examine under a powerful microscope. When McKay invited her to study a Martian meteorite, she was delighted to have something as big as one millimeter by one millimeter to work on after all those years of studying specks. Even better, she was an expert with a new type of electron microscope that could reveal the mineral composition of the carbonates locked in the meteorite. McKay and Gibson showed her the photos taken by the scanning electron microscope, and they proposed that she examine those peculiar, wormlike structures to see if they were fossils. She listened respectfully to their proposal, and when she got home that night, Kathie told her husband, "These guys are *nuts!*"

A team of researchers based at Stanford subjected chips of the meteorite to further laser tests, which yielded polycylic aromatic hydrocarbons—PAHs, for short—which are often associated with life. That finding raised more questions than it answered, for PAHs are also associated with inorganic material such as pollution and exhaust. If that were the case, the carbon in the meteorite could be the result of very recent contamination on Earth, not evidence of ancient life on Mars. Additional tests showed that the PAHs were buried inside the meteorite and probably quite old, lessening the likelihood that they were the result of exhaust. It looked like the PAHs came from Mars after all.

The team felt confident enough to announce some initial findings at the 1995 Lunar and Planetary Science Conference, held at the Johnson Space Center. In the planetary science community, the LPSC is a very big deal, a sort of scientific Super Bowl. If you don't show up for this event, scientists say, everyone assumes you've died, and when you do show up, you come to make news, if you can. On behalf of the meteorite team, Kathie Thomas-

Keprta presented a paper about the unusual and provocative features of ALH 84001 observed by the team. The paper stopped short of declaring they had found evidence of life on Mars, even very ancient, very tiny life. In fact, she adamantly denied it to a reporter from the *Houston Chronicle* who suspected she was hinting at it.

She knew other scientists would soon challenge her findings, no matter how cautiously expressed. Faulty science or clumsy handling of the situation could mar several carefully tended careers. So McKay and his colleagues ran still more tests on the meteorite with an even more powerful scanning electron microscope designed to inspect rockets for minuscule fissures; this instrument was capable of magnifying objects up to 150,000 times. McKay put a four-billion-year-old piece of Mars under the microscope, and on his monitor there appeared a bunch of wormlike forms. He printed an image and gave it to his teenage daughter.

"What does it look like to *you?*" she asked her father.

"Bacteria," he answered.

Kathie Thomas-Keprta eventually decided the guys on her team weren't nuts after all. Her conversion occurred in Building 31 at the Johnson Space Center one night when she was working late. As she examined the shapes of the nanofossils in the meteorite, she knew from experience they were of biological origin. "It was gregite, an iron sulfite present in the carbonate. It had a certain morphology known to be produced by bacteria. It was actually a biomarker, a thumbprint left by biological activity. I thought, *That's it. There's life on Mars.*

"I walked out the door to the parking lot, half-expecting to see flags waving and bands playing, but there was nothing at all out there, just a dark, empty parking lot at night."

The chain of reasoning was more or less complete. The meteorite was old enough to contain a record of Mars' early days, when water was plentiful. It had carbonaceous material; it probably had Martian rather than terrestrial PAHs; and it had gregite, a universally accepted sign of biological activity. Although each distinct link could not be taken as proof, they all added up to a fairly strong argument for ancient life on Mars.

The team, now grown to nine, approached *Science* magazine. They realized that getting the prestigious journal to accept their paper would be difficult and delicate; they might have to withstand as many as four or five anonymous critiques of their work. *Science* was tempted by the paper but reluctant to support invalid conclusions, so the publication sent out the manuscript to nine readers. The resulting article relied solely on sober observations and rigorous science, and its title reeked of compromise: "Search for Past Life on Mars: Possible Relic of Biogenic Activity in Martian Meteorite ALH 84001." The most important sentence was the summary: "Although there are alternative explanations for each of these phenomena when taken individually, when they are considered collectively . . . we conclude that they are evidence of primitive life on Mars." In other words, the meteorite offered the first scientific evidence that ours is not the only planet in the solar system where life emerged. Publication of the issue of *Science* containing the article was set for August 16, 1996.

When Jim Garvin heard about the impending *Science* paper, he felt the skin on the back of his neck prickle. "I was dumbstruck," he said. In 1990, he had looked at another meteorite, Shergotty. At the time, no one realized that particular rock had come from Mars. He borrowed a piece of it from the Smithsonian, where it is stored. "I took it up to our lab and made the measurements I'd wanted to make for impact metamorphism"—looking for evidence of shock waves, that is. "This was a passive measurement, by the way, like bouncing a laser pointer off a rock; we weren't destroying it." There he was, examining a piece of Mars without realizing it.

The force of the new paradigm—that life on other planets was probably *tiny*—spun Jim's thinking in a new direction. "We were still a few months before the launch of Pathfinder and Mars Global Surveyor, and the question was asked, 'What could be done with these ready-to-go spacecraft to look for more signs of life?' " Suddenly, Jim's Mars mission had a new reason for being. He had always believed it presumptuous to assume that life existed only on Earth, and he was sympathetic to the meteorite team's conclusions about ALH 84001. Their research science was rigorous, it was cautious, and it was consistent with the latest findings concerning extremophile life. There was something in that meteorite that could not be explained away by conventional arguments. Jim agreed with the team that the burden of proof had now shifted to those who insisted there was no life on Mars. If that was the case, he said, "an interesting explanation as to why life failed to make at least a tenuous foothold would have to be crafted."

The midsummer Martian madness started in earnest a couple of weeks before publication of the article, when Dan Goldin, the mercurial, publicity-loving head of NASA, heard that *Science* had accepted the article for publication. Next, the White House wanted to make a grand occasion out of the discovery of possible life on Mars. In preparation for the announcement, Goldin summoned David McKay and Everett Gibson to Washington. "We had thirty minutes scheduled with Dan to talk about the meteorite," Everett told me. "After an hour and a half, Dan said, 'You guys take a break, I've got some things to do, and then we'll continue.' During the break, he dictated a commencement address he was going to deliver at UCLA and handled a few other things, and then we continued for another hour and a half. I felt like I was giving an oral defense of a Ph.D. thesis. I mean, Dan went back to first principles, and he took twenty-eight pages of notes." At the end of the ordeal, Dan Goldin had one last question for the two scientists: "Can I give you a hug?" The gesture was pure Goldin. In general, NASA is not a touchy-feely place, but Goldin is a man of enthusiasms.

After that, the story began to leak everywhere. *Space News,* a weekly trade journal, hinted at the forthcoming *Science* paper about ALH 84001, and the buzz preceding an important Washington story started. Then things suddenly went awry. At the stylish Jefferson Hotel in Washington, Dick Morris, an adviser to President Clinton, told a call girl named Sherry Rowlands about the discovery, in the vain hope of impressing her. "Is it a bean?" she asked. Well, no, not really, he replied. It was, uh, more like . . . a "vegetable in a rock." When Rowlands got home and opened her diary to write about her day with Dick, she noted, "He said they found proof of life on Pluto." Scientists dread being misunderstood by the public, but who could have imagined the magnitude of misunderstanding generated by this discovery? The situation deteriorated even further when the befuddled hooker tried to peddle the story to the tabloids, which turned out to be more interested in extraterrestrial life in the form of little green men than vegetables in rocks, thank you. And her inability to recall just what planet Dick said they'd found life on—Saturn, maybe?—didn't help her credibility, either. There was no sale.

The life-on-Mars story quickly took on a life of its own. The *CBS Evening News* was making disturbing noises that it might break the news even before confirmation, according to an account that appeared in *Texas Monthly.* Other networks sensed news in the making and assigned reporters. *Science* tried to halt misunderstandings by posting the article on the Internet shortly before publication. On the first day alone, the website received over a million hits. Giving substance to an age-old dream, and terror, the article's findings excited worldwide attention. The announcement gave new impetus to America's beleaguered space program, especially its investigations of Mars. Goldin was delighted to confront a challenge of this magnitude, and the mood surrounding it recalled the great days of the space race, when Americans had an emotional investment in NASA and the nation's fortunes seemed to rise and fall with the agency. But the issue of life on Mars was more complicated to explain to the public and sell to Congress than sending people to the moon had been. There was no life-on-Mars race for politicians to exploit. National security and national pride were not at stake. Only the science really mattered. The discovery involved concepts difficult for most people, even scientists, to understand, including a meteor of unimaginable age that had traveled to Earth from an unimaginable distance, containing evidence of life that was unimaginably tiny.

NASA finally made the announcement at a flashy press conference, at which an exuberant Dan Goldin proclaimed, "What a time to be alive!" (And the head of NASA, he might have added.) Bill Clinton, campaigning for reelection, appeared on the South Lawn of the White House to hail the discovery as if it were another triumph for his administration, but he actually sounded a note of caution that went largely ignored: "If this discovery is confirmed, it will surely be one of the most stunning insights into our universe that science has ever uncovered." That was still a big *if.* And his declaration that the American space program would now "put its full intellectual power and technological prowess behind the search for further evidence of life on Mars" did not necessarily mean additional money for a beleaguered NASA. His words amounted to a mere presidential pat on the back.

The summer of Mars was under way. For a while, the names of the several NASA scientists on the meteorite team—McKay, in particular—were known to journalists and the general public. The sudden popularity threw the scientists for a loop. They naturally desired professional recognition, but not celebrity. In their line of work, being famous meant being considered suspect, a semi-charlatan, a talking head rather than a working research scientist. None of them aspired to become the next Carl Sagan, bridging the gaps among the media, the scientific community, and the public. Although their thinking was revolutionary, they weren't visionaries; they just wanted their funding, and they wanted to pursue their scientific interests. The announcement concerning ALH 84001 made it harder for them to do that, as publicity insinuated itself into the normally orderly process of disseminating scientific information. Instead of addressing specialists at conferences and publishing in specialized journals, science teams proclaimed their findings in press releases, in advance of publication. Freed of the constraints imposed in a refereed publication such as *Science,* the releases tended to make larger claims than the articles that inspired them. Conducting science by press release troubled many, including those engaged in the practice.

The announcement concerning ALH 84001 transformed NASA. For the first time, many people realized that NASA supports scientists, not just astronauts and engineers and the crews that send them into space. In its youth, NASA had accomplished one spectacular engineering feat after another: putting an astronaut in orbit, sending astronauts to the moon, keeping astronauts in orbit for months on end. These missions included science, but science was rarely the point. Flags and footprints on the moon were the point. Astronauts did collect a few hundred pounds of moon rocks for scientists to analyze, but the public had scant interest in lunar geology. Now, with the announcement of possible nanofossils in ALH 84001, NASA scientists were no longer overlooked. And with the end of the Cold War, they could participate in missions that were primarily scientific rather than political, missions that might become more significant than sending people to the moon. They suddenly had an opportunity to devise experiments exploring fundamental questions about the nature of the universe and the origins of life. Their results of their search, a NASA report concluded, "may become a turning point in the history of civilization."

The message in a bottle had arrived, but who would decipher it correctly?

Throughout the summer of 1996, David McKay expected a backlash concerning his discovery, but it was slow in coming. At first, members of the public, some of them deeply suspicious of all federal agencies, NASA included, sent him angry e-mails, most of which echoed the theme, "What kind of fools do you take us for?" One said, "Your life on Mars story is a good example of your mistaken belief that the general public is comprised of a bunch of total idiots."

Eventually, scientists joined the clamor. Some insisted that ALH 84001 proved absolutely nothing. The wormlike structures, said critics, were far too small to be bacteria; in fact, they were many times smaller than the smallest bacteria ever seen on Earth. Others insisted that if the meteorite contained evidence of biological activity, it was the result of contamination. Still others challenged the team's analysis of the PAHs. Some scientists stated flatly that McKay and his team had unfairly manipulated the evidence to support a flawed hypothesis. Everett Shock at Washington University invoked the Murchison meteorite, believed to have come from the asteroid belt, to invalidate the discovery. "It has carbonate minerals in it," he said, "and real solid evidence of water—yet there isn't anybody saying that there is life in the asteroid belt." True, no one was saying it at the time, but that situation is beginning to change as scientists have come to think of life as widely distributed throughout the solar system. Finally, the scientists attacked the reputations of McKay and his team, a tactic that took cooler heads by surprise. "It's kind of strange when scientists, who are thought to be rational, become emotional," said Marilyn Lindstrom, a curator of meteorites at the Johnson Space Center. "What bothers me most is that so many people have made up their minds before the data comes in. I mean, sometimes I'm amazed by McKay and Gibson's almost true-believer attitude."

Carl Sagan was seriously ill at the time of the announcement, with only a few months to live. During his decades with NASA, he had become familiar with both the science and the passions involved in the search for life beyond Earth, and his pronouncement on the subject was enlightening yet equivocal. "For years I've been stressing with regard to UFOs that extraordinary claims require extraordinary evidence. The evidence for life on Mars is not yet extraordinary enough. But it's a start." Although he was deeply intrigued by the meteorite team's findings, Sagan insisted that more study was required. Yet other scientists were convinced by McKay's rigorous approach. "If this is not biology," said Joseph Kirschvink of Caltech, "I am at a loss to explain what the hell is going on. I don't know of anything else that can make crystals like that."

Because McKay, Gibson, and company were cautious, even cunning, in the way they stated their findings, they made it difficult for their critics to disprove their argument. The meteorite team held that the fossils were merely possible evidence of relic life; they were not the *only* explanation for what they'd found, merely the *best* explanation. To disprove or dismiss these findings, their critics would have to understand ALH 84001 even better than the original investigators did. They would have to refute four separate, interrelated lines of argument. They would have to be familiar with geochemistry and physics and geology and of course biology. No one person knew enough about all these fields as they applied to the meteorite; it would take a team, a bigger and better team, to show McKay and his colleagues the error of their ways.

The significance of the debate transcended the meteorite itself. Even if it contained crystals that mimicked biological morphology, or contamination, the search for extraterrestrial life had undergone a sea change. Even scientists who thought ALH 84001 contained no life signs at all now found themselves thinking that if we were going to find evidence of extraterrestrial life, it would probably be tiny and ancient and carried throughout the solar system in a meteorite. McKay, Gibson, and Thomas-Keprta's real discovery was a new *paradigm*. Even if their conclusions turned out to be incorrect, their thinking was too sophisticated to dismiss. From now on, they would define the terms in the search for extraterrestrial life. Their credibility rested not so much on what they found as on *how* they found it: their precise, rigorous methodology.

Two years after the announcement, I found Kathie Thomas-Keprta in the featureless Building 31 at the Johnson Space Center, where many of the crucial discoveries concerning the meteorite had occurred. She is tall and slender, with long blond hair swept up in back. Despite the intense debate concerning her work, she didn't look embattled; she was poised, with a certain swagger and the smooth delivery of a television talk-show host, at least in one-on-one conversation. We were standing beside another Martian meteorite, EETA 79001, a cousin of the more famous ALH 84001. EETA 79001 resembles a black ice cube, about two inches by two inches. I peered carefully at this Martian specimen. There wasn't much to see except for a little hole in one side drilled by a laser to extract gases trapped within.

Her team expected a lot of debate after their discovery, she told me, although the vehemence came as a surprise. "Still, all the criticism and attacks on our findings don't bother me because I'm from Green Bay, Wisconsin, and I've been a Packers fan for thirty years, and I know what it's like to hang in there from one losing season to the next." She thought it would take five to ten years for their findings to be fully vindicated, and she couldn't wait for that day. Her case now was stronger than ever, she said. The recent discovery of microorganisms far below the Columbia River, in Washington State, gave her a lot of corroborating evidence for nano-life on Mars. No one expected to find nanobacteria a mile or more below the surface of the Earth, and no one knows how they started growing. Like their ancient Martian cousins, they live in basalt. More important, they are almost as small as the Martian nanofossils. Critics of the meteorite team insisted that the presumed nanofossils in ALH 84001 were much smaller than any organisms found on Earth—too small to be considered microorganisms. Since the Columbia River discovery, that objection lost much of its force.

I wondered what kind of energy source for life there could be found in rock a mile or two underground, where there is no sunlight, no lightning, no real heat from the Earth's core. Some scientists think the source could be as simple as water passing over the basalt, which might cause a chemical reaction. If this is the case, the answer to the Genesis Question becomes simpler all the time; it appears that the rock bottom (so it might be said) requirements for life are even more minimal than scientists believed only a few years earlier. All you need is water and an energy source for life to emerge. Water might be running through subsurface basalt everywhere; the same thing might have happened on other planets, or even on asteroids; it might be happening *now*. There might be more ways for life to emerge than we now imagine—enough to suggest that life really is an inevitable outcome of chemistry and an inevitable part of the universe, predestined, as it were, but so simple that we hardly acknowledge the phenomenon for what it is.

David McKay is tall, slender, silver-haired, professorial, imposing. As the leader of the meteorite team, he is suspicious of outsiders and chooses his words with care. His office, where we met, is capacious, even by the standards of the sprawling Johnson Space Center, and the walls are lined to the ceiling with plaques, awards, degrees, citations, and a child's squiggly drawing of a small Martian meteorite beside a large man labeled, "DAD."

"We are still getting new data," he said, as he snacked on a small bag of pretzels, eying me warily. He wasn't exactly thrilled that I'd appeared in his lair; he was sensitive to criticism and assumed I was about to add my voice to the chorus of those who angrily criticized his findings. He was about to dismiss me—or so it seemed—but he thought again and decided to test his case with me. "We are very excited about the data from the meteorite called Nakhla that fell in Egypt in 1971," he said. "The British Museum had a piece the size of a potato, covered with fusion crust, which protects it from contamination. The problem with the Allan Hills meteorite, ALH 84001, is that it may have been contaminated with carbon or terrestrial bacteria. A chunk of the Nakhla meteorite came in here, to our lab, and we had permission to break it up and pass it out to various investigators. We requested six grams. We think it's likely to have the least contamination of any Martian meteorite." I sensed he knew more, but this partial revelation was all he would risk revealing at the time.

He also wanted me to know he hadn't given up on ALH 84001 as the prime suspect in the search for life on Mars. He didn't want me to think for one second that Nakhla was a substitute for ALH 84001; rather, it offered supporting evidence. As he talked, it became apparent that he felt that all the criticisms leveled at his findings—and there had been a lot of them, more than most scientists encounter in a lifetime—had only strengthened the arguments he originally advanced. To illustrate what he meant, he invited me to sit with him before a large monitor. "Here's a new picture from the Allan Hills meteorite. We really suspect these are fossilized bacteria. They have better characteristics than what we have already seen; they are curved, segmented. If you gave this to a biologist, he'd say, 'Of course it's bacteria,' but we have to prove beyond a shadow of a doubt it's of Martian origin and fossilized. Fossilization is very common with bacteria; the organic components are replaced by mineral components such as iron oxide or silica. This can happen quickly, in a couple of weeks, and it happens when you bury the material in water. They are one hundred to two hundred nanometers long and forty to fifty nanometers wide, smaller than the big worms in the published pictures, which were five hundred nanometers long. My guess is that life is still on Mars, but it's underground, in the water system. That's where the underground organisms are living, a couple of kilometers underground. On Earth," he reminded me, "there are microbes growing four kilometers underground."

As we parted, David McKay insisted, "Our critics have proved nothing. Our research has defeated each and every one of their arguments, and the case for ancient life on Mars is now stronger than ever."

Nine months after our meeting, McKay made his latest findings public at the 1999 Lunar and Planetary Science Conference in Houston. His announcement added to the controversy and ensured that the debate surrounding fossilized Martian bacteria would continue for years. To his way of thinking, there were now *two* meteorites from Mars bearing evidence of fossilized bacteria, ALH 84001 and the newcomer, Nakhla. His detractors claimed his analysis of the newer meteorite, Nakhla, compounded the errors he had made in his analysis of the first, but his supporters insisted it offered compelling confirmation of extraterrestrial life.

Chapter Three

GROUND TRUTH

To reach the Jet Propulsion Laboratory, you take the freeway to Pasadena and get off at the Oak Grove exit, then follow Oak Grove as it winds gently toward the mountains through the luxuriant landscape. You feel the smog settle on your chest as you go. There's no suggestion of high technology in the area, just a somnolent Southern California suburb, lush, green, and slightly sullen. As you sense the end of the road approaching, you assess the looming mountains, but there's still no sign of JPL, and you begin to wonder what gives. JPL isn't exactly off-limits, but it's not easily accessible, either. It will be found only by those who put some thought into looking for it. You think you're finally there when several large white modern structures appear on the left, but as you drive up to them, you realize it's a local high school, and then, just ahead, there's a gate and a guardhouse, and that, at last, is JPL.

People arrive for work early. By 7:30 A.M., the parking lot is filled with Hondas and Fords and Nissans and Tauruses—nothing fancy, except the odd Corvette. Employees quietly fan out across the campus and go to work. The buildings at JPL are boxy, functional, crisp. Within its offices, there are the same horrible green plants you see everywhere at NASA, at headquarters or the Johnson Space Center in Houston. Once you're indoors, you can forget all about Southern California; you might as well be in Washington or Florida; it's NASA-land.

Despite its innocuous location, JPL is among the world's leading centers for spacecraft engineering and development. Started in 1936 as the Guggenheim Aeronautical Laboratory at the California Institute of Technology, JPL is now run jointly by NASA and Caltech. In the early days, there were just a few people on hand, including Frank Malina, a rocket enthusiast, and Theodore von Kármán, an influential Caltech professor. The lab barely survived the Depression, but it got a boost during World War II for experiments in rocketry. During the 1950s, JPL developed a satellite that, according to legend, could have beaten Sputnik into orbit by a few months and irrevocably changed the space race—if it had been launched. Throughout the 1960s, JPL solidified its reputation as the place for robotics—unmanned spacecraft destined for the moon and the planets—but it lacked the high profile of the Johnson Space Center in Houston or the Kennedy Space Center in Florida.

All that changed with the advent of the new Mars program in 1992, when a new generation of employees began streaming into JPL, reinvigorating the place. Unlike many of the old-timers, they hadn't come out of the military or the aerospace industry; they were just out of grad school and had grown up watching the space program on television. They were young, and they weren't burdened by the past. The men wore earrings and ponytails instead of military buzz cuts, and tie-dyed T-shirts replaced white polyester short-sleeve button-down shirts and narrow black ties. But that was just the men. Many of the new recruits were women, and among them was Jennifer Harris.

Growing up on her family's farm in Fostoria, Ohio, Jennifer never expected to explore Mars or to become a flight manager for a Mars mission. She wanted to be a concert pianist. She played the piano, the saxophone, marimbas, bassoon, trumpet, tuba; she was a one-woman band. On the other hand, she loved math and competed successfully in county-wide math competitions. Astrophysics excited her imagination, especially black holes; she loved just thinking about them. In the summer before her senior year in high school, she went to music camp, where she realized that her survival as a concert pianist would depend on her ability to practice every waking moment, and she wasn't sure that was what she wanted to do with her life. She also wanted to travel, to meet people; she was even thinking of becoming a missionary. When MIT accepted her, she went into a mild state of shock. Eventually, she chose to major in aerospace engineering—partly because it sounded like the coolest thing she could do and partly because her father had tested missiles for NASA when he was younger, and she had come of age hearing his tales of countdowns, halts, and explosions. After graduation, she went to work for the Jet Propulsion Laboratory.

Even after she arrived at JPL, Jennifer was restless. They were designing spacecraft on spec, hoping to get funding from Congress, and most projects never did. If a project actually received a green light, the lead time was awfully long. As she toiled on her subsystems, she couldn't see where her little cog fit into the machine, or if there even was a machine. She began to ask herself, "Is this all there is?"

She was single and didn't have any serious ties to Pasadena or JPL. She chose to take a leave of absence, without assurance that a job would be waiting for her when she returned, *if* she returned. She still wanted to see the world and meet people, so she decided to do missionary work in Russia. She was assigned to Sevastopol, in the Crimea, near the Black Sea, where the conditions were unbelievably grim. There was no hot water, and they lived in cement buildings that were always cold and damp. A lot of the population were flat-out atheists. The economic situation was horrendous. She was paid about $30 a week, which made her among the wealthiest citizens of the town. Everyone around her was subsisting in a barter economy, using coupons instead of cash; one Snickers bar, for instance, cost 2,000 coupons. She and her friends based everything on the cost of a Snickers bar, but that didn't help keep track of finances, because the inflation was incredible. Pretty soon that Snickers bar cost 8,000 coupons, then 16,000. People who had saved throughout their entire lives lost their fortunes overnight when the ruble crashed.

At times she wondered what kind of space program the Russians could possibly mount under these conditions. She had to wonder how they got anything done. As if the Russians' pervasive fatalism wasn't enough, there was the corruption, another thing she hadn't been exposed to back at MIT and JPL and the family farm. She knew evil when she saw it, though, and it seemed to her that Russia, or at least her speck of it, was basically run by the Mafia, the politicians, and the church, all in bed together. After a while, she wondered if she was meant to be doing missionary work, if it was really the best use of her abilities. Was this what God wanted her to do? Was this what *she* wanted to do? She had to say honestly that the answer was no; her education was going to waste there. When her tour of duty was over, she left Russia to wander around Europe.

One day, she sent a postcard to a friend at JPL to say she would be back in a few months. "Do you have any jobs?" she asked, knowing the answer was very much in doubt. The day she arrived back in Ohio, JPL called to say they had a job for her, a good job, if she wanted it, but she would have to make a decision that day or the next. The job opening was on the new

Pathfinder project, the next spacecraft to go to Mars. She said she'd take it. Jennifer was fairly skeptical about Pathfinder, but so was JPL. "A lot of people thought it would never work. There were so many things that could go wrong, especially with the Mars environment." Her new job didn't seem to have official status at JPL. Even the official Mars program people kept their distance. The development of Pathfinder struck her as a skunk works, basically. She knew what that meant: if it wasn't working, they could take it out and shoot it and bury it and no one would be the wiser.

The nature of her job changed as the mission went along. She began by working on software, "but the neat thing about Pathfinder was that once you took a job, it was sort of a 'where-do-you-fit-in?' type of thing. People didn't say, 'That's not your job, stay out of there.' They allowed you to move around, so I ended up doing more integration and testing in the early stages than operations. People were always given the opportunity to move over the borders and learn more and do more." This open-ended, go-wherever-you-fit-in approach was something very new at NASA, and at JPL, which otherwise functioned along rigid, bureaucratic lines of command. The problem with the traditional structure was that if one element was delayed, or failed, or went awry, it brought the entire system to a halt. It became accepted practice for missions to slip several years. People were confined to narrowly defined jobs, and many of their talents and interests went untapped, because they had only a single task to perform. That paradigm didn't apply to Pathfinder. Things were more flexible. It actually was faster and better and cheaper. This was all new, and very un-NASA.

Not everyone at JPL took to this open-ended approach, but Jennifer did. She became more confident in her various roles, accustomed to change. After her experiences in Russia, she knew not to overreact to situations and to plug along until she found a solution or failed miserably. In time she developed an informal network of specialists and advisers she could trust, her go-to people. The Pathfinder cradle-to-grave approach helped a lot. People came on board at the beginning, when the hardware was delivered, and they stayed all the way through to the end of operations. On the typical NASA mission, the team responsible for delivering the hardware would say, "We've delivered our hardware on time," and walk away. If the hardware happened to be a camera, and it took pictures, they felt they had achieved their goal. They didn't care if it was impossible to operate or if it didn't get the right pictures.

But if you worked on Pathfinder, you had to undergo a mental shift. If you designed your component incorrectly, if it was difficult to test or to operate, it was still your problem.

It was difficult to explain the new thinking, Jennifer realized. You had to experience it for yourself, and then it could make a huge impact. You would become committed to the ultimate goal, whatever it was. In Pathfinder, the goal was to get to Mars quickly and cheaply, and to get a rover to function on the Martian terrain. Things worked in a sort of nonsystematic way because people attacked problems where they saw them. Eventually, they generated procedures, and she wrote the documentation, but this was not a document-heavy mission, like most NASA missions. She sat down with a couple of other people, and they asked, "What are the most likely contingencies? What's our nominal plan at the big-picture level?" She realized this could be a wonderful opportunity to participate in the exploration of space, and that idea pleased her greatly. "I feel like God has blessed me in my career," she once wrote, "and I would like to glorify Him by exploring His incredible creation." So the missionary had a new mission, but even as a scientist, *especially* as a scientist, she still devoted herself to God.

The Pathfinder mission originated in a speech given by President George Bush in 1989 to commemorate the twentieth anniversary of men—American men!—landing on the moon. NASA was in the doldrums at the time; and the occasion of the speech seemed to point up how little it had done since the halcyon days of Apollo. The Challenger disaster, which occurred more than three years before the anniversary, still loomed; when people thought of NASA, they didn't visualize Neil Armstrong jumping onto the surface of the moon, they thought of the faces of the parents of Christa McAuliffe, the schoolteacher who rode aboard the space shuttle, looking in disbelief at the Y trail left in the sky by the catastrophic explosion.

Along came George Bush, discussing the future of space exploration. The demoralized NASA contingent could scarcely believe what they heard. Did the president mention "the permanent settlement of space?" Yes, he did. Did he also say it was time to travel "back to the moon, back to the future, and this time back to stay?" Indeed, he said that, as well. But surely he could not have said, "And then, a journey into tomorrow, a journey to another planet: a manned mission to Mars." Yes! The president said that, too. *Mars*. The NASA bureaucrats began to ask themselves: how much was all this going to cost? No one thought you could go back to the moon and on to Mars for under 400 billion dollars; the tasks might require twice that amount. NASA's annual budget at the time was around 13 billion. Where would the money come from? Interestingly, few doubted that the technology existed to send people to Mars, or that it could be developed quickly; if NASA had the money, they could get the job done.

George Bush's remarks evoked John Kennedy's famous speech in which he charged NASA with the duty of sending men to the moon. Without realizing it, Bush tapped into the agency's other obsession, reaching Mars, an obsession that had begun in the mind of its ace rocket engineer, Wernher von Braun, during World War II. Von Braun, a member of the Nazi party and a favorite of Hitler's, had helped to design the V-2 missile. When he became disillusioned with the Nazi war machine, the Gestapo arrested him and sent him to jail. In his cell, he turned his attention to interplanetary travel, and Mars in particular. And it was in these strange and harsh circumstances that the kernel of what would become the American effort to explore Mars was born. In May 1945, von Braun and more than a hundred other German rocket scientists surrendered to the Allies. They were swiftly transplanted to New Mexico to continue their work on rockets, this time for the United States. The German V-2 became the prototype of a new generation of American missiles, and on the strength of his engineering accomplishments for the Nazis, von Braun quickly established himself as the chief architect of the American space program's booster rockets during the 1950s and 1960s; his designs were responsible for getting American men to the moon.

Throughout his career, von Braun was mesmerized by Mars. He published his plan to send people to Mars, the one he had conceived in jail, as a long magazine article titled *"Das Marsprojekt,"* which was translated into English. In 1953, it appeared as a book in the United States: *The Mars Project*. It became a classic, but this was not science fiction; *The Mars Project* contained no inspiring rhetoric about humankind's greatest adventure. It was a how-to manual, a master plan for getting people to Mars. He used a simple slide rule to make his calculations, and its pages contained his blueprint for the actual mission, using available technology. "The logistic requirements for a large elaborate expedition to Mars are no greater than those for a minor military operation extending over a limited theater of war," he wrote. The key to reaching Mars, he believed, was sending a flotilla of spacecraft. "I believe it is time to explode once and for all the theory of the solitary space rocket and its little band of bold interplanetary travelers. No such lonesome, extra-orbital thermos bottle will ever escape earth's gravity and drift toward Mars." Instead, in von Braun's vision, "Each ship of the flotilla will be assembled in a two-hour orbital path around the earth, to which three-stage ferry rockets will deliver all the necessary components. Once the vessels are assembled, fueled, and 'in all respects ready for space,' they will leave this 'orbit of departure' and begin a voyage which will take them out of the earth's field of gravity and set them into an elliptical orbit around the sun. . . . Three of the vessels will be equipped with 'landing boats' for descent to Mars's surface. Of these three boats, two will return to the circum-Martian orbit after shedding the wings which enabled them to use the Martian atmosphere for a glider landing. The landing party will be trans-shipped to the seven interplanetary vessels, together with the crews of the three which bore the landing boats and whatever Martian materials have been gathered. The two boats and the three ships which bore them will be abandoned in the circum-Martian orbit, and the entire personnel will return to Earth orbit in the seven remaining planetary ships. From this orbit, the men will return to the earth's surface by the upper stages of the same three-stage ferry vessels which served to build and equip the space ships." It was a grand scheme, and it became the template for NASA's plans to send people to Mars, a goal von Braun thought could be accomplished by the late 1970s.

Bush's speech endowed von Braun's dormant plan with new life, but the prospect of returning to Mars raised new questions as well. If NASA planned to send people to Mars safely, scientists needed to know much more about the Red Planet. If there was life on Mars, what form did it take? Was it dangerous to humans? Could it devastate the Earth if astronauts brought samples home? How severe were the effects of radiation? And, most important, was there water on Mars? The presence of water would dramatically enhance the prospects for finding life, but more than that, water meant it would be possible to manufacture rocket fuel, oxygen, and other human essentials on Mars.

Three years after Bush's speech, in 1992, NASA announced plans to send between twelve and twenty small landers to Mars. They would fly frequently, and they would take advantage of new equipment, especially computer technology, to explore more effectively. The new program went by the name of Mars Environmental Survey—MESUR, in NASA-speak. The agency then announced another planetary program, Discovery, with similar goals; it was a nice instance of the right hand not knowing what the left was doing. Eventually, the two programs merged into one trial program: Pathfinder. It was going to be fast, it was going to be cheap, but no one knew if it would be better than previous planetary missions. Unlike most NASA missions, which are built and often operated by a private aerospace contractor such as Lockheed Martin, Pathfinder was an in-house project, designed, built, and operated by JPL. It was meant to embody JPL's prowess as NASA's robotics center, and that posed an embarrassing problem.

It had been a generation since Americans had landed a spacecraft on Mars. The old guard was gone, and few around JPL or NASA remembered exactly how that trick worked. Some scientific data had been preserved, though not completely, along with thousands of Viking images, but there was little documentation of the mission's engineering accomplishments. Rob Manning, the young leader of Pathfinder's Entry, Descent, and Landing team, sought veterans who could tell him what they had done on Viking, but many had died, and others had retired. JPL pulled a few of the old grizzlies out of retirement to help assemble a unit capable of developing a lander, and they went to work under Manning.

The idea behind Pathfinder, to develop and build a new spacecraft on a drastically reduced schedule and budget to land on the surface of Mars, sounded like a losing proposition to many at JPL, given the risks involved in getting there. Just setting their ship safely onto the surface posed difficult engineering problems. The spacecraft travels at about 17,000 miles an hour as it reaches Mars. Then it must slow to nearly zero miles an hour so that it does not vaporize in the Martian atmosphere or crash into the surface like a meteorite. The Viking solution to this problem, an expensive and cumbersome one, employed powerful, heavy thrusters capable of guiding the spacecraft gently to the surface. There was no money for that kind of extravagance with Pathfinder. Instead, Pathfinder's engineers planned to wrap the lander in a protective bubble, place the bubble inside an aerodynamic cone, and parachute it through Mars' thin atmosphere to the surface, letting the cone peel off in sections. Then Pathfinder would bounce around the surface like a big hi-tech beach ball. If all these cushioning devices worked properly, Pathfinder would still be in one piece when it came to a stop. This follow-the-bouncing-spacecraft approach was profoundly troubling to conservative NASA engineers, but Manning casually accepted the risks. "Pathfinder is just a rotating bullet with nothing controlling it. This cone shape produces some unstable results—not so unstable that it's devastating, but you live with that." When he presented his landing scheme to NASA's review board, they were, he said, "skeptical—borderline hostile, as they should be. They were paid to challenge everything. So it was a big deal when we deviated from the Viking heritage."

Even if it landed safely, Pathfinder wouldn't sit still on the surface of Mars, taking measurements, as the Viking landers had. It would carry a rover designed to roam across the surface, functioning as a twelve-inch-tall geologist. This was not a new idea; for decades, NASA had explored the possibility of sending a rover to investigate Mars. "My most persistent emotion in working with the Viking lander pictures was frustration at our immobility," Carl Sagan recalled in 1980. "I found myself unconsciously urging the spacecraft at least to stand on its tiptoes, as if this laboratory, designed for immobility, were perversely refusing to manage even a little hop. How we longed to poke that dune with the sample arm, look for life beneath the rock, see if that distant ridge was a crater rampart. . . . I know a hundred places on Mars which are far more interesting than our landing sites. The ideal tool is a roving vehicle carrying on advanced experiments, particularly in imaging, chemistry and biology." He outlined, with his usual visionary fervor, a rover-based mission very much like Pathfinder. "It is within our capability to land a rover on Mars that could scan its surroundings, see the most interesting place in its field of view and, by the same time tomorrow, be there. . . . Public interest in such a mission would be sizable. Every day a set of new vistas would arrive on our home television screens. We could trace the route, ponder the findings, suggest new destinations. . . . A billion people could participate in the exploration of another world." At the time he wrote those words, they sounded like the vaguest hyperbole, but Pathfinder and the Internet would make his outlandish prediction a reality.

Although a rover seemed like a nifty idea, it was untried. The later flights in the Apollo program had taken along a dune buggy to traverse the powdery surface of the moon. The astronauts could steer and stop the rickety lunar flivver at will. The difficulties involved in guiding Pathfinder's rover across the surface of Mars *by remote control* seemed insurmountable. What if the rover didn't emerge from the beach ball after all that bouncing? What if it got stuck on a rock or a crevice or sank into the talcum-powder-fine Martian soil? What if the beach ball landed in inhospitable terrain? What if it landed on the wrong part of Mars, where it couldn't receive signals from Earth? And yet, if it avoided all these pitfalls and worked, the rover would provide a whole new paradigm for exploring the surface of Mars, because JPL had visions of building bigger and better rovers in years to come, until they reached the size of small trucks. But most people guessed a small rover would never work, not with the two million dollars allotted for its development.

A debate sprang up over the best way to control the rover, and, given the personalities involved, it quickly escalated into a dispute over technological theology. Tony Spear, a veteran engineer at JPL, believed the most reliable and cheapest way was to tether it to the mother ship. The other approach, advocated by Donna Shirley, was to control the rover remotely, but that meant designing or finding a new radio system, one that could tolerate the extreme fluctuations in the Martian environment, including fluctuations in temperature between the rover and the lander.

Donna Shirley was a controversial figure around JPL. When her name was announced as the Pathfinder mission director, a few cheers went up, but only a few; there was also consternation. Tony Spear, the Pathfinder project manager, was nowhere to be seen during the announcement, and Donna took his absence to indicate his lack of support. She could live with that. She thought the apparent indifference had to do with the fact that she was a woman, but she was accustomed to handling that problem. Donna had been with JPL since 1966, when very few women filled responsible posts there; during her years there, she married, raised a daughter, and got a divorce. At work, she was relentlessly cheerful, almost, but not quite, to the point of bullying, and she was a world-class talker. Many bureaucrats and scientists at NASA were camera shy, but when a television crew appeared at JPL, there was Donna Shirley in her bright red dress, flashing her assertive smile, prepared to discuss in her folksy Oklahoma twang just about anything. Her appearance was perfect for television. TV producers were delighted to interview the ebullient Donna Shirley instead of a pale male attired in the gray suit, gold-rimmed glasses, and neat mustache favored by the upper echelons at NASA. But, while being interviewed, she occasionally appeared to take credit for the accomplishments of a great many NASA scientists and engineers toiling anonymously, and that did not work to her advantage.

Her detractors said she really didn't know her science well, but she made her lack of expertise into an asset because she had no scientific agenda, nothing to prove. She was content to bang heads together cheerfully and say, "Look, guys, now we are going to do it *this* way." To the increasing number of women coming out of graduate school to work for NASA, she became a symbol. These younger women liked to tell a story about the time Donna Shirley attended a launch party at Cape Canaveral. As usual in those days, she was the only woman present. A guitarist singing a bawdy song, accompanying himself on the guitar, stopped dead when he saw her. She took his guitar and completed the song herself, delighting everyone. That was great, as far as things went, but she didn't realize there was a tradition at these launch parties that a woman—a hooker, basically—was paid to show up and pull a stunt like that. One of the men assumed Donna had been hired for the occasion, maneuvered her into an alcove, and grabbed her. "I didn't exactly deck him," she said, "I just hit him on the nose."

Working on Pathfinder, she saw her team of engineers and scientists as a large family, her family. To her credit, she encouraged everyone to talk to everyone else, if only in self-defense, and she always smiled and radiated optimism. Most found it impossible to bear a grudge for long in the face of such cheerfulness; it was too exhausting to oppose her. Still, she wanted her radio-controlled rover for Pathfinder, and Tony Spear, the project manager, did not. "In his position, I wouldn't either," she said, "because he had the impossible job of landing on Mars for a fraction of what it cost the last time we landed. He had no idea how to do it, and here's this parasite coming along, giving him nothing but trouble. What I did was to convince the scientists that we really could do useful work with the rover. That was number one. Number two was to convince Tony that we really could fly without damaging his mission." When Donna presented her case to NASA's review board, one member, Jim Martin, the former Viking project manager, insisted a Mars landing could not cost less than Viking had. As for the rover, "he thought it was terrible." Donna and the rover team persisted, building better iterations of the rover and demonstrating they worked as advertised. "It became a very powerful selling tool," she realized, and eventually, to everyone's surprise, it turned into the mission's raison d'être.

If Pathfinder's engineering was, ultimately, carefully weighed, the mission's science component tended to be rushed, improvised, an afterthought. Plenty of scientists were eager to participate in the new Mars mission, but they needed time and money to formulate, conduct, and analyze experiments. Pathfinder didn't work that way. At the last minute, for instance, NASA stuck a couple of stereographic cameras on the lander and another camera on the rover. These weren't your standard television cameras; they used a technology known as a Charge Couple Device. The CCD reproduces light very accurately and is especially useful for spectroscopy, which reveals more than the naked eye can see by measuring which wavelengths of light are absorbed, and which reflected, from an object. They were useful, but they were not capable of sending back the sparkling, gorgeous images returned by Viking twenty years earlier. Pathfinder also carried an Alpha Proton X-ray spectrometer to detect the composition of Martian rocks and a weather mast to measure the Martian temperature and atmospheric conditions. Every so often, Pathfinder would collect the weather mast's data and return it to Earth, so for the first time it would be possible to obtain accurate weather reports from the surface of Mars. Everyone agreed the weather mast would be a terrific experiment, if it worked. It looked like Pathfinder had a chance to become a real mission, after all.

Manning's team conducted early Pathfinder landing tests at a NASA facility in Cleveland, Ohio, which featured a large vacuum chamber. Within, girders, lava rocks, and wood simulated the Martian surface. They dropped Pathfinder in its protective bubble onto the sharp objects and observed the result.

R-r-r-r-r-rip!

"The first time we did it, we had a tear the size of a human being," Manning said. They took it back to the lab, fixed it up, and dropped it again.

R-r-r-r-r-r-rip!

They tweaked it and tried again. *R-r-r-r-r-r-rip!* . . . *R-r-r-r-r-r-rip!* . . . *R-r-r-r-r-r-rip!*

The trials went on like that for months; they were "total disasters," said Manning, and NASA nearly canceled the mission. Late in 1995, the Pathfinder team redoubled its efforts. The engineers adjusted the spacecraft's small guidance rockets. They modified the shape of the sphere contained inside the protective beach ball. They had been imitating the Russian model, which was spherical and consequently difficult to manufacture; now they adopted a tetrahedron, which was easier to manufacture. They toyed with the air bags protecting the tetrahedron, trying one deflation strategy after another, getting incremental improvements. Gradually, they came to feel more confident about Pathfinder. They did have one advantage: because the gravity of Mars is less than half of Earth's, the spacecraft would endure less wear and tear. "We always worked in terms of the mass, and the mass kept getting bigger and bigger," Donna said. "That meant the mechanical parts had to be heavier because they were supporting all of this additional structure. The mission design people came to the rescue. They said, 'Okay, if we're going to fly into the atmosphere of Mars, there's a corridor we have to hit. If we go in too shallow, we'll just skip out of the atmosphere and keep on going. If we go in too deep, we'll burn up on entry, or we won't have enough atmosphere to slow down before we hit the surface.' So there's a narrow range of angles at which you can enter the atmosphere, and that takes some really accurate shooting by the navigators. So the navigators heard this and said, 'Okay, if we can shoot more accurately and give up some of our margin for error, we can let the spacecraft have more mass.' " Now the engineers were able to add small thrusters that would slow Pathfinder during its descent to the surface.

The mission was still alive, but the development of a decent, affordable rover still posed engineering problems. JPL had to devise a nimble mechanical creature that could scale small barriers and climb over rocks, like a little tank. To complicate matters, it would take twelve or fifteen minutes for a radio signal to travel from the Earth to Mars, which eliminated spontaneous, real-time commands. "If you're looking through the rover's eyes, and you see a cliff coming, and you say, 'Stop!' it's too late—it will be over the cliff, so it has to be smart enough to stay out of trouble," Shirley said. In addition to negotiating the Martian terrain, which was in many details unknown, the rover had to keep its solar panels in position to receive sunlight, or it would lose power and die.

Attempting to meet these requirements, JPL devised variations on a theme. They built a rover the size of a small truck, and they built one just eight inches long, nicknamed "Tooth." They built a midsize rover called "Rocky," which, when tested in the desert, actually did things required on Mars, such as scooping up soil. Rocky went through various iterations until it weighed just fifteen pounds, yet negotiated the kind of obstacles and terrain that geologists expected to find on the surface of Mars. It could perform simple experiments, and it appeared sturdy enough to withstand the rigors of landing on the Martian surface and bouncing around inside a beach ball.

The development of Pathfinder's components took place in a knowledge vacuum, because the engineers and scientists didn't know exactly where they were going on Mars or what to expect when they got there. From a spacecraft's point of view, Mars presents a landscape of treachery. The team expected to receive finely detailed studies of the surface from Mars Observer, the billion-dollar spacecraft launched on September 25, 1992. It was supposed to reach Mars the following August, when its cameras would send back pictures of the Martian surface with much higher resolution than Viking had captured in the 1970s, and those pictures were supposed to give JPL a well-informed notion of where to land their bouncing beach ball. Just when Mars Observer was to begin orbiting around the Red Planet, JPL lost the signal, and the spacecraft was never heard from again. There was speculation that a fuel line had frozen and ruptured, and the spacecraft went out of control, but nobody could say for sure—nobody, that is, but fringe elements, who concocted some fairly creative theories. There was the "Hey! That was no accident" scenario: NASA deliberately destroyed the spacecraft because it had detected signs of intelligent life on Mars. And there was the "Mad Martian" scenario: Mars Observer had been destroyed by sophisticated Martian weapons whose existence NASA conspired to conceal from the American public.

Within NASA, scientists feared they had lost their chance to return to Mars. Shortly after Mars Observer disappeared, Dan Goldin journeyed to the Goddard Space Flight Center in Greenbelt, Maryland, to rally the troops. Although Goddard is only a short commute from NASA headquarters in

Washington, D.C., the head of NASA is not in the habit of dropping in, so his presence signaled a major announcement. For many scientists, it was their first close-up look at the man whom George Bush had appointed in 1992 to run the agency. At Goddard, he reminded the scientists that NASA attempts to do difficult things, risky things, and the possibility of losing a spacecraft was an ever-present hazard, but the risk didn't mean the mission wasn't worth doing. They would continue to explore Mars. Conditioned to regard managers as antagonists, the scientists were impressed.

The loss of Mars Observer meant Pathfinder's site selection team was forced to rely on twenty-year-old Viking data. Since Pathfinder was designed to plummet to the Martian surface, it would not be able to pick and choose a landing site as Viking had. The site would have to be selected in advance, and it had better be good. A lot of the responsibility for selecting a site fell to Matt Golombek, a young geologist. If you can recall the kid in the seventh grade who always seemed a couple of steps ahead of the teacher, let alone the class, and who was wiry and agile and had a way of laughing off anything that bothered him, you have a sense of Matt Golombek. He came to the agency from Rutgers and the University of Massachusetts as one of the new generation of planetary geologists that included Maria Zuber and Jim Garvin. "You only do this because you love it. It's not like you're going to get rich or famous. You're especially not going to get rich," he says. Although he reports to work at JPL, which is a government facility, he is, like everyone else there, technically employed by Caltech. It's a peculiar arrangement, which he facetiously likens to a "money-laundering scheme to lower the number of civil servants." Matt maintains a certain skepticism concerning government work. "You know what they say about civil servants, don't you? They're like rusty old guns. They don't work, and you can't fire them."

Despite his youth, Matt brought with him long experience in Mars exploration. "I was the pre-project scientist on all the Mars missions before Pathfinder for ten years, and there was a whole string of them. I was brought in originally with something called the Mars Rover Sample Return, which was actually a politically motivated study to work with the Russians, which didn't go anywhere." This was followed by assignments on other luckless missions, including Mars Observer. "They assigned me to Pathfinder as the project scientist because I was young. Part of their thinking was, 'Well, it doesn't matter who we appoint. It's not going to mean anything.' I wasn't sure I even wanted the job, because the mission was an entry, descent, and landing demonstration that would have little or no science of benefit to anyone. *What the hell do you need a project scientist for? There's no science, right?* I mean, Pathfinder's main goal was to land safely, period."

To achieve even that limited goal, he spent two years mastering every detail of the choices before making his recommendation. The pixels in the old Viking images concealed many potential hazards. "Imagine if you looked at an image to select a potential landing site, and the smallest you see is the size of a football stadium, and you are worried about things that are the size of a meter," Matt said. "All we had was very coarse, low-resolution remote sensing information about Mars, yet we had to guesstimate that the place we would come to rest would be safe, and that the rover could travel out on it. That's a very difficult job. It was a two-and-a-half-year process. We did an exhaustive study of the options, of cost, and of the kind of science you could get at different places." He had to factor many subtle requirements into his choice. He looked for a spot where Pathfinder's solar cells would supply power, and where the antennae could communicate with Earth as often as possible. He wanted an area free of mesas, which would confuse Pathfinder's navigational system. Those and other constraints eliminated ninety percent of the surface of the planet. Geological factors eliminated a number of other tempting targets; if an area was too dusty, too cavernous, or too rocky, it was eliminated from consideration.

There was something else on his mind. What was the point in going all the way to Mars only to land in a dried-up, featureless lake bed and watch the rover go round and round? Why not use the tools they were bringing, the Alpha Proton X-ray spectrometer and the cameras? Why not make Pathfinder a *science* mission as well as an *engineering* mission? "Wait a minute," he told anyone who'd listen, *"we can actually do science."* Perhaps the mission would need a project scientist, after all. Matt saw his chance to push against the system and work with the engineers to make room for science. For Pathfinder to accomplish anything significant, it would have to land in a place with attention-grabbing rocks—rocks that would speak volumes about Martian geological history, especially the presence of water, rocks that were big, but not too big. He didn't want boulders, for instance, and he didn't want pebbles, either. He wanted, so he said, a "rock mission." He wanted a "grab bag, a smorgasbord, a potpourri of rocks." He wanted sermons in stone.

Everyone at JPL recognized that Matt was a very good scientist. Now he demonstrated that he was a very good scientific operator as well. His gift for caustic repartee concealed considerable shrewdness; depending on his purposes, he could be engagingly cynical, or firm and cool. He was persuasive with his colleagues, lacing his remarks with irony, imparting to all those around him the intoxicating sense that they were being drawn into some grand cosmic joke. Nothing intimidated him, least of all NASA's bureaucracy. It was susceptible to lobbying and influence if you knew where to apply pressure, which came down to motivating people to do something different. "The hardest part of going to Mars," Matt once told me, "was getting everyone working on Pathfinder to march in the same direction."

Unlike most scientists, he was good with the engineers; he appreciated the difficulties they faced. Scientists and engineers often develop adversarial relationships: scientists usually display scant patience for the difficulties of building and operating the instruments, and engineers tend to regard scientists as impractical, arrogant, impossible to please. Stepping into the midst of the fray, Matt pushed back on the scientists, knocking down the number of experiments, and he convinced engineers they could do things they wouldn't have thought possible. That was a formula for a very successful manager of space science. "You almost have to turn yourself into an engineer," he said, "because you have to understand what your spacecraft's doing. Your dominant job as a project scientist is to make sure they don't engineer the science off the mission. It's not that engineers are dumb—they're doing the best they can—but they don't necessarily think about science. And so you sit through interminable meetings waiting for the one silly thing that will pop up and threaten the science. I mean, it's *crazy!* The other aspect, once you get the mission going, is that you have to lead the science team. You have to show them where you're going. What's really important? How do you allocate resources? How do you keep people's egos from getting in the way? That's very tricky."

He became adept at building a consensus around the selection of a site. He led a site selection workshop at the Johnson Space Center in Houston, fielding ideas from the entire Mars community. They whittled the choices down to about ten, which Matt put on a large, complicated diagram called "The Chart from Hell." After much study, Matt, working with another geologist, Hank Moore, concentrated on a Martian basin named Chryse Planitia—Chryse Plain. Within Chryse there is an outflow channel called Ares Vallis, the geological legacy of a huge, ancient flood that deposited interesting and varied rocks on the surface. The diverse rocks were the greatest attraction, as far as he was concerned. The area's sheer size made it very appealing. Pathfinder, in addition to all its other uncertainties, could not make a carefully predetermined landing; if all went well, it would land somewhere within an ellipse sixty miles wide and twice as long. Matt fretted over the temperature range of Ares Vallis, over the distribution of rocks, and especially over the amount of dust blowing around. If you've ever come into contact with terrestrial lava dust, you immediately understand the problem. It's gritty and irritating and clings to the fingers. Martian dust, made from powdered lava, is similarly fine and gritty. It relentlessly clogs machinery and obscures solar panels.

To get a better idea of what Pathfinder might encounter if it landed in Ares Vallis, Matt used an Earth analogue—not Iceland, in this case, but the Channeled Scabland in the state of Washington. This desolate region was formed during a huge flood about 13,000 years ago; the turbulent water redeposited rocks across a flat plain, just as Matt believed had once occurred in Ares Vallis on Mars. The Channeled Scabland is much smaller than the Martian site he was considering, and tufts of grass spring from the soil, but geologically, it is remarkably similar to Ares Vallis. He took several field trips to the Channeled Scabland, and he brought along the rover to see how it would fare on the rock-strewn terrain. It ably negotiated the varied surface, and he figured he had finally found his landing site. Ares Vallis was safe, it was geologically interesting, and it was, he hoped, not too dusty.

NASA's review panel considered his choice. "You're going to go back there and kill the spacecraft—and kill your career," they said, but Matt would not back down. He had seen the results of Pathfinder tests in even worse conditions than it would encounter on Mars, and the spacecraft had survived. "We ended up making the most robust lander that's ever been designed to land on a planet. Pathfinder could land anywhere," he told the panel. In the end, Matt got his way, and his landing site on Ares Vallis, but his career would ride on Pathfinder's fortunes.

Chapter Four

FROM OUTER SPACE
TO CYBERSPACE

The Kennedy Space Center in Florida employs sixteen thousand people and covers more than a hundred thousand acres. It includes a wildlife refuge; herons, egrets, condors, crocodiles, horses, and cattle roam its expanses. From the road, you can see a six-foot-wide bald eagle's nest suspended in the branches of a tree. If you drive in the general direction of the launch pads, away from the animals, you will see the outsized Vehicle Assembly Building, a giant hangar for rocket ships, shimmering through the haze. You can pick out the odd shuttle transporter here and there; they resemble huge, primitive locomotives with giant cleats. Many of the transporters are rusting in the humidity. In recent years, NASA has built the center into a tourist attraction featuring a life-size mockup of the space shuttle, a few garish exhibits, and a souvenir emporium. On the outskirts of Titusville, the nearest town, discarded rockets litter front yards like so many abandoned cars, looking nothing like the towers of power that I remember from my youth. A heavy nostalgia for the future lingers over the place like the scent of magnolia on a humid evening, and the marquee in front of the local high school always reads, "Countdown to Graduation—Six Weeks."

This was where Pathfinder, weighing nearly a ton, arrived on a rainy day in August of 1996. There was a lot of work to do. Things were just beginning to get serious at this point. First thing, engineers wiped down the spacecraft with alcohol to prevent bacteria from Earth contaminating the surface of Mars. Then, electrical technicians wearing bunny suits to prevent dust or hair falling into the delicate machinery took the spacecraft into a large clean room and tested every circuit. They corrected software problems and installed the rover's heaters, which contained a tiny amount of plutonium. NASA wasn't eager to advertise the fact, for the use of nuclear materials in space, even for purely scientific purposes, rouses environmentalists to fury. But the plutonium was deemed necessary because sunlight on Mars is only a quarter of the strength that it is on Earth, and small solar cells alone could not generate enough power to operate even a small spacecraft and rover.

After the initial preparation, Pathfinder underwent months of additional testing at Kennedy. Often, the tests were more complicated than the actual mission would be. For a test to work correctly, the ground team had to simulate the positioning of the stars and the Sun, the amount of light, and the temperature for Pathfinder, and then program Pathfinder to respond. Nothing went exactly as planned; everything required extra effort. Work became so intense that the young engineers involved with the project didn't know what to do when they weren't testing Pathfinder; they sought any distraction available in greater Titusville. They screamed themselves hoarse in a karaoke bar, they played in mud volleyball tournaments, they surfed, they picnicked in the rain—anything to take their minds off the obstacles they faced to ready Pathfinder for space. At four o'clock one morning, they attended a shuttle launch. The rocket's glare turned night into day, and the sound of its engines was powerful enough to make observers' clothes tremble. Pathfinder's launch vehicle, a Delta II built by Boeing, was nowhere near as big as the shuttle's powerful solid state boosters, but the shuttle simply attains low Earth orbit, about 350 miles high. It circles the Earth for a few days, and then lands on a nice smooth airstrip. If you happen to be a planetary exploration zealot, shuttle missions, for all their sound and fury, can be a trifle dull. Pathfinder, in contrast, would travel 309 million miles to reach Mars, following the broad ellipse of its trajectory, and would arrive in a new world.

Whenever they could find a few minutes to spare, Pathfinder's youthful team members posted their field journals on the Internet. These were casual, subjective reflections on their lives and work, with more questions than answers. The mere act of writing made everyone self-aware. It was weird: they were doing their jobs, and simultaneously, they were watching themselves do their jobs. Anyone with access to the Internet could log on and check up on the team members' psyches. As with so much else on this mission, the plan was very cool, and very un-NASA. Taken together, the team's written observations sounded like an all-night college dormitory bull session about the meaning of God and life and truth and beauty, and that was their charm. The team members, especially the younger ones, unashamedly asked, *Who am I? Where am I going?* The field journals became confessional, a form of therapy.

The more enthusiastic correspondents realized something unusual was going on here, something that extended beyond the boundaries of the Pathfinder mission, something that the words "faster-better-cheaper" didn't begin to convey: a *transformation of consciousness*. They weren't just devising a new way to reach Mars, although that was surely foremost in their minds. They were collaborating on a new way to solve problems, to create, to communicate, to imagine.

In late October, the engineers at Kennedy mated the spacecraft with the cruise stage of the Delta rocket, and loaded fuel into the spacecraft's propulsion system. It was powered by hydrazine, nasty stuff that requires careful handling. If you touch hydrazine, it can burn your skin. If you spill it, it can start a fire. The engineers wore protective suits very much like an astronaut's while they worked. In this instance, the fueling process proceeded safely. Soon after, technicians tested the Deep Space Network (DSN), the system for communicating with Pathfinder. The DSN consists of three ground stations—one in Goldstone, California, another in Madrid, Spain, and a third in Canberra, Australia. Exquisitely sensitive, the DSN can pick up exceptionally faint signals from spacecraft as far away as Jupiter, and possibly beyond the boundaries of the solar system. In November, the team started holding practice countdowns. The launch was approaching rapidly; now it was days, not weeks, away. Back at JPL in Pasadena, a sign read, "OBJECTS ON THE CALENDAR ARE CLOSER THAN THEY APPEAR." It was a message that everyone on the Pathfinder team had learned to heed.

The closing of the spacecraft, supposedly a modest episode in the life of Pathfinder, inflated into a media event. Dan Goldin, the NASA administrator, showed up at the Kennedy Space Center and gave a rousing speech. Crowds turned out to watch Pathfinder's four petals fold shut around the rover. As everyone applauded, the engineers glimpsed a sliver of daylight between the petals. This was not a good omen: to close a spacecraft in preparation for launch, only to find that the pieces don't fit. The event had been scheduled for television broadcast, but the engineers waved away the cameras so they could study the problem. This was the first time the spacecraft had been fully loaded, and the increased weight created structural sagging, which kept the petals from sealing as designed. During the next several days, engineers worked desperately to repair the problem, and couriers bearing modified parts flew in from JPL. "OBJECTS ON THE CALENDAR ARE CLOSER THAN THEY APPEAR." One by one the new parts were mated, and the entire assembly was stacked on to the launch vehicle and trucked over to Launch Pad 17A at Cape Canaveral Air Station.

The launch was scheduled for December 2, and it proved to be an exercise in frustration. "The weather was so bad they decided to cancel the attempt for the day," Donna Shirley wrote in her field journal. "On December 3, many of us went out to the launch pad to watch the gantry roll back. This was supposed to happen at 5 P.M. but didn't actually occur until 7:30 P.M. It got colder and colder, and there was a prelaunch party scheduled. A few diehards, including me, were all that were left to see the rocket standing free of the supporting structure. It was worth waiting for, shining in the spotlights, gleaming blue and white." For a few hours that night, the countdown proceeded smoothly, but problems started to mount. "First, the winds aloft looked bad," Donna recorded. "The range sent up balloon after balloon to see what the winds were like, and gradually they began to improve. By the fourth balloon they looked acceptable, and we all began to get excited. But there was another problem. One of the ground computers was having problems. After much discussion, the launch vehicle team decided to change to a backup computer. But about two minutes before launch time, that computer also had trouble, and the launch was scrubbed. Everyone sagged. We'd been running on adrenaline, not a bit sleepy, but once there was no launch everyone went home to bed." Bridget Landry, a Pathfinder software engineer who'd been following the tortured countdown from Pasadena, also turned to her journal for consolation: "We were all so disappointed when they said we had to scrub! All that anticipation! In some ways, it's funny, all that buildup, and then nothing happens. But it's also scary: the Russian mission, Mars '96, was unable to escape Earth's gravity just a few weeks ago. Somehow, that makes us worry more about our launch. Guess scientists and engineers can be a little superstitious, just like everyone else."

On December 4, with all conditions favorable, Pathfinder finally launched from Cape Canaveral at 1:58 A.M., local time. Although the blast was visible in the night sky for miles, there were no throngs along the beach.

The event was the merest blip on the news radar; the public remained mostly oblivious to the fact that we were returning to Mars. Pathfinder was just another planetary mission, for the time being.

After launch, the engineers at JPL had a rough idea where Pathfinder was located in space, but now they needed to know precisely where it was. To get a better fix on the spacecraft, they planned to establish contact with it through the Deep Space Network. But before they could do that, the spacecraft was programmed to orient itself with a device called a sun sensor, a small disc covered with photoelectric cells. The event was supposed to occur about ninety minutes after the launch, with the spacecraft traveling at 17,000 miles an hour, but the sensor wasn't working properly, and if it failed, they would soon lose the spacecraft entirely. As everyone involved with the mission knew, if Pathfinder failed, the future of the entire Mars program, including a human mission, would be in grave doubt.

After extensive searches, the Goldstone, California, antenna of the Deep Space Network finally acquired a signal from Pathfinder, which confirmed that something was seriously wrong with the spacecraft: the sun sensor wasn't returning data. The best guess was that the sensor's photoelectric cells had been nearly blinded by exhaust from one of the launch rockets, and Pathfinder refused to pay attention to it. The fix was simple, in theory: send three new software files to command Pathfinder to tune into the sensor. The engineers transmitted the files again and again, but each time they were only partially received, and no one could say why. The situation, already very serious, deteriorated when the spacecraft began to ignore all commands. Eventually, one of the engineers realized that Pathfinder was revolving slowly and came in position to receive commands for only five seconds at a time. Now Pathfinder had two problems—the dirty sensor and the uncontrollable spinning. The double fault was likely to be fatal to the entire mission.

Working around the clock, the Pathfinder engineers compensated for the rotation, and transmitted a complete set of commands to the spacecraft. Once they did, Pathfinder paid attention to the sun sensor, oriented itself properly, and both problems disappeared. The process of resolving them, collaborative and critical, proved to be a rite of passage for the Pathfinder team. Although they were relieved to have fixed a problem that could have killed the mission, they knew that more could appear at any time. Pathfinder was single-string all the way; if a component failed, there was no backup.

In the ensuing days, a weird sense of calm descended on JPL. There wasn't much anyone could do for the next ten months besides monitor the spacecraft, whose instruments were powered down for the long cruise. "I've always known that the spirit on Pathfinder was special," Bridget Landry noted in her journal, "but when the people who worked on Apollo 11 and 13 say this project has more sense of identity and team spirit than even those two missions, you know you're involved with something extraordinary. But the feeling here, at least for me, is bittersweet, too. Now that we've launched, some people are being laid off, and even though most of us are staying, the scope of the mission means that in less than a year, all this will be over."

In the spring of 1997, as the spacecraft approached Mars, the Pathfinder team began a new series of tests to prepare for its prime mission—the weeks Pathfinder would spend on the Red Planet, roving and returning data. "Think of it as a rehearsal," Bridget Landry's journal explained. "We have a computer that simulates the spacecraft, as well as a model of the lander and a duplicate rover. We put the last two in our sandbox (a room full of sand and rocks used to simulate the surface of Mars), then close the curtains so that no one can see in, and a few people go in and rearrange the rocks. Then the operations team has to take pictures with the lander camera, determine where the rocks are, and generally do all the tasks we'll do on the first two days on Mars."

They tried to maintain a sense of humor; any joke was better than no joke. The technician who ran the sandbox—the gremlin, they called him— planted toy Martians amid the simulated rocks, and the controllers logged the little creatures as they would in an actual mission. The tests increased in frequency until they took place nearly all the time, and Bridget, for one, felt overwhelmed by the intensity. "I refer to this as 'a snake swallowing a gopher': it is an enormous amount of data/work/whatever, and it can be tracked visibly as the lump makes its way through the system."

By this point, the team members had left terrestrial time behind; they lived on Martian time. A Martian day, or sol, is forty minutes longer than an Earth day; every twenty-four hours, the team migrated forty minutes around the clock to remain in synchronization with the Red Planet. The time shift was very disorienting; after a while, they began to feel as though they were living on Mars. "I had no clue as to what time it was on Earth, although I could tell you what sol it was," Bridget recorded. "I even tried to date a check with the sol number during this time, and I dated leftover food in the refrigerator with the sol. I totally lost track of time. I couldn't tell you what planet I was on, never mind what day it was. I was often surprised to walk out of JPL and find the sun up." While she worked the graveyard shift, she found it "strange to be driving home under rosy skies, pulling the pillow over my head to shut out the morning light. Stranger still to hear meetings called for midnight or 2 A.M., and having to ask whether an event scheduled for 6 means A.M. or P.M." One night, while driving home on the freeway, Bridget fell asleep at the wheel. There was no accident—no one was hurt, fortunately—but the mishap terrified her. After that, she kept a sleeping bag and pillow in her office so she could sleep there, if she needed.

Bridget's role on Pathfinder changed along with her schedule, and throughout the project, she never knew what to expect. Her first job was straightforward: adapting sequencing software to Pathfinder's commands and instruments, but then she worked with the imaging team in a job so idiosyncratic there wasn't even a name for it. For official purposes, she called herself an uplink system engineer, but the title didn't begin to describe the nuances of the position. After a while, she came to see the benefits of this elastic, improvisational approach; it was liberating, as long as she could keep up with it. "I think this is the main reason we succeeded—there was so little redundancy in staff and hardware that you *had* to understand not only your little piece of the puzzle, but how it all fit together. There was no one else to do it. If people couldn't think this way, they left." She managed to hang on, but found her new job staggeringly intricate and pressured. If just one system on which she worked malfunctioned, the entire mission could go haywire. "I always mean to go back and clean up the code I write, but there is only so much time and money and energy—you do the things that matter most, until it's good enough. Striving for perfection is a good and worthwhile effort. Expecting actually to attain perfection can kill you."

What started out as a cool new job—but still, just a job—became much more than that to Bridget and to the other team members. Most were young—this was their first big mission, in some cases their first job at NASA—and while they were confident of their technical skills, which was the reason they were hired in the first place, they were less certain they could manage the emotional strain. That was something they didn't teach you back at Caltech or MIT: how to deal with failure and uncertainty. "This is a nerve-wracking experience," Rob Manning confided in his journal. "It is really tough to go to another planet." Normally a jovial soul, Rob endured nightmares. In one, "I launched my dog Scooter up in the spacecraft, and it landed on Mars. I realized, too late, how was I going to get him home? I was really upset for letting him go!"

On Monday, June 23, David Mittman, the new Pathfinder flight controller, awoke at three o'clock in the morning and showered in the dark so as not to disturb his sleeping family. He had managed a few minutes extra sleep by shaving the night before. He dressed, and drove to work at JPL. The place was already alive; June 23 was going to be a crucial day in the life of Pathfinder, which would land on Mars on July 4. "All the flight engineers have been developing 'sequences'—collections of commands executed in order by the spacecraft—in preparation for landing," he wrote. "These sequences have gone through many reviews and revisions. Some sequences have been changed as many as twenty times as we've learned more about how to operate Mars Pathfinder." Mittman and his team were about to send nearly 300 sequences to Pathfinder via the Deep Space Network's transmitter in Madrid. The task was time-consuming, because each sequence, traveling at the speed of light, as fast as anything can go in the universe, required nine minutes and forty seconds to reach Pathfinder as it sped toward Mars. The transmission proceeded smoothly for nearly twelve hours, until glitches in terrestrial equipment halted the proceedings.

The next day, the Pathfinder team agonized over the rover. They knew that getting the twenty-five pound, six-wheel rover down the ramp from the mothership and onto the Martian surface would be fraught with hazard. "First, we'll look at the tilt of the ramp to make sure it's not too steep for the rover to drive down," team member Matt Wallace wrote in his journal. "Then, we'll look to make sure the ramp isn't twisted, because if it is, the rover could fall off the sides. We'll check to make sure the air bag material has properly retracted. If it puffs up and gets around the sides of the ramp, there's a potential for the rover's wheels to snag on the material. We'll have to look for rocks down at the base of the ramp to make sure that once the rover actually gets down on the surface, it can go somewhere." They ran simulations in the JPL sandbox, and for once, things went according to plan. The team began to gain confidence that, despite all obstacles, they could actually pull this mission off.

The primary mission, consisting of their top-priority, do-or-die tasks on Mars, was scheduled to last just seven days; after that, the Pathfinder team could claim success, technically, but they prayed the mission would last much longer than that. If they could get to seven days, they figured they could get to thirty, which would be very cool—thirty days with a rover on Mars. After that, they expected to encounter severe battery problems. Pathfinder's batteries were not rechargeable—again, to save weight. Heating the rover sufficiently to survive longer than a month on Mars would be problematic, but if by chance it endured the cold, there were plans for a full year's operation. At this point, no one knew how long the mission would last.

On June 25, the team prepared Pathfinder for its fourth and final trajectory correction. They pinned down the spacecraft's location and velocity as it sped toward Mars. They calculated distance by ranging—sending a signal to the spacecraft and seeing how long it took to return. For velocity, they relied on the Doppler shift. "Think of the spacecraft as a train whistle that's producing a tone," Rob Manning explained. "As the train comes toward you, you hear a higher pitch. After the train passes by you, it produces a lower pitch. This is called the Doppler shift. We can use the Doppler shift to tell us how fast the spacecraft is moving away from us. With the combination of Doppler and ranging—ranging tells us how far, Doppler tells how fast—we can figure out exactly where the spacecraft is."

It was more difficult to determine precisely where Mars would be when the spacecraft was scheduled to land on its surface at 10 A.M. Pasadena time on July 4. The calculations were complicated by Jupiter's gravity, which distorts Mars' orbit. The gravitational pull of one celestial body on another is called perturbation; astronomers rely on it to infer the existence of celestial bodies. But perturbation plays havoc with exact calculations of orbits and trajectories. There can be as much as eighteen kilometers' leeway in Mars' orbit, for example. Depending on the effects of perturbation, Pathfinder would land as much as a hundred kilometers in one direction or another within the ellipse selected by Matt Golombek as the landing site.

The trajectory maneuver involved less than .03 meters per second change in velocity. "Maneuvers are performed in one of two modes: turn and burn, or vector," Pieter Kallemeyn, a Pathfinder navigator, explained in his journal. "In the turn and burn mode, the spacecraft is turned in the direction we want the velocity change to occur. A pair of thrusters is fired for a predetermined period, which results in a velocity change. A vector mode is performed in two parts: axial, which is along the direction of the spin axis, and lateral, which is roughly 90 degrees from that." After the maneuvers, the team awaited the result from the Deep Space Network. They had worked perfectly, and Pathfinder's luck continued.

Hours before landing, the navigation team concluded that Pathfinder was on target, traveling at about 15,200 miles an hour, and heading for the center of the ellipse. Its angle of entry was off by only three-quarters of a degree; the amount was so small they decided to forgo another burn. "This is the equivalent of playing a round of golf in which the teeoff is in Pasadena and the pin is in Houston," Kallemeyn said. "We're basically hitting a hole in one."

They took last-minute bets on the odds of the mission's success. Rob Manning, the most optimistic team member, said he was ninety percent convinced it would work, but Jennifer Harris, who knew Pathfinder as well as anyone did, wasn't so sure. Simply maneuvering Pathfinder into the correct trajectory would not guarantee a successful landing on Mars, she knew. During the four-and-a-half-minute entry, descent, and landing sequence, the spacecraft was programmed to execute forty-one intricate pyrotechnic events designed to blow the hatches and bolts and then release the parachutes. Each event had to occur at precisely the right moment, or the mission would fail.

On the eve of Pathfinder's descent, the team members at JPL—a hundred scientists, engineers, and managers—each contributed a dollar to a pool. Everyone selected a place on Mars where they thought Pathfinder would land and marked the spot with an orange sticker on a large black-and-white map pinned to the wall in Matt Golombek's office. Orange dots ranged across the entire map; one guess was as good as another.

The same day, Dan Goldin, chastened by the uncertainties surrounding the mission, warned, "People have to be grown-up enough to understand that bold things, like Pathfinder, run risks. I want my people to try, and if they fail, learn from their mistakes and try again."

Bridget Landry wrote in her journal: "I . . . really . . . hate . . . waiting."

The following morning, July 4, 1997, Pathfinder plunged through the Martian atmosphere as planned, the heat shield fell away as planned, the parachute deployed as planned, the forty-one pyrotechnic events fired as planned, the air bag cushioned Pathfinder as it bounced thirty feet high and continued to bounce about twenty times across the Martian landscape until it came to rest right side up—all exactly as planned. Pathfinder's petals opened, and the transmissions began. Just like that, the thing worked. The surprised and ecstatic engineers and scientists at JPL learned about the local conditions they would face during the mission, and the news was promising, better than expected, in fact. Pathfinder was tilted at an angle of only two degrees, with the high-gain antenna pointing almost exactly at Earth. None of the team members guessed the exact spot where the spacecraft finally came to rest on Mars, but it was well within its large landing ellipse, in the midst of a field of varied medium-sized rocks, each with its own geologic story to relate. This was exactly the setting Matt Golombek had been predicting all along, and he savored his vindication. "We've got a mission!" he cried. "The landing site choice consumed me for two and a half years," he said later. "That was, without a doubt, my greatest satisfaction. We just nailed that site." The exhausted young engineers at JPL remained at their consoles receiving their first data from Mars: no pictures yet, just simple confirmation that Pathfinder was healthy, and that we were back on Mars.

Within eight hours of landing, Pathfinder deployed its three-foot-high weather mast and took its first detailed measurements of Martian atmospheric conditions. The spacecraft's weather station consisted of a cylinder surrounded by six wires, some of which were warmed by electricity from the spacecraft. They detected wind speed, wind direction, and Mars' extreme temperature fluctuations. In a single Martian sol, the temperature fluctuates as much as it does on Earth during the course of a year, and daily pressure drops are equivalent to those occurring during a terrestrial hurricane. At the moment, it was −76° F at the Pathfinder site, with steady light winds blowing from the southeast. For the following day, the prediction was that the temperature would reach 10° F during the day, −105° F at night. There's a lot of bizarre weather on Mars.

The next day, the rover whirred to life, and a new crisis erupted.

The protective air bags remained puffed up around the petal on which the rover sat. If the rover rolled down the ramp toward the surface, it would become fouled in the air bags and shade the solar panels, which generated the spacecraft's power. If the rover became tangled this way, the Pathfinder mission would die. The JPL engineers had encountered this scenario during their sandbox tests. At the time, they'd been annoyed about having to build files containing instructions to handle the emergency; they knew it would never happen in a million years. Now that it had occurred, they used computer files compiled during the tests to fix the problem and boasted of their foresight.

JPL sent the instructions to raise the petal, reel in the air bags, and drop the petal back down, pushing the air bags into position. With the impediment removed, the rover rolled slowly down a ramp onto the surface of Mars and started taking pictures as it explored the alien surface. Its six wheels left little tracks in the Martian fine material as it obtained what geologists call ground truth, the most reliable of all data. It was one thing to see a feature from above, from a spacecraft passing overhead, but quite another to approach the feature, pick it up, and photograph it at close range. That's ground truth, and that's what the rover, the twelve-inch-tall geologist, could get as it bumped up against a rock to take data. "What a rush!" Bridget Landry wrote. "Could it possibly have gone any better? Right up to the end, there were people (some of them at NASA headquarters) who honestly believed we'd be a smoking hole in the ground by now!"

At JPL, Jennifer Harris took over from Rob Manning as the flight director. The move had been planned in advance, because Manning knew he would be exhausted from lack of sleep by this time, and someone with a clear head was needed to keep the mission on its demanding course. Jennifer loved her new position, even if she aged about two years during the first few days. She had never experienced this level of confusion, except, perhaps, as a missionary in Russia. Somehow, she had to make sense of everything, to make decisions, to operate on several levels simultaneously. She was not saving souls this time, she was fixing problems, but it felt similar. Maybe this was what God meant for her to do, after all. She defined her job as "coordinating the chaos," and it was stressful. Things frequently went wrong, there were dead spots when they didn't hear from the spacecraft, and she would have to commence an "anomaly investigation." When this happened, she would call a meeting with fifteen or twenty team members, and they would try to figure out what was wrong and how to fix it. Each time they met, the team managed to get the spacecraft up and running again. Usually, it was a question of rescuing Pathfinder's computer from a crash, which was easy enough, once they identified the problem.

The enormity of what they'd all accomplished hit her several days after the successful landing, as she was driving home in her car. She had a picture of Mars taken by Pathfinder posted on her steering wheel so she could look at it as she drove along the freeway. Glancing at the image, she realized, *This isn't the rover in the sandbox at JPL that I'm looking at. This is the real thing. We're really on Mars.* The thought brought tears to her eyes. They really were on Mars.

When JPL held a press conference to reveal the first images returned by Pathfinder, Peter Smith, who designed the camera, turned the briefing into a spectacle. "The eyes of the camera are our eyes," he said to hundreds of reporters. "We are there together. You might say that the people of Earth are the soul of this robot. So for the first presentation of images, forget about the engineering and scientific aspects. Open your mind to the experience and beauty of landing on Mars." He described the journey from the camera's point of view—"You start to raise your head, and this is what you see"— and he revealed a striking mosaic of an oddly familiar, yet distinctly alien panorama, our world transformed into a dreamscape. If you're a planetary scientist, you *live* for this moment. It's a vindication, a revelation. The memory of it will carry you through some very bad, underfunded days. Now, in these images—some in color, others in black and white—Mars silently reasserted its fascination, its claim on the imagination. To see Mars again was like seeing a photograph of a ghost; the spectral became corporeal. It was time to behold, to realize that the Viking images of Mars had become a well-tended memory, a landscape saturated with garish, tawdry red. The Pathfinder images, especially those in color, were far more subtle. The surface of Mars appeared to be ocher, burnt umber, caramel, butterscotch, tarnished silver, rust, and a number of other subtle shadings, but not red. It was pitiless and moody, a desert, an ancient biblical landscape. The scene was powerful enough to make you realize that in the beginning, God also created the Red Planet. The solar system seemed simultaneously bigger and smaller—bigger because the images offered additional testimony of a new world, and smaller because of the unexpected immediacy of the scenes conveyed through space to this auditorium. The new images invited you to walk across the landscape. You could practically hear the stones crunch under foot and feel the faint, rapid Martian breeze. It was the moment to rediscover and reimagine Mars. Soon there would be time to analyze, hypothesize, debate, and quantify, but this was the moment to appreciate, to stop thinking, to indulge in the mystery of this planet.

After Smith finished his presentation, there was a silence, followed by applause in the JPL auditorium. Now everyone knew we were back on Mars.

In the following days, NASA posted Pathfinder images, data, and weather readings on the Internet. They expected to attract perhaps 25 million hits a day, but by July 7, three days after landing, the Pathfinder websites had received 100 million hits. On July 8 alone, they got 47 million hits, and over the next month, the sites received more than half a billion hits, more than any website in the history of the Internet. The whole world was watching.

An instant mini-culture sprung up around Pathfinder. JPL renamed the spacecraft the "Carl Sagan Memorial," in honor of the visionary scientist. The rover became better known to the public under the name "Sojourner," after Sojourner Truth, the itinerant abolitionist, an African American, and a woman. Displaying an exact feel for publicizing the mission, NASA had solicited suggestions for names from schoolchildren and had chosen this one. The geological formations Sojourner examined acquired even more whimsical names—Barnacle Bill, Half Dome, Shark, Scooby Doo, Yogi, Wedge, Cradle, Flat Top, and dozens of other occupants of what became known as the Rock Garden. Everyone embraced the nicknames and the wonderful illusion that Mars was as close as the nearest desktop computer, familiar yet exotic, and there for the taking.

NASA asked Jim Garvin to make a few public remarks about Pathfinder on the very day it landed on Mars. His talk took place where he works: the Goddard Space Flight Center in Greenbelt, Maryland. It was the Fourth of July, and the house was packed. *What was going on here?* This was something he'd never seen. His wife, Cindy, was there, and so were his son and his new-born daughter, who wailed throughout his talk.

Later, a television crew from a local channel interviewed Garvin, and the correspondent, an attractive young woman, sat with him in front of a bank of monitors, while he provided commentary on the features of Mars being seen for the first time. He gave her an abbreviated course in planetary geology and discussed craters and dunes and rock formations. He drew subtle analogies to Iceland. When the interview ran on television that night, Cindy became infuriated; it was practically all interviewer and no Jim. There she was, that attractive young interviewer, offering *her* commentary on Mars, *her* geologic analogies to Iceland. "Jim," Cindy asked, "does she even know where Iceland is?" He didn't think so, because he had to keep reminding her of its location.

After the first wave of euphoria subsided that evening, the magic phrase occurred to him, *"We're back.* After twenty-one years, we're back. . . ." He stumbled around his house in a daze, repeating the phrase, as though he were in a science fiction movie. The phone kept ringing. Friends were calling. Mars was all over the news. New Zealand television, of all things, insisted on an interview with him; would he please come to their studio in Washington, D.C., right away? He brought Mars to the people of New Zealand by virtue of analogy, talking about their volcanoes, some of which are Mars-like. When he arrived home late that night, he did another interview, this time for the *Baltimore Sun,* and by now the excitement of returning to Mars had intensified. "Maybe 'epiphany' is the word for what I felt when Pathfinder arrived safely on Mars. My wife wonders why I'm so emotional," Jim said, "but the successful landing brought me to tears."

As the prime mission continued, the rover performed flawlessly, and the whole summer seemed charmed, the summer of Pathfinder, the summer of Mars. NASA's decision to place the daily operations of Pathfinder and its data on the Internet made the mission participatory. Suddenly everyone with access to a computer was on equal terms with the Red Planet. The mission was the ideal Internet event: timely, technical, visual, and data-rich. It appealed to kids in college, in high school; for them, Pathfinder was phat. If you knew your way around NASA websites, you could, without much trouble, find the e-mail addresses of most everyone on Pathfinder and send them questions. A high school student in Keokuk, Iowa, who went on-line had access to the same information as the Pathfinder team members. She could view and download the latest pictures from Mars, the weather data, the constantly updated reports on the spacecraft's health. She knew as much about the mission at a given moment as the project scientists did. She could travel instantly from cyberspace to outer space.

The two types of space merged to create a new environment, in which discoveries were offered to the public at the moment they occurred. "We brought space close to people," Jennifer Harris said. "This wasn't some big, complex, I-can-never-be-part-of-it thing. The public got to go along, to be explorers in space. The Internet helped, but so did the casual way we allowed people to look at the data." The Pathfinder mission became the most public, the most popular, the most carefully watched science experiment in the agency's history, and the public came to appreciate a new facet of NASA— its scientists and engineers, not just its astronauts. They were so young! There were dozens of these bright, enthusiastic kids exploring Mars, and the Internet made them appear suddenly familiar and accessible. NASA had become cool all of a sudden, and Pathfinder team members became Internet celebrities, glamorous geeks whose names and deeds were known internationally. "Very soon, we'll have a virtual human presence throughout the solar system," NASA predicted. "A merger between cyberspace and outer space. This will be a new universe for us to explore, built around real astronomical and planetary phenomena. And you'll never have to leave the ground to get there."

Whenever the pressure of working on the now hugely successful Pathfinder mission threatened to overwhelm her, Bridget Landry fled to masquerading. She was lured into the masquerading subculture, which can be described roughly as "Star Trek" meets vaudeville, for the opportunity to wear outrageous clothes. "One of my earliest costumes was an old 'Star Trek' uniform, which was made from a pattern smuggled off the set by a friend of a friend's father." She later became a "historical costumer," and realized, as she put it, "there was a real need for someone to make historical foundation garments—in a word, corsets." She made up some corsets for bridal boutiques, "the historical costuming community," and the Goth market. In and around her work on Pathfinder she managed to make seven new corsets and to attend a costume convention. The exotic costumes gave her, she said, "the chance to be someone else. As I get older, I understand that is just an exploration of different parts of myself, to which I previously had little or no access. I get a lot of ego boost from wearing a costume before an appreciative audience, and I have to admit, I love messing with people's minds: wearing something sexy or slinky or ditzy—I have one costume where, as soon as I put it on, my brains dribble out of my ears—and then telling people what I do for a living. Stretching the envelope of what technical women should be."

She showed me a picture of herself in one of her costumes. There was a large round disc coming out of her head, yards of multicolored chiffon falling to the floor, and a golden asp slithering up her leg. She had three-inch-long fingernails, lots of bare midriff, and her face was painted bright green. She looked like a cross between Princess Leia of *Star Wars* and Cleopatra, but she told me she represented the Whore of the Five Universes. "I believe her attributes are corruption and lust, but I won't swear to that," Bridget said. In an e-mail, she asked me, "Don't I look like your standard rocket scientist? :-)"

Her résumé was almost as distinctive. It contained the expected credentials, including her master's in planetary science from Caltech, her impressive experience at NASA and JPL, her awards and publications, but the most significant entry appeared at the top of the page:

GOAL

To be a truly outstanding second banana

She defined herself this way: "Rocket scientist, dancer, master costumer. Whatever your box is, I DON'T fit into it."

Bridget wasn't the only Trekkie on the Pathfinder team. "Star Trek" came up whenever the team members discussed their influences and inspiration, and eventually I realized there was a straight line from "Star Trek" to Pathfinder. The cultural norm and inner life of Pathfinder was pure "Star Trek," an endlessly unfolding fantasy of space exploration. Early in the project, LeVar Burton, who was starring in the latest "Star Trek" movie, stopped

by JPL and received a hero's welcome from the Pathfinder team. He epito-
mized the spirit of everything they were working toward, to the point where
he had to remind them that he was only acting, while they were exploring
space for real. "I think Pathfinder and all the 'Star Trek' shows can be seen
as similar phenomena," Bridget says. "There's a positive feel to both, a valu-
ing of intelligence and competence, as well as a drive to do things not just
faster, better, and cheaper, but *right.*"

It is possible that Bridget represents a new modality at NASA. Whenever I talked with her, I felt as though I was holding a conversation with
someone who lived about a hundred years in the future. You have to wonder what Wernher von Braun would have made of Bridget Landry and
Matt Golombek and Jennifer Harris and Rob Manning and all the other innocent kids in blue jeans and T-shirts executing his vision of reaching Mars.
What an unlikely set of protégés he spawned. You have to wonder what
strange alchemy of *Peenemünde* and "Star Trek" led to this spectacle taking
place simultaneously in a pleasant suburb of Los Angeles, on Mars, and in cyberspace.

Near the end of the mission, exhaustion set in. Matt Golombek, the project's
chief scientist, who had also become its chief spokesman, showed signs of fatigue when he appeared on television. He had five-year-old twins at home,
and his wife worked late; these kids were *hungry,* and they didn't want to wait
for Pathfinder or anything else. As soon as he walked in the door, Matt
turned from a rocket scientist into a short-order cook, and when he left, he
turned into a rocket scientist once again. Pathfinder ran people ragged in this
way, yet the excitement, the adrenaline rush of retrieving data from Mars was
addictive. How often would you have your own spacecraft on Mars, while
the eyes of the world were on you, and your site was getting more hits than
any other site in the entire history of the Internet? Maybe once in a lifetime,
if you were exceptionally lucky. "I don't know what else I'll be doing in my
career," Matt said, "but I'm sure it will never add up to this."

Pathfinder was the Hail Mary pass of planetary missions, but, inevitably,
the mission began to wind down. On September 24, the first signs of serious trouble appeared. The spacecraft returned its last images of the day and
went to sleep, as planned. The next day, engineers attempted to wake it to

photograph the morning sky over Mars, but it remained unresponsive. "We assume that the spacecraft has dropped into a contingency mode," Bridget reported. In the following weeks, Pathfinder succumbed to the harsh Martian environment. Its batteries fell victim to the intense cold and started to fail. Signals from the spacecraft flickered and winked out for unnerving intervals. Ambitious plans for the rover to roam far beyond the rock garden, out of sight of Pathfinder's camera, were canceled. The prime mission finally ended, and although it had lasted ninety days, far beyond the required seven, the realization that it would not continue for a full Earth year came as a disappointment nevertheless. Depression settled over the team, or what was left of it. "It's almost spooky around here these days," Bridget wrote. "So many people have left the project that the floor feels abandoned. I have this weird feeling that six months from now one of the flight software guys and I will be the only ones left and we will be alternating signatures on all the required paperwork. This feels different from the quiet of the night shift; that had a sleeping feeling to it, the natural need for rest between busy days. This feels empty and alone, hollow."

In late September, Jennifer Harris took her turn as mission director, "the guy in charge," as she put it. She felt ambivalent about assuming the role, since it meant she would preside over the last days of Pathfinder, which she knew would be difficult to accept. Donna Shirley was still there, too, encouraging the team members as they tried to diagnose Pathfinder's problems, which, everyone conceded, were terminal.

They made endless attempts to revive the spacecraft. They tried to turn on the main transmitter. Nothing. They tried to turn on the auxiliary transmitter. Nothing. They tried it again and again. They tried it still another time, when the Sun was high in the Martian sky, and the weather was warm. Each time they sent a sequence, they had to wait twenty minutes to evaluate the results, which were always the same: nothing. Jennifer took more drastic steps. She declared a "spacecraft emergency," which gave Pathfinder top priority on the Deep Space Network, used by dozens of other spacecraft. That request ticked off the controllers of the Galileo planetary mission, who told Jennifer and her team: "Give up." She knew the Galileo team had problems with their own spacecraft, but she did not appreciate that message in her hour of need, and she would not give up.

On October 7, she convened a core group of team members to consider

Pathfinder's dwindling options. "It's the middle of the night, and we're all so depressed," she wrote. Then, from the next room, they heard someone shouting. He was listening for the spacecraft on the Deep Space Network, and he heard it. They jumped up and ran into his room and studied the readout. It was not exactly what they had hoped to see—not a robust signal, merely a feeble pulse from Pathfinder's auxiliary transmitter. They did the only thing they could, which was to send sequences like mad, instructions to turn this subsystem off and that subsystem on, and hope *something* worked. All the while, Jennifer suspected they might not be receiving Pathfinder's output. This signal was so faint it might just be background noise, but the Deep Space Network technicians tracking Pathfinder assured her this *was* the signal. She argued back, no, it's not, but they finally convinced her it was Pathfinder.

The spacecraft was asleep on the surface of Mars, and death was creeping up on it. In its feeble condition, the spacecraft automatically awoke and shut down, unable to accept commands from Earth. The engineers called it "The Kevorkian Syndrome." There was still a chance they could rescue it, but it was a long shot, at best. It was midmorning on Mars, and they hoped the Sun's warmth would help restore Pathfinder. As they waited, the signal actually strengthened. The computer on board the spacecraft was working, at least a little, so they sent a bunch of commands and waited. Nothing.

For the next four days, they tried to revive Pathfinder. Their receivers searched a broad range of frequencies on the assumption that the Martian chill caused the spacecraft's radio to vary its operational range, but they found nothing. They tried alternating the main and auxiliary transmitters, communicating through Goldstone or Madrid—it was all getting to be a blur. No matter what they did, they failed to detect a signal from Pathfinder. At one point, they sent a command to switch on Pathfinder's transmitter, and then they sent another command to switch it off. The next day, they hoped to wake Pathfinder, but the crippled spacecraft failed to respond. They tried again the following day, and the day after. Nothing.

Jennifer suffered pangs of self-recrimination. Why had she agreed to send a command to switch off Pathfinder's transmitter? They should have left it on! Perhaps the mission would have turned out differently if they had. She would never know.

"Still no word from the spacecraft. This is getting scary," Bridget Landry wrote in her journal on October 15. "Theories abound, of course. One is that because the transmitters and the spacecraft have been off for so long, the spacecraft is cooling down to temperatures below where it was tested. This may have changed the wavelengths at which the spacecraft is transmitting and receiving. Another is that the switch between the low-gain and high-gain antennas may be stuck in the high-gain position. Also, if the battery has died, the spacecraft may not be able to track time correctly. I have this image of the spacecraft clock flashing '12:00' like my VCR does when there's been a power outage. But the fact remains that we haven't had any data from the spacecraft since September, we can't reliably command the spacecraft, and we don't know what's wrong. This may be the end of the project. Sad and scary both."

On November 4, four months to the day after Pathfinder's landing, they finally gave up. Jennifer hated to think of leaving the spacecraft stranded. Even now, it was just possible that something aboard Pathfinder was not completely dead, that some subsystem had enough power to click and to whir and to send faint messages. In case of a miraculous resurrection, they continued to transmit a command once a week, and later, once a month, but nothing ever came back. The spacecraft and rover slowly froze in the frigid Martian climate, in Ares Vallis, where it will remain for billions of years as a monument to the team, the technology, and the Pathfinder mission.

A lot of folks at NASA rushed to take credit for Pathfinder's success and to send up the cry, *"Faster! Better! Cheaper!"* As the project scientist, Matt Golombek reveled in the data Pathfinder returned. It was time to begin thinking of Mars in a new way, as a planet far more like Earth than scientists had believed. "The rover discovered rocks that are generally similar in composition to the continental crust on Earth," he wrote. "If these rocks are representative of the ancient cratered terrain, they suggest more Earth-like processes of crustal formation of Mars than previously believed. The rover also found rocks that appear sedimentary in origin and some that may be conglomerates, which suggests that liquid water was stable for a long period of time (required for their formation), and that the climate was warmer and wetter than at present." The geologic evidence strongly suggested the presence of water on Mars, a great deal of water, which meant a great deal of potential for life on Mars.

Coming off the success of Pathfinder, members of the Pathfinder team received plum assignments at JPL, but they were, in the end, just jobs. Pathfinder by comparison had been more than a job, more than a mission; it was an experiment in improvising new solutions and doing the impossible. When they did their jobs well enough, the mission itself became the least interesting part of it; the interactions among the team members mattered most—that and the exhilarating sense of cutting loose from NASA's slow and steady approach to planetary exploration. For a time, Jennifer Harris, who had given so much of herself to Pathfinder, was bereft without the mission to sustain her. "At this point, I'm twenty-nine years old, and my career has peaked. Nothing will ever be like this." Of all the challenges she faced during the mission, this was the most difficult, seeing it end.

There were consolations, of course. The team members participated in beer-soaked Pathfinder parties, held an uproarious Pathfinder wake, and read about themselves in the glossy pages of the popular press. *Glamour* and *Ms.* selected Donna Shirley as their "Woman of the Year." Annie Leibovitz photographed key members of the team, including Jennifer Harris, for *Vanity Fair's* "Hall of Fame." Leibovitz posed them in the rover's sandbox, setting up the shot so that it looked like a Pathfinder mosaic. The effect was so hip, so postmodern, it was unbelievable! Who would have imagined a year before that *Vanity Fair,* of all places, would glorify JPL? The magazine even compared Pathfinder's renderings of the Red Planet to the "opening vista of a Sergio Leone film witnessed through a bloodshot eye." Martian chic was born.

"Was it worth it? What did we get for our $170 million?" an interviewer for National Public Radio asked Jim Garvin. *This is ludicrous,* Jim thought, *ludicrous!*

"We went to a planet one hundred million miles away," he carefully explained, "dropped a rover controlled by a RadioShack-type walkie-talkie on its surface, wandered around for a hundred yards, measured for the first time what the rocks on Mars are made of, which we did not do with Viking back in the seventies, sniffed the Martian atmosphere, measured the weather, and did all that on a spacecraft that cost one-tenth of what it cost twenty years ago. And, on top of that, the mission captured the imagination of the American public during the doldrums of the summer of ninety-seven. Perhaps I'm biased, but we energized our thinking about our role in the solar system and catalyzed the search for life on Mars. That's what we got for our taxpayer dollars."

For planetary scientists like Jim Garvin, the Pathfinder mission would continue for years beyond its official conclusion. "If you look at all the things Pathfinder had to do to deliver a lander to the surface of Mars," he told me, "it is a remarkable engineering achievement. The science legacy can't be described, yet. We will sift through the data, and the mission generated *tons* of data, textbooks full. I'm talking about the broad scientific community, not just the people on the mission's science team. The knee-jerk reactions have been done, but the follow-up will take years." When he was working on Viking, he measured every pit in every rock he studied—hundreds of rocks and thousands of pits—to find indications of what might have happened to those rocks. They hadn't even begun to do that for Pathfinder.

Garvin applauded Pathfinder's success but lamented the mission's missed opportunities. If only it had carried more science, better instruments, a more durable power system. There were dozens of other experiments Pathfinder could have performed if NASA had spent a little more time and money on the project. It seemed a shame to send a rover all the way to Mars equipped with such low-resolution cameras, for instance. As he saw things, the real problem with going to Mars wasn't the *cost,* which, in the overall scheme of NASA's budget, was insignificant, but the *time* involved in planning the mission and traveling to the Red Planet. There weren't many opportunities to go to Mars; when you got one, you wanted to make the most of it.

Part Two

CODE S

Chapter Five

SHOOTOUT AT CALTECH

"Science is not an art—though some may feel it resembles art in certain respects. Science may not even be a science."

—HENRY S. F. COOPER, JR.

Jim Garvin finally had his chance to explore the Red Planet when Mars Global Surveyor was launched on November 7, 1996. He planned to be present at Cape Canaveral Air Station, "basking in the glow of the Delta rocket," as he put it, but his infant son needed to see a doctor on launch day, his wife was pregnant and not feeling well, and he was juggling three separate NASA assignments. "When MGS launched," he told me, "my mind was surging with conflicts—the hopes of a great launch so the instrument would do its stuff, and the wish to be at the Cape, combined with the deep emotions of needing to be there for my son." The "instrument" was a laser altimeter designed to map Mars, and he had worked on it for a decade. It was, in some ways, his claim to fame. The Mars Orbiter Laser Altimeter (MOLA) belonged to a suite of five science instruments aboard MGS; the others were a thermal emission spectrometer, a magnetometer, a camera, and an ultra-stable oscillator. If they all worked properly, the data they returned would rewrite the history of Mars, including the planet's *biological* history, if any. They were all crammed into a spacecraft about the size of a refrigerator.

Seconds after liftoff, the Delta launch vehicle soared out of sight, and within minutes, the spacecraft was safely on its trajectory to Mars. Going to the Red Planet is considerably more complicated than sending a spacecraft into Earth orbit or even to the moon. Because both planets travel around the Sun in elliptical orbits, chasing each other, so to speak, the most direct route from Earth to Mars requires a tremendous amount of thrust, which translates into fuel and weight—too much, really, to be practical. Instead, engineers rely on Hohmann transfer orbits, named after Walter Hohmann, a

German mathematician who calculated trajectories for interplanetary travel balancing distance, gravity, and planetary motion. Hohmann transfers use less fuel, which means they are less costly. A fuel efficient transfer to Mars begins with firing a rocket from Earth in the direction of the planet's rotation (to get an extra kick), then following a course that gently curves beyond Earth's orbit around the Sun, gradually catching up to Mars' more elliptical orbit, and finally intersecting the Red Planet. By that time, Mars and the spacecraft have both traveled halfway around the Sun, more than 300 million miles.

The precision required to complete a transfer to Mars, the unpredictability of rockets, and the fragility of spacecraft have created a lot of frustration over the years. The Russians started hurling objects at the Red Planet in 1960, and they endured at least five humiliating failures in a row. More prone to fatalism than their Western counterparts, Russian rocket scientists wondered if an unknown force or alien intelligence was bent on destroying their spacecraft, and after the failure of the first American spacecraft aimed at Mars, even NASA engineers started to banter about a Great Galactic Ghoul lying in wait somewhere between Earth and Mars.

Finally, in November of 1964, Mariner 4, an American spacecraft, returned twenty-one photographs of Mars, taken with a single black-and-white television camera at a distance of about 6,000 miles. The stark images revealed no Martians, no canals, no cities, no vegetation, no signs of life in any form, but there were abundant craters, and at the time, they appeared to be an ominous sign. "We were all shocked by seeing such large, lunar-like craters," said Bruce Murray, a Caltech geologist, of this first modern look at the planet. "It meant that Mars had not recycled its surface the way Earth does. There must have been no rainfall, no weathering, in any way comparable to that of Earth's for billions of years, in order for Mars to resemble the moon." The multitude of craters suggested that Mars had practically no atmosphere and perhaps never had one. And it looked lifeless, a barren and unforgiving piece of celestial real estate that negated decades of scientific speculation and fantasy concerning life on Mars. At the time, scientists thought that only impact craters—created by meteors, asteroids, and comets bombarding the surface—existed on Mars. If scientists had realized that at least some of the craters were volcanic, they would have altered their conclusions, for volcanic activity means heat, and heat means at least the possibility that conditions favorable to life once existed.

Between 1969 and 1972, the American Mariner series continued, occasionally meeting with modest success, and the Russians repeatedly attempted to reach the Red Planet, often failing to get off the launch pad and never attaining their goal. In 1972, Mariner 9, one of the few spacecraft that slipped through the clutches of the Great Galactic Ghoul, altered the stark impressions of the planet gathered by its predecessors. The focal point of this mission was an intriguing white spot on Mars then known as Nix Olympica, the Snows of Olympus. Mariner 9's camera showed this was actually the largest shield (that is, gently sloping) volcano ever seen, much larger than any terrestrial counterpart. It was formed somewhat differently than shield volcanoes on Earth, owing to the lesser Martian gravity; it was seventeen miles high and nearly four hundred miles across. Nix Olympica was renamed Olympus Mons, and suddenly Mars was not the dormant, moon-like planet scientists had imagined only a few years earlier. Michael Carr, a Mariner geologist, observed, "Olympus Mons and some of the other large Martian volcanoes may still be active. Eruptions are probably widely spaced in time, occurring every few thousand years." With the revelation of volcanism on Mars, the possibility of finding life gradually resurfaced as a subject of respectable scientific inquiry. Mars had the same geologic language as Earth, which was reassuring, but it contained an exotic and nuanced vocabulary that needed years of study to comprehend fully and interpret correctly.

During the ten-month-long cruise of MGS, Jim pursued his other NASA responsibilities. Cindy, his wife, gave birth to their daughter, Danica, whom they named after a television actress they had admired in the days when they had time to watch television; only later did they discover that her name means "morning star." By the time the spacecraft finally reached Mars on September 11, 1997, he was almost a year older and the father of two small children.

As the time of the maneuver designed to insert Mars Global Surveyor into orbit around the Red Planet approached, Jim focused all his attention on the mission. He knew that no spacecraft, Russian or American, had been successfully inserted into Martian orbit since the two Viking orbiters a generation earlier, in 1976. As Martian gravity began to exert its pull on Mars Global Surveyor that day in September 1997, a group of small guidance rockets commenced a twenty-two-minute-long burn designed to slow the speed of the spacecraft from 11,386 miles per hour to precisely 9,842 miles per hour, easing it into orbit around Mars rather than letting it streak into space or crash-land on the surface. There was no margin for error. Trying to insert a spacecraft into orbit around Mars can be compared to firing a bullet from the moon and hitting a particular window on the Empire State Building. Trying to maneuver the orbiting spacecraft as it dips lower and lower into Mars' thin but turbulent upper atmosphere can be compared to guiding a radio-controlled toy in New York City from Los Angeles with enough precision that it stays on the sidewalk.

A few minutes after the burn commenced, Mars Global Surveyor disappeared behind the Red Planet, an event known as occultation. The mood at JPL turned exceptionally taut, for everyone remembered the same moment in 1993 when Mars Observer had slipped behind Mars and never resumed contact with Earth. Now there was just one question in everyone's mind: *Do we have a mission?*

In the midst of the background murmur came an alarming howl . . .

"*. . . SHUUUTTTT UUUPPP!!!!!*"

Everyone looked around to see who had roared. It was Wayne Lee, the mission planner, normally a patient and polite young man, a poster boy for the new NASA. But now Wayne was wired and desperate for news from Mars. He stunned everyone at JPL into silence for about thirty seconds, and then the anxious hum resumed.

Do we have a mission?

The press was milling about, most of it foreign and television, ignoring an overwhelming backdrop of trajectories and data.

Do we have a mission?

And then communication with Mars Global Surveyor, 158 million miles away, was reestablished. The burn was successful.

YES, WE HAVE A MISSION!

For the first time in twenty-one years, a spacecraft from planet Earth was orbiting Mars. Cheers erupted from the ranks of the scientists, their managers, and even from the foreign press. Men and women who normally shrink from physical contact lifted each other off the ground in great bear hugs as the euphoria spread, and everyone grinned a grin that said, *"We're back."* T-shirts, pins, and baseball caps with the MGS insignia suddenly appeared. NASA, badly in need of a victory, finally came up with a big one.

Amid the hoopla, the men and women of the press groped to understand the reason for the cheering. *MOI . . . Mars Orbital Insertion . . . what the hell is that?* They could understand a spacecraft landing on Mars, but the concept of a planet's gravity gently capturing a spacecraft and holding it in a carefully calculated orbit didn't register. Of what use was an orbiter? NASA explained that a *lander* was stuck in one place on the surface, while an *orbiter* circled the entire planet, like a spy satellite. It was much more versatile than a lander; it could do all sorts of acrobatics in the Martian atmosphere; it could go anywhere and see anything. The distinction was lost on most of the media. They didn't want an aerial ballet. What they really wanted was an artificial object plummeting to the surface of Mars with a great big crash.

"We on the Mars Orbiter Laser Altimeter team are very excited since we realize a ten-year-old dream today, as we turn on our laser," Jim Garvin told me four days after the successful insertion. "If all goes well, MOLA will peel back the veils of uncertainty about the landscapes of the Red Planet."

All did not go well.

"There is exciting and frightening news," Jim said later. "Just after the successful launch last November, one of the solar panels that powers the spacecraft failed to fully deploy and latch open. This caused NO problem in the cruise to Mars, but since Mars Global Surveyor was designed to be 'cheap,' the spacecraft must rely upon using the Mars atmospheric drag on these panels to place itself into the proper orbital vantage point for making global observations. After making orbit, the intrepid MGS has been dipping within seventy miles of the surface of Mars to slow itself down and ever so slowly place itself into the desired mapping orbit—circular, polar, and with a two-hour orbital period." That's known as aerobraking—using the atmosphere to slow the spacecraft's velocity. Aerobraking is a risky maneuver, but NASA engineers have used it successfully, notably on the Magellan mission to Venus in 1990. It is also time-consuming. Mars' atmosphere is less than one-hundredth the density of Earth's, and aerobraking would require well over a year to complete.

After fourteen aerobraking passes, the spacecraft began behaving oddly, and ground controllers became terrified that the malfunctioning panel would break. It provided half of the power needed by the spacecraft to function, not to mention aerodynamic stability. So they suddenly had a major issue to contend with. All the while, they were receiving data from Mars Global Surveyor, but for only twenty-two minutes per orbit, and at this point, the orbits lasted about thirty-five hours each. If the spacecraft remained in this weird elliptical orbit, they would get only about two percent of the expected data for the northern half of Mars, even after two years of orbiting the planet. So it was essential to modify the orbit.

The setback stimulated Jim Garvin. He saw it as a test of mental agility, and he felt prepared for any eventuality, except the loss of another spacecraft. He told himself the Mars missions were like the long journeys made by the mariners of antiquity. Problems were inevitable. The question was, what could they do to fix them? During the search for a solution to the problems afflicting Mars Global Surveyor, Jim decided a delay in the mission might not be such a bad thing, after all. The spacecraft could collect and transmit data while it was aerobraking, even before it reached its ideal orbit. Sure, the data were incomplete and subject to correction, but they provided his first chance to see the news from Mars. And when the MGS science teams, including Garvin's, all gathered at Caltech to exchange their first returns from the Red Planet, he would be there to spread the news throughout the Mars community.

Caltech enjoys an intimidating reputation, to put it mildly. Students at rival institutions look on their Caltech counterparts with a mixture of awe and scorn, and outsiders are wary of the place. To hear them talk, you'd think Caltech was the West Point of rocket science, a place where scientific macho and bravado are de rigueur. The Caltech approach to defending a research paper or a thesis resembles the Marines' approach to character building: you tear it down, and you build it up again as a product worthy of the institution. Legends circulate about the informal requirement that any Caltech student must be able to disassemble a car and then reassemble it indoors, make it run, and then disassemble it again, cart it piecemeal to the street, reassemble it once more, and make it run on the street.

Caltech has been home to a number of Nobel Prize laureates—Linus Pauling, William Shockley, Richard Feynman, and Murray Gell-Mann among them. The institution rose to preeminence from humble California roots, having been founded in 1891 as Throop University, the successor to a small arts and crafts school. Early on, the United States government looked to Throop as a source of engineers for projects of national significance. In 1911, President Theodore Roosevelt came in search of engineers for the Panama Canal. Eventually, Throop changed its name to the California Institute of Technology and took over the management of high profile scientific projects such as Mount Wilson Observatory. During the Second World War, Caltech moved closer to the government, devoting resources to weapons research and science programs. By the time NASA took over space flight from the military, Caltech had become a prime developer of unmanned spacecraft for the agency in conjunction with the Jet Propulsion Laboratory, located a few miles away. Through war and peace, Caltech has retained its unique character, scholarly yet state of the art, the cradle of American rocket science.

In February 1998, Jim Garvin flew from Washington to Caltech to help present the earliest data from the laser altimeter. He was one of a hundred scientists converging on the area to take the measure of Mars, and he was psyched. He arrived in Los Angeles on a stormy Saturday evening with *El Niño* at its most extreme, flooding highways, disrupting traffic, jolting Southern California out of meteorological complacency. He managed to navigate his rental car through the downpour to Pasadena, where his team—the mighty MOLA science team—would convene first thing in the morning before presenting the data to all the other scientists on the following day.

MOLA is a small, very accurate laser capable of firing billions of impulses at the surface of Mars and measuring the minute variations in the time it takes for the impulses to return to the spacecraft. The laser impulses last only eight billionths of a second. Some of the geologic features the laser altimeter will measure are larger than anything on Earth. "MOLA must measure the precise depth of this abyss from hundreds of miles away, while flying in a spacecraft cruising at nearly three miles per second," Jim said. "MOLA is the most precise altimeter ever built for studying another world. It can detect changes in relief at levels of a few feet over hundreds of miles. It can measure the shape of landscapes at what I call 'human scales.' " And it can help to detect places on Mars where life might have existed, places where water now exists. "The only known reservoir of water is the polar caps, and only the Northern polar cap has been shown to have a water ice component." I should explain that Mars' polar caps contain two kinds of ice—"dry" ice from carbon dioxide, and ice that is simply frozen water. It's the water ice that's relevant to life and attracts Garvin's attention. "We don't know anything about how thick the ice cap is and hence its volume. MOLA can measure its apparent thickness, and from a sequence of MOLA transects, we can model its shape and compute its volume. We can test hypotheses about the history of the water and how it might be recycled annually and seasonally on Mars." All this from an instrument the size of a microwave oven, circling a planet a hundred million miles away.

The MOLA team to which Jim belonged consisted of a dozen or so scientists, plus a few part-time members. They had been meeting, incredibly enough, for a decade. Until now, their deliberations had been closed; I was the first outsider allowed to observe their deliberations at close range. "I'd better warn you up front," Jim said to me right before the meeting, "things might not be what you expect. We all have a lot of respect for each other, but things can get pretty heated and personal. It's a lot like an episode from 'Seinfeld.' After all, this is the first time that we have actual data, so everyone's keyed up."

"Where exactly does 'Seinfeld' come in?"

"You'll see."

The morning dawns cool and tranquil; the Caltech campus exudes Spanish Mission elegance and repose. After *El Niño*'s violence subsides, tendrils of mist hover low over the sparkling, lush grass, until the sun's angle of incidence reaches them, and they glisten and disperse into the soft, gardenia-scented air. The place resembles a monastery nestled in a comfortable suburb, and on this placid morning, you expect to see contemplative monks traversing the hushed quadrangles. Instead, ten or twelve members of the MOLA team find their way to a conference room in the Arms Laboratory,

where the walls are papered with charts and graphs and pictures and calculations, all displaying their first, very incomplete data from Mars. This should
be a moment of triumph for the scientists, now that they have the information for which they've waited a decade. Instead, they are preoccupied with
funding in the new NASA. In the current mantra, "faster, better, cheaper,"
it's the word *cheaper* that keeps coming up in discussion. The entire MGS
project costs $187 million, including, the team is quick to add, the "launch
vehicle." The rocket—a reliable Delta II built by Boeing—sucks up over half
the budget, but you can't get to Mars without one. The science component
comes to about $90 million, which includes the cost of designing, develop-
ing, building, testing, and operating the instruments on board the spacecraft,
each of which is intended to explore some aspect of Mars, and to coexist
peacefully with the other instruments. If NASA were building many copies
of each experimental device, the cost per item would fall off dramatically,
but it doesn't work that way. Each one is handcrafted and costly.

Nevertheless, NASA has found ways to make the instruments into relative bargains. MGS's failed predecessor, Mars Observer, cost nearly a billion
dollars, five times more than this mission. There was no backup spacecraft,
no redundancy. The failure reinforced the notion that the beleaguered, demoralized agency had become the gang that couldn't shoot straight. On his
late-night television show, David Letterman reviewed his "Top Ten Reasons
for Losing Mars Observer." He started with "Mars probe? What Mars
probe?" and moved on to "Forgot to use the 'The Club,' " and on down the
list of excuses to number one: "Space Monkeys." Nobody at NASA was
laughing.

Now that Mars Global Surveyor is finally returning data, everyone's
main concern is that NASA will continue to pay for the mission. It hasn't
occurred to me that NASA would actually pull the plug on a brand-new,
fully functional mission, but under the new "faster-better-cheaper" regimen anything is possible. The funding universe at NASA shifts almost every
week; these changes cannot not be detected by scientific instruments, only
by gossip and guesswork. "Whenever the budget changes, people's careers are
put on the line," Greg Neumann, the custodian of the data acquired from the
spacecraft, tells me. A change in the budget is generally a euphemism for a
cut, and cuts create complications. Mars Global Surveyor, for example, is a
rebuild of the old Mars Observer. It's lighter and cheaper and carries fewer

experiments, but otherwise, it's pretty much the same. To save money, NASA refused to make changes in the original specs, and some elements of the Mars Global Surveyor date back to 1988. The spacecraft's CPU, its computer processor unit, is an 8086 model, with 32 kilobytes of memory. It is more than ten years out of date and lags many generations behind current processors. The CPU inside my mail-order desktop is many times more powerful than the one inside the spacecraft. Relying on an old chip has created new problems, forcing NASA engineers to scrounge around for old equipment compatible with it.

At the start of the meeting, the MOLA team's leader, Dave Smith, listens, nods, and paces. With a full head of hair, a mustache, and a distinct burr in his accent, courtesy of his English birth and education, he looks as though he would be driving racing cars along European speedways if he were not doing science at NASA. Dave's field is planetary geodesy, which concerns the shape, makeup, and physics of planets, and how they interact. During the MGS mission, he has been putting his considerable energies toward functioning as a science manager—a tough job, since scientists tend to be loners and explorers who resist being managed. So Dave is invariably charming, but he is also a relentless coaxer and goader, in short, a coach. When he talks, his arms shoot off in all directions, and the pitch of his voice fluctuates wildly. When he becomes very excited, he jumps to his feet and dances a little jig. He is often too energized to sit still, so he stands. He paces. He struts. One of his tasks is to bring his brilliant, argumentative team together to make recommendations based on their individual and collective findings. That, in NASA speak, is called convergence, and that exalted state can be elusive. Spontaneous convergence occasionally occurs in the midst of a meeting, but more often, when he tires of all the discussion, Dave bashes a few scientific heads together into a kind of willed convergence so they can get on with things.

I was once startled by a photograph of the team taken when they were just beginning to meet, never suspecting what a marathon their association would become. Dave looks particularly bold and athletic. In the years since the photo was taken, his hair has gone gray, he has ripened, matured, persisted, and the effort to stay the course has, I suspect, taken its toll on him. How much simpler life was for everyone on the team in the beginning, when there was no spacecraft or data, just a schedule and expectations. In those days, they devoted their meetings to data from Viking, and they anticipated the returns from Mars Observer. That mission originally was set for 1990; it slipped a year ("slipped" being a NASA euphemism for inevitable delays) and finally launched late in 1991. Dave still glows with the memory of the spacecraft; it was beautiful, a work of engineering and scientific art. Excitement ran high, and then, suddenly, they lost it. One moment there was a signal, and the next there was none. It was a very depressing time, those months after the loss. Dave, I've heard, was rarely seen around NASA for a period. When he reemerged, Jim and some others talked him into backing another, much cheaper spacecraft with a laser altimeter aboard: Mars Global Surveyor. It took some doing, but eventually Dave decided to go along with the plan.

Maria Zuber serves as the team's co-principal investigator; she's the Co-PI, in NASA-speak. She is about forty, with an athletic physique and short, light brown hair. These days, she is the Griswold professor of geophysics and planetary science at MIT, and she holds a number of appointments at other institutions. She sits on many scientific committees and advisory boards, in addition to her teaching duties at MIT, her participation on the MOLA science team, and her research interests in theoretical modeling of geophysical processes and in the analysis of altimetry, gravity, and tectonics as they affect the Earth and the rocky planets. All this, and neither of her parents attended college. Somehow, she found time to marry (to a financial executive) and raise two kids.

Ever since she was a four-year-old in Pennsylvania, looking up at the night sky, Maria wanted to explore space. As a slightly older, precocious child, she started building telescopes; she ground the lenses herself, partly because there wasn't money to buy a telescope. "The first several weren't that good, but I was determined to get it right, so I kept at it. One of my favorites was one where I used a stove pipe as a tube, and it won grand prize in a school science fair when I was in fifth grade. Despite my best grinding efforts, it had a slight chromatic aberration, but I guess the judges didn't notice. I think my parents still have that one. When I looked through telescopes, I wasn't partial to any particular kind of object. I was driven to astronomy because I wanted to figure out what was up there. I managed to get a pretty good familiarity with the night sky from years of observing, but that didn't really cure my curiosity because once I started seeing *what* was there, I started to think about *how* it got there. I'm still working on that part."

After receiving her bachelor's degree from the University of Pennsylvania, Maria did graduate work in geophysics at Brown, where she played basketball with a vengeance and became the best player on the *men's* geology team. She was the only one who could shoot worth a damn, she says. Eventually, other teams realized the lone woman on the team was the high scorer, and they started guarding her. She still has a competitive streak; once, when I wondered aloud how she managed to compile an impressive shooting record even though she isn't especially tall, her gaze hardened, she looked at me as though I'd missed the point, and the conversation skidded to a halt. She's learned to be tough, for scientists can get down-and-dirty, especially among themselves. At the start of her career, she appeared before an intimidating crowd of scientists at Cornell. Everyone was there, including Carl Sagan. She delivered an elegant paper, well-written, smoothly delivered, and when she was done, a Caltech professor shot to his feet and started to tear into it. Jim Garvin, who was also in attendance, was outraged on her behalf. He'd seen other scientists break down and shuffle off the stage after being subjected to this sort of treatment, but when the Caltech professor was finished with his diatribe, Maria politely answered every objection, showing no fear or anxiety.

On the MOLA team, she works very closely with Dave; he speaks for her, and she speaks for him. If you tell Dave something, you can be fairly certain Maria will hear it, and if you discuss the matter with Dave, that's as good as telling Maria, too. She's more easily exasperated than Dave, who seldom utters a discouraging word, and she's more sensitive to the strains eating away at her team—the jet lag, the professional anxieties. Now, at the Caltech meeting, she wears a pink Oxford cloth shirt, blue jeans, bright pink woolen socks, and loafers. At various times, she slumps, she sprawls, she hides under her bulky sweater, concealing her hands, and lets her head droop almost to her chest, yet even when she seems transported into a private realm, nothing escapes her notice.

"Exploring space seems very glamorous," she tells me by way of introduction to the MOLA team, "but politics, budgets, and other forces beyond our control drive the system. I sometimes wonder if people knew when they got into this how hard it was going to be and how long it was going to take. I wonder if they would have started it in the first place. When students enter the field, they usually walk into a project where data is coming in and they analyze it for a thesis and it seems like a pretty good way to make a living. Then, you work for ten years on the next project before you get anything, and it takes real conviction to persist. But I think everybody appreciates what we are getting now more than they would have if the missions had taken only a couple of years." Any group that meets regularly over a decade inevitably develops its own idiosyncratic culture, and the MOLA science team has its version of tough love that bonds its members even as their careers take them in different directions. "The personal dynamics have certainly evolved," she says. "I consider some of the people on my team like big brothers because we have lived through experiences no one outside the team can understand. Either there are crises or you pass a milestone and life couldn't be better."

Now, at the Sunday morning meeting in the Arms Laboratory at Caltech, life is marvelous. Dave Smith reports on the spacecraft's health and well-being. He announces that Mars Global Surveyor is in the midst of aerobraking, the slow-motion dipping into the Martian atmosphere. He describes the complication this created, the one Garvin told me about: one of the spacecraft's large, wing-like solar panels buckled on deployment. But now there is a solution. After jiggering and rejiggering the computer commands, JPL's engineers redeployed the panels, and while they are still not in their optimal configuration, they're at least sturdy enough to permit aerobraking to continue.

As any rocket scientist knows, there's a faster way to perfect an orbit: include thrusters and a large supply of rocket fuel, then fire the thrusters to change the orbit as needed. These little bursts of fire are called delta-v burns and take only minutes to accomplish instead of the months required for aerobraking, but the necessary fuel adds weight to the spacecraft, as do the navigational rockets, and a heavier spacecraft requires a much more powerful—and expensive—launch vehicle. These days, NASA relies on the Delta II, built by Lockheed Martin, to launch its planetary spacecraft. It runs about a hundred million dollars per launch, and it's not reusable. If the spacecraft were to include navigational rockets, which require additional fuel, NASA would be forced to use the much bigger Titan missile, also built by Lockheed Martin, and the Titan costs $400 million. At present, the sole customer for the Titan is the CIA, which uses it to launch spy satellites. So there is considerable money to be saved by eliminating weight and resorting to aerobraking, enough to fund four times as many missions.

The meeting picks up momentum when Maria Zuber startles her colleagues by reporting that Mike Malin, who processes and controls the photographs taken by MGS, suggested shutting down the laser altimeter until spring. *Mike Malin!* Merely mentioning the name makes everyone jumpy. Malin can do this drastic thing, if he chooses, because his instrument, the Mars Orbiter Camera, is a "primary experiment" and takes precedence over the laser altimeter, which is considered secondary. A groan of astonishment goes up. He might be jesting, or testing—you never know with Malin. They all acknowledge the man is a genius when it comes to his cameras, but he can be maddeningly elusive. He's a member of the MOLA team but rarely attends meetings; he's missing this one, which comes as no surprise to anyone. Maria complains out loud that he hasn't answered her e-mail in six months. In a culture where everyone rushes to release data to affirm findings, Malin holds on to data, and not just any data, but the precious images of Mars.

Malin doesn't play by the same rules as the other NASA scientists; he is the Great Exception. He won a McArthur "genius" award while doing research at Arizona State University, which supports a distinguished program in planetary science, but rather than stay at Arizona, which, it is said, he didn't consider prestigious enough, he decided to start "Malin Space Science Systems," located in San Diego. This arrangement makes him the only businessperson on the Mars Global Surveyor science teams. Malin founded his company in 1990, and at first, he tried to do everything by himself. Eventually, though, he realized he needed to delegate responsibility. Then he had to deal with NASA as his major client. The agency could be slow to pay, and the prospect of cutbacks constantly loomed. Nevertheless, his colleagues tend to forgive him everything once they finally see his spectacular images. His Mars Orbiter Camera (MOC), a scientific version of a spy satellite, can zoom in to record images with resolution fine enough to make out people walking across the surface of Mars (if there were any), and it can zoom out to make a portrait of the entire planet in a single frame. If deployed in Earth orbit, the camera system would be illegal, because its fantastic resolution exceeds the limit set for spy satellites.

Then Jim Head, Tim Mutch's former colleague from Brown, inadvertently stirs up more trouble with his report. He begins calmly enough, discussing methods for selecting a landing site on Mars for a new spacecraft, due to arrive at the Red Planet in 2001. The general idea is organize their latest data into validated data sets useful for site selection. A data set is derived from raw information gathered by the spacecraft's instruments, and it is heavily massaged by various specialists on the team who note correlations to and departures from previous findings. Interpreting and correcting raw numbers while allowing for various errors and glitches requires considerable experience and familiarity with the limitations of the instruments generating them. Two equally adept scientists treating the same raw numbers will often arrive at different conclusions; there are lots of judgment calls.

Head has barely begun his presentation when an edgy voice from the back of the room calls out, *"That's bullshit!"*

Silence.

"This is a total waste of our money! They are designing a billion-dollar mission without key information!"

The voice belongs to Sean Solomon, reclining in blue jeans on an old couch. Sean is tall and rangy. A professor at MIT for more than twenty years, Sean was president of the American Geophysical Union, and he is now the head of the Division of Territorial Magnetism for the Carnegie Institution of Washington. Both the name and the Institution are 1904 vintage; these days, his division is engaged in geoplanetary studies. We're talking astrophysics. Cosmochemistry. Geochemistry. Planetary physics. Seismology. He has worked on a number of important NASA planetary missions over the years; Venus and the moon figure prominently in his past. Sean has the credentials; he knows his way around the solar system and around NASA. He wields considerable power and influence, and everyone else in the room is faintly awed by him.

"Speak up, Sean," says Head, trying for some irony. Nobody laughs.

Dave Smith appeals to Jim Garvin for enlightenment on the question of site selection, but Garvin only deepens the confusion when he displays a plot of the laser altimeter's "hits"—or individual measurements—on the surface of Mars and explains how they yield a coherent map. This is where individual interpretation comes in, and Sean isn't buying Garvin's. Sean wonders aloud how Jim can confidently depict arcs and parabolas and other geologic features when the plot looks like footprints made by a deranged chicken staggering across a sheet of paper. The hits seem to be completely random.

Jim tries to defend his interpretation, but Sean interrupts him. "I'm sorry, your baud rate is just overwhelming me." Jim falls silent. The team has just experienced what they call a "Garvin minute"—that's when Jim asks to speak for *just a minute,* and a quarter of an hour later, he's still holding forth. Even Jim is aware of his tendency to talk a lot; he feels it may be genetic. Sean persists. "If you want to tie our stuff to a crappy United States Geological Survey map, you are not upholding the honor of our team." Sean isn't smiling.

Before MOLA, the best available maps of Mars were based on various sources. Photographs and stereo imaging were useful for determining the topography of Mars. Measurements of the density of carbon dioxide in the Martian atmosphere helped to determine how much atmosphere there was between the spacecraft taking the measurement and the ground. The higher the atmospheric pressure, the lower the altitude, and vice versa. Maps relying on this data are inaccurate up to one vertical mile. No wonder Sean has a horror of depending on them. In disgust, he adds, "Listen, David, this isn't a paper in *Science.*" What Sean means is that they aren't discussing a mere model, an academic exercise. This is the real thing. They have sent a real spacecraft to Mars, it cost American taxpayers several hundred million dollars, and NASA had better find a site that will actually work, and not just on paper.

Garvin, I notice, never flinches under Sean's scrutiny and mockery. In fact, he later expresses tremendous admiration for Sean—"a phenomenal scientist with tons of experience and a born leader." You have to talk one-on-one with Sean to understand what Jim means. Sean is a relentless critic, but he is also an enthusiastic teacher who enjoys imparting information and understanding. When I take a few minutes to ask him several elementary questions about planetary motion, he carefully sketches diagrams to illustrate what he means and makes sure that I'm following him; for all his ferocity, it's perfectly fine to say to him, "I don't understand." That's the hardest thing to say to Sean, or to anyone else in a MOLA meeting, but it's also the most important thing to say. A lot of scientists don't understand each other, even the scientists who sit side by side on the team. In his role as chief gadfly for MOLA, Sean wants to make sure they do, and he wants to make sure that the data they put out is completely accurate and reliable. There are so many ways it can be tainted or misunderstood—by faulty interpretations or faulty instruments or faulty application of standards. And if the MOLA team acquires a reputation for putting out sloppy data, well, it will be a disaster, because that's what they're all about—bringing back good, clean, reliable data from Mars. Sean wants to make sure things stay that way.

The data will eventually appear in various forms, intended for various audiences, which makes their job more complicated. There will be carefully edited, circumspect articles in professional journals such as *Science* and *Nature*. There will be various data dumps on the mission's websites. Some of those will be highly interpreted, simple enough for a child to understand, while others will consist entirely of the raw numbers. There will even be a series of compact discs that anyone can purchase for ten dollars apiece. The CDs will contain images, text, and a staggering number of numerical measurements. One of the CDs has already been released. When Neumann describes various technical errors on it, Dave becomes so agitated he feigns a heart attack. Although the CD is actually filled with useful and reliable data, Dave utters dark suspicions that its shortcomings can be attributed to flawed decisions made years ago at NASA, when Mars Global Surveyor was in the early planning stages.

Then there is the problem of when to release the data. That is almost as crucial to the team's reputation as the data's accuracy. They want to get the data out and impress their colleagues—that's their job—but they don't want to lose control of their ability to refine and manipulate it. It is theirs and theirs alone, for a while, at any rate. What a glorious feeling it is to know something about Mars that no one else knows, something big. There will be many such moments ahead, deliciously private frissons of iconoclastic insight into the dynamics of the Red Planet, but those moments won't last long. Sooner or later, the scientists on the team will have to get the data out.

Seeking wisdom on the subject, the team turns to Peter Ford, who helps to assemble the raw, fragmented images of Mars into a coherent, accurate map; this is painstaking work. Peter, who teaches at MIT, is also an expert on the topography and cartography of Venus. Tall, donnish, with swirls of salt-and-pepper hair, he stands out in this crowd. His voice—high, thin, and precise—requires close attention, and his accent suggests the patient deliberations of the British Upper Class. He speaks haltingly, nearly drawling, giving an impression of being above the fray as he enunciates some general principles governing the release of scientific data. The MOLA team, he says, should have all the data to themselves for three months: a month to study it, a month to pass it back and forth among themselves, and then, after they have attached their names to it, a month to go public with it in a forum such as *Science*. It all makes eminently good sense.

Sean Solomon has a better idea. "Set an example," he urges Dave Smith. "Be the most generous principal investigator on Mars Global Surveyor. Share all the data immediately, since it becomes public knowledge on a CD in a month, anyway." An intense rivalry exists among the various science teams, even though they share the Mars Global Surveyor, and feelings run especially high between the MOLA people in this room and the team behind the magnetometer, designed to measure the magnetic field of Mars. "They're like vultures," says a voice in the back of the room, referring to scientists outside the team. "Nice vultures."

Dave declares he does indeed want to share data with the other MGS teams for the sake of building goodwill, but he points out that the others won't be so eager to return the favor. In fact, he's certain they won't. Nevertheless, he wants to set an example. He wants to get the data out as soon as they can—just as Sean urges. He briskly cajoles the MOLA team into accepting this procedure.

The discussion is just a prelude to a consideration of the MOLA team's first paper to appear in *Science*. The team has a lot at stake here. If the paper is received well, if it becomes a cover story, it will begin to vindicate a decade's work—not to mention several hundred million dollars of taxpayers' money. "The good news is we are guaranteed to be accepted by *Science*," Maria announces, "so we can do whatever we want. We did not offend any reviewers based on not referencing their work."

Peer review is a fact of life at *Science*; articles submitted to the publication are evaluated by a committee of scientists. No one on the MOLA team knows for certain who they are, but guesses abound. The peer reviewers accept or reject the submissions and suggest changes; they urge caution concerning the implications of the findings. Such comments can drive an impatient or insecure scientist to distraction. The peer-review process gives *Science* a remarkable amount of clout. No other journal, with the possible exception of *Nature,* enjoys quite the same reputation; if an article withstands the scrutiny of its anonymous reviewers and appears in its pages, it is automatically considered important.

In the case of the MOLA paper, the changes *Science* wants are small but critical. "I can't see the justification for the statement . . . that the MOLA data indicate less dust and more ice in the layered terrains than formerly thought," reads one comment. "I also find the argument for a thickness of 5 km of ice at the poles very weak. You assume ice, and assume a state of equilibrium profile. Why?"

"That's *exactly* what we would have written if they'd given us the space!" says one team member. Another cautions that they should be wary of these editors, who are just interested in selling magazines. As the discussion continues, the MOLA scientists boast that they know the identity of their anonymous critics and are confident they can answer their concerns. For the first time, I see another side of the MOLA team. Yes, they are agonizers. Yes, they enjoy savaging each other. But they have a certain swagger. They aren't about to let the anonymous gnomes of *Science* intimidate them—well, perhaps a little, but they will deal with the objections. In the end, the team members believe they are smarter than their critics.

The meeting goes on like this for hours, and it's late in the afternoon when Jim Garvin finally makes his main presentation. Earlier, he simply commented (in his "Garvin minute") on map methodology; now Jim is supposed to discuss his preliminary findings in his principal area of expertise, the craters of Mars. I can only imagine what Solomon will do to him this time around.

When Jim analyzes Martian craters, he isn't just looking at distant holes, he examines subtle and complex clues to the geologic history of the Red Planet. He seeks evidence, or at least suggestions, of how the solar system formed and sites where the conditions necessary for life might have started. To Jim, cratering is "the most ubiquitous process in the solar system. It formed the planets, bombarded them in their infancy, and left behind holes the size of continents, like Hellas on Mars. Cratering has even affected the history of life on Earth." Even better, cratering is exciting and explosive. "Impact cratering involves the collision of materials with planetary surfaces at 'cosmic velocities'—as much as ten miles per second, which is very fast. Simple physics shows that the amount of energy released is huge, and the collision process almost always leads to the annihilation of the impacting object. It basically blows up, forming what we call a cratering flow field that excavates a cavity—a hole—and forms a rim—a ring of hills. It also dumps stuff in a region beyond the rim called the ejecta blanket."

As Garvin would be the first to tell you, craters vary greatly from planet to planet. On Earth, we tend to notice only recently formed, relatively small craters; they're less than two miles in diameter, and they aren't very deep, at least compared to those on Mars. Cratering is heavily influenced by gravity, and since Mars has about sixty percent less gravity than Earth, cratering is "more efficient" on the Red Planet. "Cratering efficiency is simple. It measures how easy it is on a given planet to make a crater relative to the energy required. On planets with a lower 'g'—or gravity—it's intrinsically more efficient." This seems counterintuitive—if Mars has less gravity, wouldn't Martian craters be *shallower* instead of *deeper* than terrestrial craters?—but as Garvin explains, our intuition fails to take into account how explosions work in a gravity field. The effect of gravity is *acceleration,* and Newton "showed us that force works to retard the flow field development while increasing the time available to melt the target and fill the crater from within." So on Earth, with its high "g," craters fill up faster and tend to be shallower than on Mars.

The available Martian cratering record is much older than Earth's; with all the geologic changes in the surface of this planet, our cratering record stretches back merely hundreds of millions of years, but on Mars, the evidence of cratering has survived for billions of years, all the way back to the earliest days of the solar system. Martian craters, therefore, are far more numerous than terrestrial craters, and potentially more revealing about the geologic history of the Red Planet and the solar system. With his laser altimeter, Jim will measure these craters, inside and out, in more detail than anyone has ever measured them. "Since cratering events generate lots of energy, some of which takes the form of heat, I believe there is a high likelihood that temporary lakes or 'ice rinks' once formed in some Martian craters. Young craters that liberated the frozen groundwater and converted it to liquid water could be life-bearing oases on Mars." In his search for potential oases, Jim plans to catalog as many as five hundred young craters on Mars, more than he ever hoped to see. Perhaps some will be fertile; perhaps *one*.

The Martian landers, including Pathfinder, carefully avoided craters, boulders, or anything else that was really interesting in geological terms for the sake of landing on a smooth surface and not falling off the side of a mountain into a canyon and sinking into the Martian dust. This cautious approach made for safe landings but dull prospecting. The Pathfinder's rover and the Viking's landers reported on their rather quiet neighborhoods, and that was about all. The orbiting laser altimeter, by way of contrast, goes anywhere, and returns again and again. It will gather a comprehensive view of the planet; nothing on the surface will escape its notice. It is, simply, an eye in the Martian sky.

With an overhead projector's ghostly image of laser altimeter data looming on the screen behind him, Jim, in a state of high excitement, begins to describe his initial observations. The craters are far more varied than anyone suspected; they have many puzzling characteristics, some of them may be volcanic in origin; there is no way to explain some of things they are beginning to see. Jim gathers speed, and his presentation tumbles out in a hyperventilated rush. Meanwhile, Sean Solomon begins a litany of comments calculated to slow Jim's baud rate: "I don't understand . . . I think you're confusing two issues, and you are confusing me on the issue of ejecta volume . . . You are multiplying your errors . . ." Jim persists until he runs out of breath. At the end of his presentation, Sean cackles derisively and calls out to Dave, "I think you'd better tell *Science* we're not ready!"

Now I see how the MOLA scientists get caught in the cross fire of these science meetings: Dave Smith builds 'em up, and Sean Solomon tears 'em down.

Dave holds his head in his hands, then lifts it slowly as he addresses Jim, who has patiently endured Sean's onslaught. "I see you are being guided by the tradition of an outdated model," Dave says diplomatically. "It's a distorted data set. What are the error bounds?"

Jim answers as best he can, faltering occasionally, and apologizes for speaking so long.

"If you go on for another minute, it'll be an hour," one of the scientists says, bludgeoning Jim into silence. "Have you ever seen him speechless before?"

Jim awaits another volley from the back of the room, where Sean Solomon sits, twitching slightly and smiling enigmatically. All Sean says this time is, "Very promising." Jim leans forward and prepares to launch into an impassioned rebuttal when he realizes he is no longer under attack, and a defense won't be necessary. A contemplative hush settles over the room. This isn't the satisfaction of convergence, exactly; it's the collective realization that Jim Garvin has ventured so deeply into the subject of Martian craters that no one else present can challenge him effectively. There is nothing that Sean can say and nothing anyone else can add. Mars is getting closer all the time, but for now, it still exceeds the limits of science.

Chapter Six

THE HONOR OF THE TEAM

On Monday, the day after their marathon closed session, the mighty MOLA science team presents its data to the plenary Mars Global Surveyor Geoscience Workshop at Jet Propulsion Laboratory, a short drive from Caltech. The point of the workshop, Jim tells me, is "to review the new data being received from Mars Global Surveyor and what implication they might have for the evolution of the solid planet." He tells me that the meeting will be open to all MGS scientists—but not to the public, and certainly not to the press—and it will likely be organized by experiment, with each investigation allotted an amount of time to be used as each principal investigator sees fit. The idea is to bring everyone up to date on the new Mars that is emerging, so that the teams can cooperate, and interpret their own data in light of their colleagues' findings.

That's the theory, and it will be honored, to an extent. In practice, though, this workshop is an opportunity for each science team to see how the competition is faring. Everything said today, the scientists understand, is preliminary, subject to revision, qualification, and debate. Yet they are itching to reveal whatever data they have, however incomplete. Various motives are at work here: the urge to enlighten, the desire to show off, and the need to justify funding for these experiments. Although it may sound cynical to say so, the ideal experiment, from a funding standpoint, yields tantalizing and provocative results, *yet requires further study* (and therefore, more funding). A wholly conclusive result isn't as desirable from a funding perspective, for it means the work is done, and no more money is necessary. The Mars experiments, inconclusive and provocative as they are, approach the funding ideal.

The workshop commences with Mario Acuna's touting his new findings about Mars' magnetic field. Mario is heavyset, with large, almost outsize features, and a low voice infused with an Argentine accent. Because his experiment has been wildly successful (but needing further study!) and offers the scientific equivalent of instant gratification, he attracts the envy of the MOLA team, which still awaits its payoff. From their perspective, Mario and his magnetometer team seem like close relatives who have suddenly and unaccountably struck it rich. Mario's good fortune came about this way: when it became necessary for Mars Global Surveyor to go into an extended aerobraking mode, the spacecraft's elliptical orbit around Mars took it lower than originally designed, so low that it dipped below Mars' ionosphere. On the unexpectedly low swings around the Red Planet, the magnetometer detected signs of an ancient magnetic field. If not for the extra aerobraking, the discovery would not have been made. It was an instance of serendipity, and it was beautiful.

Mario shows some slides illustrating his findings, and he grins. His experiment, the magnetometer circling Mars, has quickly overturned accepted notions of the Martian magnetic field. Currents caused by cooling and rotation in a planet's liquid metallic core create a magnetic field, and before Mars Global Surveyor, the prevailing wisdom was that Mars had none. No one knew why, or what its absence might mean. Now, Mario and his magic magnetometer have discovered that Mars does, in fact, have magnetic "anomalies"—intense magnetic fields frozen in the crust from an earlier time when the Martian core was still warm and churning. But Mario's findings raise new questions. Why did the Martian magnetic field fade? Is there a connection between Earth's magnetic field and the presence of abundant life on this planet? Persuading the people of Earth that magnetic fields on Mars matter is a difficult task, but Mario believes his findings have tremendous implications. "These locally magnetized areas on Mars could not form without the presence of an overall magnetic field that was perhaps as strong as Earth's is today," he says publicly soon after the JPL convocation. "Since the internal dynamo that powered the global field is extinct, these local magnetic fields act as fossils, preserving a record of the geologic and thermal history of Mars."

Mario brashly describes his findings as a scientific "bonanza." He insists they are deserving of more study (and, everyone understands, more funding). He even tells the scientists gathered at JPL that the loss of the billion-dollar Mars Observer in 1993 has turned out to be a good thing in the long run, at least for *his team,* because otherwise they wouldn't have been able to fly their instrument around Mars on the new MGS and make this "wonderful discovery."

Not everyone is as overjoyed as Mario.

"This is the same data we were shown five months ago," Greg Neumann mutters to me as Mario sings his own praises. "There's nothing new here, it all comes from a single orbit, and the data may not be valid." You have to know Greg to understand where he's coming from; more conscience than professional jealousy is at work here. He is balding and bearded and looks like a daguerreotype come to life. He has an impressive reserve that harks back to an earlier century. He wears sandals and a moss green triangular felt hat with three coils of cord wrapped around its crown. Gleaming pins commemorating planetary missions decorate the hat, which Greg is always careful to remove indoors. When he speaks, he reveals himself to be gentle and considerate, essentially a very intelligent, conscience-stricken hippie, the kind of person who was unlikely to have been part of NASA in the old, Cold War days. He tells me that both of his parents were historians, and when he was in college, he appeared headed in that direction himself. Then he spent a summer in Iceland, studying the terrain, and became engrossed in geology. He is not the first geologist who went out on a field trip with no idea what to expect and returned with a career. Still, there were hurdles for him to overcome at NASA. He took no astronomy or geology courses in college, and he refused to work on anything classified because he believes that all scientific information should be shared. So it's satisfying for him to be responsible for releasing the MOLA team's data to other scientists and to the public, and he has retained his sense of independence in this role.

Greg isn't about to shape his data to fit a NASA mold. As far as he's concerned, the mold broke at the end of the Cold War. Until then, a Chinese wall separated the NASA and the Department of Defense, which created a tremendous amount of duplication in the two agencies. They each had their own separate hardware, technology, and facilities. Occasionally, some piece of hardware—usually a booster rocket—migrated from the military to the civilian side, but that was a time-consuming and infrequent occurrence. The worst part of the separation was that NASA was unable to use classified technology belonging to the Department of Defense—technology that could have advanced the cause of peaceful space exploration. "Nothing much had happened at NASA since the end of the Apollo program," Greg says. "Many guys were just sitting around, marking time. There were a lot of buyouts and

early retirements; then the new guys came in, and the place changed dramatically."

A school of thought holds that without a space race to motivate NASA,
and to motivate Congress to fund NASA, the American space effort will
wither or eventually become privatized, like the airlines or the shipping industry. That may yet happen, but in the short run, the end of the Cold War
has invigorated NASA's engineers, scientists, and bureaucrats. The first real
stirrings of a NASA commitment to science, as opposed to winning a space
race, became apparent even before the end of the Cold War, with the Viking
missions to Mars in the mid-1970s. Viking was a hugely expensive project;
it would cost $5 billion to replicate it today. That's a lot of money to spend
on a pure science mission, far more than NASA would consider spending
these days. All that money bought better equipment than current missions
use: higher resolution cameras, more durable hardware, and the luxury of redundancy. There were *two* identical Viking missions, each one consisting of
two components, an orbiter and a lander. If you're going to explore Mars ro-
botically, that's the way to go, if you can afford it.

Greg is an encyclopedia of thoughts about NASA, its purpose, and its
evolution. Talking with him, I sense that he considers it part of his professional responsibilities to keep the agency on the side of the angels. To Greg,
this means distancing it from the military, and on that score, he is hopeful.
He describes a moment in 1992 when Al Gore, the newly elected vice president, suggested that the Department of Defense (DOD) start collaborating
with NASA, which came as a pleasant shock to people like Greg Neumann.
"I asked guys in Naval Research if they could do something, and they would
say, 'Let us think it over and get back to you in a few hours,' and they would
come back and say, 'Yeah, we think we can do that.' At NASA, everything
was heavily bureaucratized and time consuming and worked around 'command blocks,' but all that began to change when the DOD started working
with nonclassified information. The end of the Cold War not only ended
Soviet-American rivalry but also brought down barriers within the U.S.
Defense establishment. Suddenly, separate organizations, with distinct cultures, could share information, and things began to happen very rapidly."
Greg points to the Clementine spacecraft as an example. Originally, this was
a spy satellite designed to look for missiles on Earth. It was converted to map

the moon's surface and, in the process, discovered that water may exist on the moon. The work on Clementine was all done on a last-minute basis. Engineers sketched out its earliest configuration on cocktail napkins late one evening over beers, and Maria Zuber was programming codes for it while it was on the way to the moon. "After Mars Observer was lost in 1993," Greg reminds me, "Clementine was the only game in town."

At the end of Mario's presentation, Greg falls silent, and the MOLA team manages to conceal its mixture of contempt and envy of the magnetic scientists, but just barely. Only Sean Solomon, who enjoys diplomatic immunity within the scientific community, crosses the gulf in the auditorium and huddles with the opposition, as if he were an emissary.

"When Russians have their meetings, Americans often can't go because it's Christmas in the West," says Anton Ivanov as we sit in JPL's sunny cafeteria, eating an abbreviated lunch between sessions. At twenty-seven, the tall, somewhat shy young Russian is a Caltech graduate student, working for Duane "Dewey" Muhleman, a professor emeritus of planetary science at Caltech and a member of the MOLA team's 'murderers' row' of prominent scientists. Dewey is old enough to be Anton's father, or perhaps his grandfather. Anton's actual father, Boris, is a celebrated planetary scientist in Russia, known for his exploration of Venus. As a result of its successful Venus missions, Russia has been identified with Venus in an unofficial way, just as the United States is now unofficially identified with Mars. By international treaty, no country can claim Mars, or any part of Mars, or the moon, for that matter. The situation is roughly similar to that of Antarctica, which does not belong to any one nation. Nevertheless, the scientific colonization of space has already begun, and the political will inevitably follow.

Even now, the people who work at NASA remain highly respectful of the Russian rocket engineers and scientists. No matter what fiascos Russia suffers these days, NASA veterans have vivid recollections of Sergei Pavlovich Korolev, the brilliant and mysterious architect of the Russian space program, who died suddenly and strangely after surgery in 1966, at the height of the race for the moon. Known only as "The Chief Designer," Korolev was responsible for the success of Sputnik, for putting the first human being in space (Yuri Gagarin), and for other triumphs of the Soviet program, and not a few people wonder if the space race might have turned out differently had Korolev survived.

Some of this old Russian rocket glamour has trickled down to Anton, who came of age in the glory days of the Soviet space program, and who now participates in the MOLA team's meetings under Dewey's auspices. When he was very young, Anton began reading sci-fi and scientific books about the planets. His next-door neighbor and family friend, Sasha Basilevsky, was a major player in the Soviet space science program, and he dropped by from time to time, telling tales of the moon and Venus. Later, Anton's father introduced the boy to computers, an integral part of planetary research, and at fourteen, Anton wrote his first FORTRAN program. His father compiled it in his office and brought home the results. "I corrected the errors, and he compiled it again the next day, and so on," he says. "Then we bought a programmable calculator, and you could write a simple program with it in codes."

Anton attended the Moscow Engineering Physics Institute, where he majored in applied math and computer science, and finally became involved in space research when he entered the space science department of the Institute for Applied Mathematics. At a science conference, he met Dewey, who became his American mentor and brought him to Caltech as a special student. "My first impression as I arrived at Los Angeles International Airport was, *Wow!* Look how many lights there are! But Caltech wasn't terribly friendly at first, and I felt quite lonely in the first couple of months," he says, although he came to feel at home in Pasadena.

Russian rocket scientists who make it to the West quickly become enthralled with the extent of American resources and the open-ended way Americans transact scientific business. "Russian decisions are frequently made behind the scenes," Anton tells me. "You've got to have a powerful principal investigator. Ours was said to be Lenin's grandson. At one meeting, he said a certain radar experiment we were planning was not flying. End of discussion. Dewey Muhleman, who was there, was going nuts. There was a Russian general shouting. I was telling Dewey, 'Be calm.' No one stands up at the meetings. Everyone works behind the scenes. In the United States, the challenging is all in public."

That afternoon, Maria Zuber speaks on behalf of the MOLA team. Mario Acuna and his amazing magnetometer make for a tough act to follow, but Maria displays a sure feel for the data; she is MOLA's advocate, and this is the MOLA team's day in court. She painstakingly synthesizes masses of data and gracefully sidesteps the debates that sundered her team the day before. One way or another, she keeps telling the assembled scientists, "These are data you can trust." She amply demonstrates that the laser altimeter works just fine and will eventually yield an extraordinarily accurate map of the Red Planet, its polar caps and craters, and even its possible oases of life. From her smooth presentation, you would never guess that only the day before Dave Smith was feigning a heart attack over the data on the preliminary CD and that Sean Solomon was savaging Jim Garvin's graph.

At the end of her presentation, Arden Albee stands. As the project scientist for the entire MGS mission, Arden looms large. From a distance, he appears to be gruff and intimidating. A NASA veteran, he holds several degrees in geology from Harvard, and although he occupies such a high level on the organizational chart that he's not really involved in the daily life of the experiments, Arden is the one to whom everyone must finally report. Now he thanks Maria and the rest of the MOLA science team for revealing their "wild ideas and hypotheses" to their colleagues and rivals and lets his little joke float over the assembled scientists. Speculation, exaggerated claims, and oohing and ahhing over partial data rub him the wrong way. He cautions everyone not to discuss their findings publicly for a month, a request that many will find difficult to honor. "This will revolutionize all our thinking about Mars the way the study of plate tectonics completely changed our understanding of Earth's geology," says one scientist to another, who tugs on his woolly beard in reply.

Only toward the end of the meeting does the subject of astrobiology—life beyond Earth—finally come up. The scientists have shied away from discussing the issue until this moment, when Bruce Jakosky, a tall, bearded Caltech-trained geology professor at the University of Colorado, weighs the pros and cons. He lists the "high-level requirements" for extraterrestrial life—liquid water, access to biogenic elements, and a source of energy—and on that basis he declares that the latest results from Mars Global Surveyor are still inconclusive. Yet the possibility of life on Mars cannot be ruled out. No one takes issue with him, despite the controversy surrounding the subject. Astrobiology seems way too hot to handle at the moment. Greg explains that "life is the Holy Grail" of all Mars research, and the search for it must be carried out correctly, the methods above reproach, the conclusions unassailable. That isn't going to happen here. Arden Albee, the boss of bosses, insists that the evidence presented today offers a devastating argument *against* life on Mars. "No life on Mars!" he proclaims, as if by fiat.

Unmoved by the pronouncement, Bruce Jakosky says simply that Albee's conclusion applies only to life on the *surface* of Mars, or within the last 3.2 billion years. What about *below* the surface? What about *early* in the life of the planet, during its first billion years? Jakosky takes a long view of the possibilities of extraterrestrial life. He has just completed a thorough overview of the subject, *The Search for Life on Other Planets,* which argues that the discovery of life-forms beyond Earth may be traumatic and difficult for society to accept. Think about the near-havoc Darwin wreaked on science and society, Jakosky says in this book. Consider how jarring to humanity were the discoveries of Nicolaus Copernicus in the sixteenth century. Think about it.

From the time of the ancient Greeks through the Dark Ages, most astronomers believed that the Sun revolved around the Earth. The most influential exponent of the geocentric system was Ptolemy—Claudius Ptolemaeus—a Greco-Egyptian mathematician who lived in Alexandria in the second century A.D. In this model, the Sun, moon, Mercury, Venus, Mars, Jupiter, and Saturn revolved around the Earth, each in its own crystalline sphere, traveling along a perfectly circular orbit, nested one inside the other. The stars occupied the outermost sphere of all, and as the entire system rotated, it generated the music of the spheres. One outstanding problem with this harmonious system was that the planets occasionally seemed to adopt a retrograde motion, as if backtracking. In his *Almagest* (Arabic for "The Greatest"), a thirteen-volume study of astronomy and mathematics, Ptolemy explained the planets' peculiar motions across the heavens on the basis of epicycles, essentially little orbits within their larger orbits. The geocentric system remained dogma throughout western Christendom for the next 1,300 years.

A profound cosmic disillusionment occurred in 1543, when a Polish physician and lawyer named Nicolaus Copernicus issued his *De Revolutionibus Orbium Coelestium (On the Revolutions of the Celestial Spheres)*, which he dedicated to Pope Paul III. In this work, Copernicus proposed that the Earth revolved around the Sun, not the other way around, as the system codified by Ptolemy insisted. Copernicus died only a month after his book was published, and his ideas caught on slowly, without official opposition from the Church. They occasioned a number of theological debates, including one on the nature of extraterrestrial life, arguments that contain distant but distinct echoes of contemporary astrobiology. If the Earth is not the center of the universe, but merely another planet, as Copernicus proposed, could life exist on many planets? And if it did, what form did it take? Had these extraterrestrials also fallen from grace? Was Jesus Christ their Savior?

As the years passed, the Copernican view became more influential, and it attracted controversy and controversial supporters such as Galileo. Brilliant, abrasive, and quick to offend, Galileo was forbidden by the Church to teach Copernican theories, even though the Church had not officially condemned them. Galileo inflamed his critics by calling Ptolemy's supporters "dumb idiots." After a trial in 1633, Galileo was ordered to spend the remainder of his life in seclusion, under house arrest. Undeterred by the Church's condemnation, he insisted there was no contradiction between belief in God and belief in a Sun-centered, Copernican universe. He expressed the relationship between science and the Supreme Being this way: "I do not feel obliged to believe that the same God who has endowed us with sense, reason, and intellect has intended us to forgo their use."

In 1859, Charles Darwin launched another scientific revolution, this one concerned with the evolution of life. By emphasizing natural selection as the engine of evolution, Darwin demonstrated that life had evolved in a series of more or less logical, if random, steps from the most primitive forms of life on Earth. Taken together, these two revolutions were humbling. Darwin, for instance, suggested that life, even human life on our little planet, was not the miraculous, unique event of our fond imaginings, but the outcome of random and routine biological processes. Darwin endured—and continues to endure—his share of slings and arrows, but his work, like Copernicus', permanently altered our world view. If our planet is not the center of the universe, if human beings evolved naturally from simpler life forms, and if life exists elsewhere, then we are not alone; neither are we particularly significant in the cosmic scheme of things. Whatever significance humanity may have becomes an entirely new and different question than has been assumed for thousands of years. In his book, Jakosky writes:

> There has been a gradual recognition that life probably formed on Earth through chemical processes acting within the ambient environment, that a similar environment on another planet could result in the formation of life there, that there are several planets and satellites within our own solar system that have the potential to harbor life, and that planets with appropriate environments may be widespread throughout the universe. Of course, expecting that life will be found on other planets is different from actually finding it.
>
> Finding non-terrestrial life would be the final act in the change in our view of how life on Earth fits into the larger perspective of the universe. We would have to realize that life on Earth was not a special occurrence, that the universe and all of the events within it were natural consequences of physical and chemical laws, and that humans are the result of a long series of random events.

Jakosky's assumption is not the premise of escapist entertainment, but a terrifying revelation of the accidental nature of our existence and our place, or lack of it, in the universe.

With all these questions and implications left hanging in the air, the convocation at the Jet Propulsion Laboratory concludes, and the scientists disperse to waiting airplanes, crowded classrooms, shrinking budgets, and baffling data. Shutting down her laptop, Maria looks at me plaintively. "Can you *believe* this is how we explore outer space? Can you believe it is done this way? It makes no sense! It's amazing anything comes out of this." And I am amazed, as well, by the complete lack of consensus concerning Mars. Given all the disagreements I have heard today, I wonder aloud if anything has changed regarding the prospect of life on Mars since the Viking era.

"Everything has!" Maria shoots back. "In the late seventies, the data were either negative or inconclusive. The two Viking landers we sent to Mars were looking for life as we know it, breathing oxygen, and so on.

Since then, the Mars rock has changed everything to do with people's attitudes and perceptions." The meteorite from Mars bearing nanofossils is a good start, but still not widely accepted as proof that life once existed on the Red Planet.

Maria emphasizes that a number of other recent advances have pointed to the possibility of life beyond Earth. In astrophysics, organic molecules have been identified in molecular clouds, which are found throughout galaxies. She tells me that from analyzing RNA in cells, "we now know that the most primitive life-forms are not oxygen-breathing creatures that exist in temperate environments"—which was what the Viking life-detection experiments were designed to seek—"but methane breathing, sulfur-munching 'little monsters,' single-celled, extremely simple life-forms." She mentions their more complex descendants who exist on Earth today, hideous-looking tubelike creatures called black smokers living on the seafloor at vents where volcanic gases are released into the ocean. "Black smokers thrive in two-hundred-degree water. No light. No oxygen. And there are extremophile life-forms under the ice in Antarctica, and recently discovered organisms in a mine shaft in California that thrive in a pure acid environment. Maybe we need to be a lot more open-minded," Maria says. "The one thing these extremophiles have in common is that they all like water. The old formula for life was chlorophyll, oxygen, water, and sunlight, but now they're finding that water and an energy source are all that's necessary." I've heard this before, but it bears repeating: finding evidence of liquid water on Mars is tantamount to finding evidence of life.

Maria adds, "I think the obituary for life on Mars is quite premature. Evidence presented against life includes the significant ultraviolet radiation at the surface due to the thin atmosphere and a variety of observations suggesting that water didn't last on the surface long enough to chemically react with rocks. But it should be apparent even to a non-scientist that no one understands the data, yet. Currently, I am wondering how you liberate a lot of water on the surface, yet it does not persist. I don't know where the water went, but I know it was there. And it may *still* be there, trapped beneath the surface. Understanding the history of water is the key to finding out whether life could have developed on Mars. When you are in a mode where the data is flowing for the first time in twenty years, it is important to keep a flexible point of view."

That night, after nearly ten hours of scientific debate, the MOLA team locates a trendy Pasadena microbrewery. A sense of calm settles over the members as they eat and drink and slowly unwind from the rigors of science meetings, science arguments, and science anxieties, all of which have a way of disappearing after a beer or two. Jim Garvin is there, in his Indiana Jones leather jacket; Sean Solomon is there, unexpectedly quiet and affable, in his leather jacket; and in a corner, Dave Smith, the team's Energizer Bunny, is still going strong.

When the check arrives, silence descends. The prospect of settling the bill causes each scientist to wince a little, mostly for show. They study the statement carefully; it's not a mammoth amount, just $130 or so for ten people. The scientists painfully withdraw crumpled ones and fives from their wallets; the money looks as if it's been pried from piggy banks. They try to figure out who owes what, as they count and recount the money three, four, five times. I wonder why they take the bill so seriously, and debate it so strenuously, and then I realize that it's more raw data requiring interpretation and substantiation. Only then will it be considered valid. When I offer to pay a bit more than my small share, they tell me I can't, it's not proper, there are strict rules governing these matters, and they are so serious that I let them go back to dividing the bill among themselves. "My wife says I'm financially challenged," Jim mutters, shaking his head. "I can run a hundred-million-dollar mission, but when you work for the government . . ." He speaks darkly about looking at private schools for his children, realizing he might not be able to afford the tuition on his NASA salary. Then, after a long day of pondering fundamental questions about the nature of the solar system and the universe, Jim and the others scatter into the fragrant California night.

The MOLA team's new data appear in the March 13, 1998, issue of *Science* under the title "Topography of the Northern Hemisphere of Mars from the Mars Orbiter Laser Altimeter." It is part of a series of reports from Mars Global Surveyor in the issue, one for each team, plus an overview by Arden

Albee of the entire mission. The cover features a moody image of the Red Planet taken by the Mars Global Surveyor's camera; it reveals the surface of Mars with supernal clarity. It looks close enough to walk on. Inviting enough to touch. Familiar enough to inhabit. The MOLA article is very much a bulletin from the front. This is not the team's definitive word on Mars; it's just a first impression, but it's startling. The introduction states, "The first 18 tracks of laser altimeter data across the northern hemisphere of Mars from the Mars Global Surveyor spacecraft show that the planet at latitudes north of 50° is exceptionally flat; slopes and surface roughness increase toward the equator." The next observation is more tantalizing, since it advertises the presence of water on Mars and, indirectly, the possibility of life-forms: "The polar layered terrain appears to be a thick ice-rich formation."

These provocative observations start Jim thinking: if Mars has active, or recently active, volcanoes—say in the last million years or so—and if any of these volcanoes are near the ice, it is possible that they could melt some of it and produce conditions favorable to the generation of life. It will be a tough sell, he knows, and he keeps the idea to himself for a while, as he begins to sift through the MGS data for evidence of volcanism that might fit his hypothesis. At Goddard, his chaotic desk disappears under stacks of glossy reprints of the issue, and he gives them away to interested parties at every opportunity.

During the weeks following the JPL and Caltech meetings, events threaten to overwhelm Jim Garvin. He endures chronic anxiety over the Mars Global Surveyor's risky aerobraking maneuvers. The technique has never been tried in the Martian atmosphere, and now, because of the buckling of the panels, it will take a year longer than planned.

In late February, his father, retired from IBM and living in Arizona, suffers a heart attack. Jim realizes that he must make the cross-country trip to visit, though it means that he will probably miss the turn-on of MOLA, when the data from his instrument will finally begin to pour into the Goddard Space Flight Center. For a decade, he has anticipated his first glimpse of the instrument's rich, error-free data stream; instead of this scientific nirvana, he finds himself on an airplane, immersed in a family crisis. In the end, he compromises by visiting his father briefly and flying home to Washington, exhausted, but just in time to catch MOLA's awakening from deep space hibernation.

Subject: MOLA TURN ON AND PLAYBACK!
Date: Sat, Feb 28 1998 11:56:09
From: Jim Garvin <jgarvin@nasa.gov>
To: Laurence Bergreen <bergreen@NYCnet.net>

Got up at 1 AM to be at Goddard Space Flight Center for the data playback. It was stressful since I didn't want to wake the kids. Got there in time to see the playback from MARS! What an exhilarating feeling to see the cross section of the huge polar cap—at least 2 kilometers in relief—and clouds! MOLA was able to range off of clouds, which tell us about atmospheric water vapor and aerosols, which affect climate.

My emotions were high since I learned the evening before that my Dad was home from the hospital after his heart attack. God and Mars are good.

Jim

Chapter Seven

GODDARD

Jim Garvin comes straight at me as I disembark from a train near Greenbelt, Maryland. Six weeks have passed since the Caltech conference. It is now a splendid day in early April; soft breezes blow, trees burst into blossom, and shoots of tender grass peek from the ground. He's wearing a leather jacket, a loose-fitting blue button-down shirt, and a tie, given to him by his mother, depicting Snoopy in outer space. It's held in place by his NASA identification tag, which dangles from a metallic chain. He wears the tie partly to amuse his son, Zachary, "three point seven years old," as Jim puts it; and, I imagine, partly to indulge his own whimsy. He looks like a scientific Pied Piper, here to lead the children of Earth to Mars. His pace comes as a jolt; I've forgotten how fast his baud rate can get. I've learned to call this type of behavior "the Garvin Effect," meaning an overwhelming amount of energy and enthusiasm. Jim Head, who taught Garvin at Brown, once told me that "it takes a few days for the air to settle after Garvin visits," and I knew exactly what he meant.

At the moment, Jim is waving a large, multicolored graph, excitedly showing it off. He tells me it exploits recent data from Mars to plot the topography of a large volcanic crater, Uranius Patera, which has never before been mapped in this much detail. Uranius Patera is in the northern hemisphere of Mars, and as geological monuments go, it's very young. The climatic conditions there are utterly different from those responsible for this mild spring morning in Maryland. The atmosphere at Uranius Patera is frigid, unbreathable, polar. Fierce winds drive the fine Martian grit into the ice. The *patera*, or volcano, stands as mute testimony to ancient geological mayhem, and it may bear witness to water. There are some interesting anomalies about this crater that Jim's mind has fastened onto, and he can't wait to return to the lab to analyze them.

In the weeks since I last saw Jim, he's been difficult to reach, no question; his e-mails were cryptic. Occasionally, he fired off midnight bulletins telling of time-consuming, off-site review panels on which he sat, of articles he co-wrote about MOLA's latest findings for *Science* and *Nature* and the *Journal of Geophysics Review.* He kept me current on the health of Mars Global Surveyor as it continued aerobraking itself into circular orbit around Mars. Then, in one very recent e-mail, he finally came across:

Subject: MOLA TURN ON AND PLAYBACK!

Date: Sat, 28 Mar 98 05:05:03 EST

From: Jim Garvin <jgarvin@nasa.gov>

To: Laurence Bergreen <bergreen@NYCnet.net>

Good Morning, Larry!

We are here at Goddard Space Flight Center awaiting playback of our first MOLA pass over the north polar cap of Mars. We got the first ever cross section of the north polar cap and saw consistent clouds with MOLA! A great pass—it came down by 3 AM and we have been delightedly scrutinizing it. Looks like the cap may have 1.6 miles of relief.

Jim

After receiving a series of such communications, I sensed he was too excited, too wound up to sleep, even in the best of circumstances. No matter how busy he says he was—and no doubt he's doing the work of two people, or maybe more—he seemed to be holding his breath. And now is the time to exhale.

"Any surprises so far?" I inquire as we stride through the parking lot.

"We have a lot of information in the shape of a polar cap now. We knew so little about it that 'surprise' is not the right word to use. We're getting excited about things we've never *seen* before. Give me one of your bags."

"That's okay, they're light."

"Are you sure? We think we may have a crater in the ice cap, which would be great because it would puncture through the whole thing."

I glance at the chart he brought to the station. "Does this show liquid water or ice?" Water, of course, is the indicator of potential life.

"We can't tell, yet. The good thing is we're detecting abundant clouds up around the polar region. Now, the question is: are they water, or not? If they are, then they have a water cycle that's putting that water into the atmosphere. We have about a hundred-yard walk here, I'm sorry," Jim says, as we try to locate his aging but immaculate silver BMW bearing "MARS 7" plates.

We zip off to the Goddard Space Flight Center, his place of work. The center is named for Robert Hutchings Goddard, the original rocket scientist—and a Mars fanatic. As an innovator and inventor, he's as important as the Wright brothers, but not nearly as well known. Born in 1882, Goddard, like so many other scientists and astronomers, grew up under the sway of the science fiction of his day, especially H. G. Wells' tale of Martian invaders, *The War of the Worlds,* which Goddard read in serial form as a child. Soon after reading Wells, he sat in the branches of a cherry tree and began to daydream. "I imagined how wonderful it would be to make some device which had even the small possibility of ascending to Mars, and how it would look on a small scale, if sent up from the meadow at my feet," he wrote some years later. "I was a different boy when I descended the tree from when I ascended, for existence at last seemed very purposive."

Goddard blossomed into an extroverted and popular student at Worcester Polytechnic Institute, where he wrote both the music and lyrics for the school song. In the other hemisphere of his brain, he envisioned a vacuum tube railway running from New York to Boston. It was a marvel of complexity and hugely impractical. Later, at Clark University, he resumed his work on rocketry but abruptly stopped when doctors diagnosed him with tuberculosis. The experience of illness proved to be a turning point in his life, focusing his intellectual powers. While convalescing, he applied for patents essential to modern rocket science: one for a combustion chamber and nozzle, another for feeding propellant (liquid or solid) into the chamber, and a third for discarding successive stages of a rocket as they consumed fuel. He eventually received all three, which became the cornerstones of American rocket technology.

As he recovered from tuberculosis, he published scientific papers. His monograph, *A Method of Reaching Extreme Altitudes,* attracted little notice at first, but he managed to obtain funding from the Smithsonian Institution for further rocket experiments. The Smithsonian publicized his work, and to his surprise, his obscure theories, in corrupted form, inspired a front-page story in the *New York Times* on January 12, 1920. The next day, an editorial in the paper accused him of making the same mistake Jules Verne had made. As everyone knew, a rocket couldn't burn in the vacuum of space. Goddard published a polite rebuttal indicating that further study was needed. Meanwhile, the notion of human space travel achieved popular currency. A suffragette received a lot of publicity when she volunteered to go to Mars in a rocket. In this feverish atmosphere, Goddard attempted to clarify the misunderstandings about his work—for one thing, he wasn't interested in sending people into space, only spacecraft—and then gave up and retreated into honorable silence, which led to more publicity, as he became known in the press as the Mystery Professor.

His ideas were taken more seriously abroad, especially in Germany. The president of Clark University ominously remarked, "German and U. S. scientists are in a rocket race. . . . They are using the principle Dr. Goddard has worked out and which has been published." His theories helped to inform the development of the V-2 missile and stimulated the imagination of Werner von Braun, who often said he had been inspired by Robert H. Goddard.

On March 26, 1926, Goddard launched a small liquid propellant rocket from a farm in Auburn, Massachusetts. It was a simple affair; the launch tower was a modified windmill ordered from the Sears Roebuck catalog, and the rocket reached a height of only 184 feet, but it was the first liquid propellant rocket to fly successfully. As engineering milestones go, the launch was equivalent to the Wright brothers' celebrated flight at Kitty Hawk, North Carolina, twenty-three years earlier. Goddard was still guarding his privacy, and many months passed before word of his accomplishment reached the newspapers. When the story finally broke, he attracted the attention of Charles Lindbergh, who used his prestige to obtain $50,000 from the philanthropist Harry Guggenheim for Goddard's rocket experiments.

With his first solid financial backing, Goddard and a small team of assistants packed up their rockets and traveled to Roswell, New Mexico, of all places. The irony in his choice of location is extreme, for these days, of course, cultists believe that Roswell is the center of secret and nefarious government-sponsored UFO activities. In those days, it was mostly an endless prairie. There Goddard and his team launched ever more sophisticated rockets, each one nicknamed "Nell." By the time various Nells were able to reach an altitude of 2,000 feet, Goddard's funds ran out, and he returned to Massachusetts to resume teaching and to continue his work until his death in 1945. Awards and recognition, including a Congressional Medal, came posthumously to the Mystery Professor of American rocket science.

The NASA center bearing his name is a sprawling, campus-like complex set in the hills of suburban Greenbelt, Maryland, about a thirty-minute drive from Washington, D.C. It could easily be mistaken for the campus of a large state university, were it not for the impressive security surrounding it. Nearly twelve thousand are employed here. About a quarter of them are civil servants, like Jim; the rest work on site for various private contractors such as Lockheed Martin, Allied Signal, TRW, Raytheon, and other major aerospace industry players. This is also the home base of many of NASA's leading scientists. In fact, Goddard claims to have the largest number of scientists anywhere. The science department here—Code S in NASA speak—includes Dave Smith, Greg Neumann, and Mario Acuna. Goddard covers more than a thousand acres and has more than thirty capacious buildings; new ones go up all the time. The place is so big you generally drive from one anonymous building to another. They have numbers, not names, and it takes several visits to learn to distinguish one from another. To help myself tell them apart, I have given them descriptive names: "Red and Beige Building at Right Angles to the Curving Building," "Building with the Cafeteria," "Building that Curves Weirdly to the Right," and the memorable "Building with the Large White Tank of Nitrogen in Back."

I sign in, receive credentials, and duck back into Garvin's waiting BMW. "They'll need to see this as we go through the gate," Jim says, pointing to my badge. The guard waves us through, and we drive onto the site. Jim parks, and we trot toward one of the featureless buildings. Inside, this building, like the rest of Goddard, has a strictly government-issue, cinder-block feel. The scientists who populate its warren-like offices and corridors look like refugees from academia, and many are. They wear khaki pants and flannel shirts and hiking boots; most everything about them is environmentally correct. A significant number are women. Little Martian trolls adorn computers and windowsills, and crumpled cardboard rockets reminiscent of Buck Rogers and Tom Swift teeter on ledges. I get the sense that the more retro the toy, the greater its value; a rusty, dented rocket ship from the 1950s has more cachet than, say, a recent *Star Wars* figure. These decorations give the place the feeling of a large kindergarten.

A conspiracy nut could get lots of ideas by reading the news items posted in the hallways. Taped to an office door, a yellowing newspaper article of doubtful provenance reveals the untold story behind the loss of Mars

Observer: the spacecraft discovered evidence of current Martian life, and NASA destroyed it in order to keep the discovery a secret. There are astounding stories on display about the military's armed conflict with the Red Planet's current inhabitants, all of it hush-hush. An article from a supermarket tabloid offers photographic evidence of a top-secret manned base now operating on Mars. To my eyes, it looks just like an Arctic base, right down to the little windows and idling Jeeps. I think the waiting dogsled has been airbrushed out of the picture.

And then there's the "Face on Mars," the round, Sphinx-like visage staring dully into space from the Cydonia region of Mars. It's been causing controversy since Viking imaged it in 1976. There it was, "scientific" evidence of an ancient Martian civilization, according to some people, and NASA has been concealing it from the world! The Face amounts to a kind of Rorschach test. If you study it from one angle, it looks an awful lot like a rock. From another, it bears an uncanny resemblance to Senator Edward Kennedy—at least, that's what some people think. Others claim it's a likeness of Carl Sagan. Or it might just be a portrait of a Martian druid, and if that's the case, what are the Martians trying to tell us?

Another posted article proclaims:

MARTIANS TRASH GODDARD
MOLA MEANS WAR,
ALIENS SAY

Greenbelt, MD: Martians destroyed the Goddard Space Flight Center today. The attack came in revenge for zapping their planet with laser beams from the altimeter aboard the Mars Global Surveyor spacecraft. . . . Dave E. Smith, Goddard's lead scientist for MOLA, said, "This could have been avoided if the Martians had only sold us a map!"

"I'm a conspiracy theorist myself," says Greg Neumann, the MOLA team's data expert. "I know where the conspiracies are and are not, and they are not here. People shouldn't confuse their well-founded suspicions of government with NASA, which is a civilian agency and does not do classified research." He's sitting in a large conference room stacked with computers, silently contemplating the latest information from the Mars Global Surveyor; he wears a black fleece jacket, a blue Oxford cloth shirt, blue jeans, socks, and sandals. There's a low electric hum in the background; you can feel ambient energy surging. "Let me give you an example of what I mean. NASA's use of fissionable, or nuclear, material in outer space is sensitive. Most scientists oppose the use of nuclear material in Earth orbit, as do I, but there are some who advocate that NASA should be allowed to put the material on the Galileo spacecraft, which is now circling Jupiter, because it's the best use of available technology. The same situation applies to the Cassini spacecraft, which is on the way to Saturn; it's so far from the Sun they have to use nuclear energy to power it. The Cassini spacecraft had an environmental impact statement this thick." Greg puts four feet between his outstretched hands. "Still, there was a sizable protest against the launch of Cassini, and some of the people who spoke at the rallies were notable scientists, but it seemed to me the level of debate was vague and totally emotional. Sad to say, a lot of things are dangerous. Every day we ship natural gas across state lines, for example. On the other hand, we aren't killing trees to get our data out, we're using the World Wide Web, where the space is free."

Greg takes his role as data guru with complete seriousness and ruminates endlessly on his product. "Data are never perfect. They are always contaminated with errors. The art is in how you handle these errors. If it were all perfect, we could be theoreticians, plug the data into a computer, and have the computer solve it, but life is not like that. We run it through sanity checks. If the Northern hemisphere of Mars has a corrugated shape, I'm going to wonder about it. When I came on board here, in the summer of 1993, to analyze the data from Clementine, the moon spacecraft, we had the noisiest data ever: ten percent data, ninety percent no data, or false data. The detectors hadn't been calibrated accurately to track topography. On the other hand, working with MOLA is relatively easy, and everybody's thrilled with it. Right now, it's performing at its theoretical limit."

He'd prefer to hold on to this material and refine it to his satisfaction. "But we can't just live in our own world. There are a lot of people in Congress who would like to see us on the street. The days of the gentleman scientist, if they ever existed, are long gone. The idea is to get the data out. The scientists would like to have a jump on everyone else, of course. I've had requests from kids to see these data. We would like to do cutting-edge science *and* have people share the excitement."

Jim is eager to show me the latest data from the Mars Global Surveyor, but the server at JPL, where the data are collected before being sent to Goddard, is down. He sees the glitch as symptomatic of progress. "Viking's memory couldn't hold a whole image; it was only sixteen kilobytes. It would take part of a picture, radio it back to Earth, take some more, radio it back. Then you had to go to JPL in Pasadena, and wait, and they gave you *tapes*. You got your tape, ran back to your favorite computer, mounted it, and worked on your data. These days, it all goes into computer banks. It's great stuff."

"When will their server be up again?"

"We think today. We were able to look at data from yesterday even though our network was down, in engineering mode, and that's where this picture I will show you is from." Even at Mars' closest approach, it takes well over ten minutes for a radio signal to travel between the two planets, and the situation rules out real-time communication between the spacecraft and Earth. "Unfortunately, there's a speed limit in space," Jim reminds me: the speed of light.

We step into his office. It has a geology of its own, peaks and valleys made of teetering piles of reports and raw data. His bulletin board displays an old photo of Bobby Orr, the famous Boston Bruins goalie, diving across the ice to block a puck. Jim, a veteran of high school and college hockey leagues, holds out his hands to me, pointing to various fingers. "I have injuries from hockey here, and here, and here, and this finger will always be crooked from an injury. And I have one here—" he points to an eyebrow "—which is probably what knocked all the brains out of my head!"

Waiting for the server to wake up, Jim and I wander into a room, where he points to a replica of Mars the size of a basketball, and pulls out his laser pointer. "Mars, Earth, Venus, the moon, and Mercury are the big rocky planets," he reminds me, as he waves the pointer at a map of the solar system. "Then there are two big asteroids out beyond Mars—Ceres and Vesta— that could be included, but we won't; they're even smaller than the moon. The planets in our neighborhood are *rocky*. They're made of stuff, but other planets, such as Neptune and Jupiter, consist largely of gases coupled to dense interiors that are made of metallic hydrogen." If you compressed hydrogen in gaseous form enough, it would become a metal; this process is difficult to simulate on Earth. You would have to bore deep into the Earth's core to find pressure strong enough to convert hydrogen into its metallic form. Anyway, unlike the outer planets of our solar system, Mars, Earth, and Venus were all probably formed from the same basic materials. Jim tells me that planets are either hot as they form, and they stay hot, or they're cold and they heat up because of internal processes. "In the case of Earth, there's a crust. It's like a sheet of ice covering a pond, and it's very thin relative to the size of our planet."

Mars is also hot inside, even though its surface is often ice-cold. "The first law of thermodynamics says one simple thing: everything in the universe likes to achieve a state of relative equilibrium," Jim notes. "Planets that are hot want to cool down. They want to remove that heat, transfer it away. There are different ways of doing it. Big planets like Earth and Mars and Venus have vigorous internal engines to transfer the hot energy out to the surface, and they're called volcanoes." To Jim, there is beauty in this mechanism: volcanoes as giant, mesmerizing energy transfers.

One other thing about Mars that attracts Jim's attention: with its low atmospheric pressure and internal heat, the Red Planet might have had water in all three physical states—liquid, solid, and gas—if not now, then early in its history. "Earth and Mars both have a rocky outer shell that is partly mobile and is affected by the action of *liquid* water in the system, the action of *frozen* water in the system, and the action of *gases* in an atmosphere that interacts with the surface." Taken together, they constitute the planet's water cycle. It's similar to Earth's cycle, but there are some crucial differences. Water moves around Mars in unusual ways, perhaps because of its giant volcanoes and dust storms.

Jim displays another, very recent image of Mars. It's a crater about thirty kilometers across showing what he says is strong evidence of floods of "mud-like substances." There's a story behind this crater. It formed a hole, excavated it, threw materials out of the hole, and then piled them up in a "ramp-like deposit" around the perimeter. He theorizes that this impact melted the frozen ground and triggered mud flows—"we call them fluid ejectas"—from which he deduces that something extra was needed to make this crater in just this way. We don't see anything quite like this crater on the moon, Jim says, and only indirect evidence of them on Earth, yet we see them on Mars. He asked himself, "What do they suggest about what's *under* these craters?"

He found an answer while playing in the mud, as good a way as any to do planetary geology. "I did an experiment in the Reis Crater in Germany with a bunch of school kids. There was a big puddle of well-mixed mud. I took a rock and went *whoosh!* as hard as I could. I don't have a ninety-mile-an-hour fastball, but it's reasonably hard, in the sixty-mile-an-hour range. I produced this beautiful little crater with residual patterns of mud not unlike this one on Mars. But it was not a hypervelocity impact. The craters on Mars were made by something coming in faster than your literal speeding bullet; something like thirty kilometers a second. On Mars, there are *thousands* of these craters. What are they saying? Are they telling us that ground ice has been liberated by the energies of impact? We don't know. We have theories."

To test his theories, Garvin and a few colleagues undertook an expedition to a crater in central Kazakhstan known as Zhamanshin. "It's thought to have formed about eight hundred thousand years ago, and the impact landform was extremely shallow, similar in shape to some impacts on the Red Planet," he says. Just getting there proved hazardous. "This was in the last days of the Soviet Union, and the only way we were able to undertake this scientific expedition was thanks to a Soviet academician who had an interest in geology." They drove along a bulldozed road for more than three hundred miles to reach the crater. "I was the first American the Kazakh nomads who wandered through there had ever seen. They touched me to see if I was real or a ghost." Before Jim's expedition could return with special equipment, "the country went crazy, and we haven't been able to get back since."

In the absence of new data, Jim returned to old data from the Viking missions to satisfy his curiosity about Mars. "Viking left us a legacy of many things, of sunsets and sunrises and views of the moons, summits of volcanoes, clouds and canyons. Viking was so darned good! I can't say strongly enough how amazing the mission was. Viking was designed to last thirty days in its nominal mission, but it worked for five years. It was still running in 1982, and every two days a picture would appear at the National Air and Space Museum in Washington, direct from Mars, in real time."

The current orbiter, Mars Global Surveyor, circles Mars at a distance of about two hundred kilometers (one hundred and twenty-four miles), and in the absence of a thick atmosphere, the images taken from this altitude dazzle with their clarity. "Imagine this kind of data for thousands of places on

Mars, not one or two, which is all we have now, but everywhere! You take a Viking image of a big crater, say thirty miles across." Jim serves one up; it looks as blurry and quaint as a snapshot in an old vacation photo album. "There's a big lumpy ejecta blanket all around it. It looks like a fried egg." In this image, the Martian landforms *are* vaguely reminiscent of one egg, sunny-side up, but I wouldn't have noticed the resemblance unless Jim suggested it. "Now here is the track that the laser altimeter made of the same crater." Jim shows me the latest image from Mars Global Surveyor, and the same crater suddenly leaps out in all its spooky three dimensionality. Unconsciously, I extend my hand toward the screen to touch it; at this moment, I can practically *feel* Mars. "You can see how the crater goes down and comes back up, and has a little mountain in the middle. It's about twelve hundred feet high." I feel a sudden urge to walk across the crater, to leave some footprints on the little mountain in the middle. After our day trekking across Surtsey's Mars-like terrain, I can imagine what it would be like to walk over the hills and dales of Mars.

Jim theorizes that craters on Earth gave early warriors the inspiration to build ramparts as protection against invaders. He believes the circular shape of many castles derives directly from craters, and, I suppose, if you look at a crater long enough, it does come to seem like a natural fortress. "When you think about Mars," Jim says, "it's important to remember the planet in the context of Earth and Venus, just for scale. Remember this: Earth has two topographies, continents and ocean floors. And Venus has high places and low places. But Mars has hemispheric boundaries between the smooth northern plains and the rough, cratered southern upland, so the question on Mars has always been: what's the source of that dichotomy? Is it age? What if all the low plains of Mars were flooded? That is not unrealistic, given some of the new emerging evidence of the role of water on Mars, but we don't know."

"I bet you have theories."

"You can't preclude Elvis!" He displays a picture. "I call this, 'Headlights on the Horizon of Mars.'" And that's what it looks like, or possibly the twinkling lights of Martian cities. Finally, I ask him what it is, really. "Actually, it's a sunset from Viking 1, but when it was first shown to some of those Elvis folk, they said, 'Oh, my God, there are *lights on the horizon!*' But that's only the Sun twinkling behind big rocks on the horizon."

"So you haven't found those Martian cities, yet."

"Not yet, Larry, but we're still looking. You'll be among the first to know."

On a large monitor, Jim pulls up another crisp new image of the Martian surface. It's black-and-white, very sculptural and detailed. There appears to be a profusion of hills and gullies. It was acquired, Jim tells me, by the Mars Global Surveyor camera system, run by Mike Malin, who analyzed it and presented it to scientists at the most recent Lunar and Planetary Science Conference in Houston. "Mike stood up in Houston and said, 'Look, I'm going to show an image. I will not release it to the public yet, because I know what they'll say.' " Jim points to a Martian hillside; a silt deposit seems to drape down one side. "Malin said, 'People are going to say this is a sea and there are all kinds of microbes growing there, but I don't know what it is.' " Jim concurs. It could be volcanic. It could be "strange chemistry." It could be an oasis for Martian life-forms. There are many tantalizing suggestions of what this area might be, but nothing definite.

By superimposing Malin's images over the measurements taken by the laser altimeter, Jim can decipher Martian relief. He switches from the image to a MOLA readout, which is impenetrably numerical to most people, but to Garvin, these numbers reveal patterns the eye cannot see. "This is right off the engineering computer," he says. "These are TIU counts." Suddenly, I begin to feel an attack of math anxiety coming on, but Jim swiftly walks me through the demonstration. "TIU means Time Interval Units. Each of these marks represents a laser 'hit' on the surface of Mars as MOLA travels overhead, bounces an impulse off the surface, and records the difference in time on the return trip to figure out the height of the surface. It's an extremely precise measurement." This readout, he says, reveals a caldera a mile deep. A caldera—the name is Spanish in origin—is essentially a large hole in a volcano.

Jim wrote the computer program that analyzes the shapes traced by the laser altimeter. He can subject all sorts of things to his formulas; at the request of the nursing school at Johns Hopkins, he once measured the shapes of wounds to better understand their healing patterns. Now we're looking at Martian contours. "There's a little cloud," Jim says, "and if I track along here, you see the edge of the polar ice cap, right there. These are dunes." Spikes in the readout depict a "cloud sitting in a little trough." I can see it;

I imagine I'm flying through it. "These troughs are produced by the way the sun preferentially melts one side of the ice cap, and dust accumulates because of wind. Now, these are cross sections. We've already built up ten of these across the north pole and will continue to do so for the next month. . . ."

Claire Parkinson, who has been working nearby, creeps in silently and joins us. Claire is a climatologist with an interest in astrobiology. She is tall and lanky, with million-dollar cheekbones and piercing blue eyes, but she has little interest in appearances. Science commands all her energy and attention. She seems shy until she gets going on a subject that engages her interest, and then she comes to life.

Jim glances up, acknowledges Claire, and continues studying his data.

"Is that below the surface of the ice?"

"Well, Claire, we don't know. That's the problem. We think we've punctured all the way through to the base level, giving us a probe of the sub-ice topography, which is a great thing."

Claire says, "To put it in Earth terms, this mile thickness of ice on Mars would be like two miles of ice in Antarctica."

"Here we have a crater, our old friend Uranius Patera," says Jim. It's the one he showed me at the railway platform. "It's a low-relief volcano, and rather large, bigger than Mauna Loa in Hawaii. If we were walking across it, you would see the kinds of lava we saw as we traversed Surtsey, only it would go on for tens of miles. Then we would come to a hole, not unlike the one on Surtsey, but it would be thirty miles across, not two hundred yards. It would look as though we were peering down a natural piston of Mother Nature's volcano engine. The region inside supplied the liquid rock, which we call magma, and it foundered and collapsed probably hundreds of millions of years ago." Jim makes some rapid keystrokes. "What I'm now doing is selecting the feature and saying, 'Why don't I measure it?' This is what I spend my summer vacation doing. How deep is it? Let's take a guess." He makes a few more keystrokes, and the computer displays the answer. "It's a thousand meters deep. There's a little terrace here. We can measure its width. There's a little break and slope there, five kilometers. And then I can show all the numbers I calculated and print it out—it's printing now on that machine—and there's a feature on Mars that may tell us about the thickness of that ice cap."

He prints a MOLA interpretation of Uranius Patera, taken yesterday, more than a hundred million miles away from the Terrestrial Physics Laboratory at the Goddard Space Flight Center, where we sit. It's a graph tracing the contours of what is unmistakably a crater; for Jim and his colleagues, it is a precious artifact, proof that they can map a distant celestial object with great accuracy. As I examine the graph, something on the computer screen catches Jim's eye—"there's a little bank of clouds! That is a little bank of low-lying clouds sitting above the surface."

The image makes Mars seem very close. The illusion is pleasant. "Every one of these points is about a couple of football fields apart. There is a caldera. So what does that thing look like? Let's measure this volcano." A few keystrokes, and the slopes appear on the screen.

"Look how steep those sides are," Claire marvels.

"Very steep, Claire. This volcano is pretty big. Let's first measure the crater. It's very steep—eighty kilometers across. That's fifty-five miles."

Claire whispers, "Wow." Then she corrects him: "Fifty miles."

"Still pretty big. How deep is it? Let's guess the average depth is two kilometers. How wide is the volcano? Well, let's see. The volcano is two hundred and sixty-one kilometers across. We can show all the numbers, and they tell us the volcano is very steep. What else have we here? Ah! There are some little craters and some more little volcanoes." Jim's eyes alight on a graben, which, he says, is a place where the surface ripped apart and dropped. Then he spies another trough. "It's very steep and eight kilometers wide. It's only one hundred meters deep. See that little black line? That's a little cleft. You wouldn't see that in existing maps of Mars. Those maps are good to one hundred meters. We're good to forty centimeters." About sixteen inches. "One of the things I can do with these measurements is figure out what the craters tell us about water on the surface of Mars. I measure where the crater crosses the original ground. Then I measure the width of the ejecta blanket. And by looking at the numbers I can tell whether an object may have impacted into frozen or wet ground, making the crater look different from one that did not."

Jim gives me one last example of MOLA's extremely precise measurements. "There's the floor of a volcano." He points to a rising bulge of ticks, each corresponding to points on Mars just fifty meters apart. "The floor domes up; it's tumescent. It's arched up because the volcano formed, and the floor dropped down inside the crater after it flooded the lava out, and then it re-domed. So this may show evidence that this particular volcano had a slight resurgence. Whether it's active now, we don't know. We're not measuring the gases."

"Are there active volcanoes on Mars now?"

"We don't know. We are looking for them."

"What about hot spots?" I ask, thinking of potential breeding grounds for life.

"People get nervous if you call them that. There are biological connotations." He prefers to call them "thermal anomalies." Jim checks his watch. "Let's go."

On the way out, he recounts a conversation he had the other day with his next-door neighbor. "He told me, out of the blue, 'You are so damn lucky.' I asked him why. I told him I can barely afford the house we live in, and I work eighteen hours a day. And do you know what he said? 'But you like what you do.' " Jim pauses for a nanosecond of reflection. "You know, he doesn't like what he does and never has. I think most of us in this field feel so fortunate. People ask, 'What's your most lasting impact on planet Earth?' It's your children, of course. But being the person that helps a new thing happen is enough for me." He hurries off to a meeting, promising to come find me when it's over.

Each morning, Claire Parkinson rises at four and walks or jogs to Goddard—winter and summer, rain or shine. "It's only about a mile," she said one morning after she arrived for work soaking wet, "but it was raining and *hailing,* and, well, you can get pretty wet in one mile." No matter. She comes into her office seven days a week, and she particularly enjoys Goddard on the weekends because no one disturbs her then. Holidays barely intrude. Over one Thanksgiving, she worked in her small, immaculate office until late afternoon, when she finally broke away to join a couple of other rocket scientists in Washington for a turkey dinner. I might wish she had more of a social life, but Claire doesn't see things that way; as far as she's concerned, there aren't enough hours in the day to do all the things that need to get done. I think of her as the archetypal NASA scientist.

Claire's idea of a terrific adventure is an expedition to the Pole—North or South. On her return from a recent journey, she sent an e-mail that opened this way: "Hi Larry—Just got back Sunday night from a North Pole expedition (it was neat)." Because the North Pole is over water, unlike the South Pole, which is on terra firma, its actual location can be difficult for the traveler to pin down, she explained; the ice floes over the North Pole constantly shift. They made sure they were over the Pole by riding a dog team across the ice, taking precise measurements as they mushed to their destination. The point of this particular trip was to test new NASA communications gear, so they made the first-ever live web-cast from the North Pole, and the first-ever phone call from the North Pole to the South Pole.

The hardships of a Polar trip—the lack of showers and privacy, the wretched food, the isolation—scarcely mattered; the cold invigorated her. She endured temperatures of −20° F, worse with the wind chill, a truly Martian cold, but all she could think about was how happy she was to be working at the North Pole. When she got home, she told me she had suffered only a little patch of frostbite on the tip of her nose; it hadn't even turned black, and she thought she would be okay, unlike some members of the dogsled team, whose fingers and toes had suffered badly in the frigid conditions. It sounded as though she had just returned from a week at the beach with only a little sunburn to complain about.

Claire has published a book about satellite observations of the Earth, as well as a comprehensive history of scientific landmarks from the thirteenth century to the present entitled *Breakthroughs: A Chronology of Great Achievements in Science and Mathematics.* This labor of love took thirteen years to complete. Of the entire range of scientific endeavor, her first love, if it can be isolated, is math. "It's so powerful. Once you learn a technique, you can apply it to so many things." She entered Wellesley in 1966 and majored in math, but during her junior year, she began to doubt its relevance to issues such as the Vietnam War and civil rights. She even felt slightly guilty about majoring in it. At the same time, she developed an interest in astronomy and obtained access to Wellesley's observatory. After she graduated Phi Beta Kappa, she received her doctorate in climatology from Ohio State.

Trying to reconcile her spiritual beliefs with her work became a major issue for her at Wellesley, especially when she took a required course in biblical history. "Prior to college, I had believed in and clung to the basic Christian teachings that I had grown up with. In the course, I was exposed to huge numbers of seeming contradictions in the Bible, plus many explanations of

the motives of different writers of the Bible, and this was all far more traumatic to my religious beliefs than anything connected to science." At NASA,
assumptions in science are challenged continuously. "There is nothing in-
consistent about engaging in scientific research and believing in God," she
says. "In fact, the search for understanding the universe can easily be interpreted as a search for understanding and hence glorifying the magnificence
of God's creation." As a student of scientific breakthroughs, she was naturally
interested in the discovery of possible nanofossils in a Martian meteorite, but
not at all surprised. "It's not revolutionary for me, because I've believed it all
my life. I believe there are much bigger life forms out there right now." If
life is found on Mars, or elsewhere, a very reasonable response could be awe
in what God has created. "On the other hand," she explains, "science certainly does sometimes come in conflict with specific religious beliefs such as
the archaic claim that the Sun circles the Earth, or the claim that human beings, in the form of Adam and Eve, emerged full-blown during the first week
of creation. In cases like this, scientists are inclined either to stay away from
the controversial topics or to examine the evidence and then, if need be, ad-
just religious beliefs accordingly, such as saying that the seven 'days' of creation are really seven epochs, or that the Adam and Eve story is really an
allegory."

As a part-time historian of science, she adds, "Historically, science can
be viewed as having altered the role of the Supreme Being rather than elim-
inating the Supreme Being. Without science, a society could easily view the
Supreme Being as all powerful, making the Sun rise each morning, bringing rains and storms, and arbitrarily inflicting disease. As science has developed and furnished explanations of why the Sun appears to rise in the
morning and what causes eclipses and how storm systems move across the
planet and how diseases emerge and spread, specific occurrences ascribed to
the Supreme Being are no longer ascribed in the same way. Instead, the role
of the Supreme Being became one of starting the system and establishing the
rules by which it would run—so-called laws of nature. Many scientists, like
many non-scientists, also believe that the Supreme Being can do whatever
He wants. If God wants to step in and create a gigantic thunderbolt out of
nowhere to destroy a community, He can do that, and if He wants to cure
someone of 'incurable' cancer, He can certainly do that, also."

Some years ago, she looked through her young niece's science textbook. Her niece attended a school with a Baptist affiliation, and the textbook had a definite point of view. Claire recalls a section that reviewed theories of formation of the moon. The section began by stating that science is based on observable phenomena, which is certainly true enough, and went on to argue that since no one was around to observe the formation of the moon, science alone could not explain it. That much is also true; there isn't widespread agreement in the scientific community about how the moon was actually formed. The textbook went on to mention three scientific hypotheses concerning how the moon came to be. In one, a massive body struck the surface of the Earth billions of years ago, probably where the Pacific Ocean is now, and blew a big chunk of the planet into orbit. In another scenario, the Earth's gravity captured a massive body wandering through the solar system. And in the third, smaller bodies accreted in orbit around the Earth to form what we now call the moon. At the end of each scientific hypothesis, the book asked the same question: "Now isn't that ridiculous?" According to the textbook, the correct answer concerning the origin of the moon could be found in the Bible.

Claire has a question of her own: "What will happen when my niece takes the SAT?"

The use of technology to observe the heavens is a very recent phenomenon. At the beginning of the seventeenth century, lens grinders in Holland, who had long manufactured eyeglasses, produced a simple magnification device known as a *perspicillum,* an early name for the telescope. By 1608, the device was on display in Venice and was advertised as useful for "seeing stars which are not ordinarily in view, because of their smallness." The following year, Galileo, having heard of the invention, constructed his own version, and presented it to the senate in Venice, successfully demonstrating the advantages of the perspicillum to detect distant ships at sea.

Galileo then turned his telescope to the heavens and made a series of astonishing observations: the moon was not a perfect sphere, as many believed, but was pockmarked with craters; the Sun had spots; and the Milky Way consisted of a multitude of stars. His most provocative discovery occurred when he sighted three (and later, four) little moons circling the planet Jupiter. If his observations were correct, and they were, the prevailing Ptolemaic doctrine that all celestial objects, including the Sun, circled the Earth was gravely flawed. Delighted, excited, and eager to claim credit for these discoveries before anyone else, he quickly published his findings in a book titled *Sidereus Nuncius,* usually translated as *Starry Messenger.* With the invention of the telescope and the publication of Galileo's findings, it was as though humanity, blind from birth, could suddenly see.

In 1636, Francisco Fontana, an Italian, sketched the first drawing of the Red Planet. It was a crude illustration that might have confused flaws in his telescope with actual Martian features. Twenty-three years later, Christiaan Huygens, the Dutch astronomer, produced a far more sophisticated rendering that suggested a dark "sea"—a Martian landmark now known as Syrtis Major. Huygens' telescope was sufficiently sophisticated to reveal that Mars rotated on its axis, as the Earth does; he was even able to determine that the Martian sol lasted about the same as a terrestrial day. Only seven years after these discoveries, Giovanni Cassini refined Huygens' estimate of a sol to twenty-four hours, forty minutes, just a few minutes longer than our current understanding. Both Cassini and Huygens noted the prominent Martian polar caps and other major geological features. The more astronomers studied Mars through their rapidly improving telescopes, the more Earth-like the Red Planet appeared.

In the 1780s, King George III's astronomer, William Herschel, turned his attention to Mars. Born in Germany, Herschel was a prominent musician in England, where he composed, taught, played the organ, and functioned as director of music for the city of Bath. In his spare time, he built and sold telescopes and pursued his passion for astronomy, which eventually led him to manufacture instruments as long as forty feet; on occasion, Herschel composed music to commemorate his enormous telescopes. In 1781, he discovered the planet Uranus and later tried to map the entire universe in three dimensions. He also made a number of interesting observations about Mars. He found, for instance, that its angle of inclination to the Sun was about the same as Earth's; it was natural to infer that Mars enjoyed four seasons, like Earth. He supposed its polar caps contained ice and snow—another inference that has stood the test of time. Herschel also guessed that the Red Planet was covered by a "modest atmosphere" as well. One of the few observations he could not interpret concerned "occasional changes in partial bright belts; and also once on a darkish one." Perhaps the changes were caused by clouds; more likely, they were the result of the violent dust storms we now know sweep the planet. He declared its "inhabitants enjoy a situation similar to our own." This was the conventional wisdom of Herschel's era: Mars *was* inhabited.

Mapping the features of the Red Planet began in earnest more than 150 years before Mars Global Surveyor turned on its cameras and laser altimeter. During the 1830s, two Germans—Wilhelm Beer, a banker, and Johann von Mädler, an astronomer—attempted a complete map of Mars. They were the first to assign Martian latitudes and longitudes, which would become crucial for all future navigating around the Red Planet. Zero latitude necessarily falls on the equator, just as on Earth. Zero longitude was another, more arbitrary matter. On Earth, zero longitude, otherwise known as the prime meridian, is fixed to pass through England's Royal Observatory in Greenwich. Beer and Mädler decided that the Martian prime meridian should pass through a prominent dark area they mapped on the Martian Southern Hemisphere. (Scientists later refined the origin of longitude to pass through the center of a crater named Airy-O. Navigation engineers and geophysicists number the degrees of longitude increasing toward the east from the prime meridian, while geologists measure longitude increasing toward the west.) Camille Flammarion, a French astronomer, corrected accumulated inaccuracies in his two-volume work, *La Planète Mars et ses Conditions d'Habitabilitié,* while other astronomers embellished the Martian map and named Martian features after themselves and their colleagues, a practice that occasioned debate among competing countries, who naturally favored their own scientists over foreigners.

In the summer of 1877, Mars' elliptical orbit brought it close enough to Earth to inspire revolutionary assessments. Giovanni Schiaparelli, who oversaw Italy's Brera Observatory in Milan, turned his eyes to the skies night after night and mapped the Red Planet in greater detail than ever before. Schiaparelli fixed the latitude and longitude for more than sixty Martian features and replaced antique place names with more sophisticated and descriptive Latin terms, many of which survive to the present. He described what he called *canali*—canals, or, more accurately, channels running across the surface of Mars. Schiaparelli conceived of the channels as natural Martian rivers connecting Martian oceans. To be visible on Earth, millions of miles away, they would have to be several miles wide, at least. The following year, Schiaparelli delivered a report about his findings to the Italian Society of Spectroscopists. In his report, Schiaparelli wrote, "There are no large continuous masses on Mars, as the entire surface of the planet is divided by many *canali* into an enormous number of islands." At about the same time, other astronomers, using improved telescopes, also thought they were seeing linear features of some kind, but the *canali* were still identified with Schiaparelli. His findings remained a subject of discussion among scientists for years to come, especially when Mars moved into a favorable position for viewing from Earth, a time known as opposition.

During the opposition of 1894, an American named Percival Lowell turned his telescope on Mars, and in the next decade used his findings, beliefs, and charisma to make Mars into a scientific and popular obsession. He rose to fame by proselytizing on behalf of intelligent life on Mars, but his reputation as a scientist survived even after his theories were debunked. More than anyone else of his era, he established Mars in the popular consciousness and positioned it for serious scientific investigation.

Percival Lowell personified that late-Victorian phenomenon: the gentleman scientist. This was a well-educated, upper-class man of means attracted to the sciences and able to support his own research, equipment, expeditions, and assistants. All of this was true of Lowell. He possessed an impeccable if suffocating pedigree; the Lowells ranked among the most talented, energetic, and wealthiest of old New England families. Percival, born in 1855, demonstrated the prodigious and erratic Lowellian intellect early in life when he memorialized the destruction of a toy boat in a hundred lines of Latin hexameter. He was eleven at the time. At Harvard, he distinguished himself both in mathematics and in English composition, claiming the prestigious Bowdoin Prize in composition. Soon after, his celebrated cousin James Russell Lowell transferred the title of "most brilliant man in Cambridge" to him. After graduation in 1876, however, Percival Lowell spent years searching for himself and his place in the world. He turned down a teaching position in mathematics at Harvard, he traveled, he worked in the family business, rising to the position of treasurer, but he spurned that too, and he spent much of his thirties voyaging repeatedly to the East. Later, he participated in a trade mission to Korea and published a series of books about his experiences in the East, concluding with *Occult Japan,* a study of Shinto trances, derived from his research into séances. It was a strange and unlikely apprenticeship for an astronomer and student of Mars, but it does explain a great deal about his ravenous intellect and his attraction to exotic subjects.

As Lowell's interest in the East waned, he gradually turned his attention to stargazing. He bought a telescope and struck up a correspondence with the astronomer William H. Pickering, but Lowell's interest languished until his aunt presented him with a copy of Flammarion's *La Planète Mars*. Nearly forty, he read the treatise rapidly and wrote one word on its pages: "Hurry." Percival decided that Mars offered a major scientific opportunity for him, but more than that, he needed to hurry because a favorable opposition to the Red Planet was about to occur, and he wanted to capitalize on it. His intellect had finally found a worthy object. Within weeks, he arranged to go on an expedition to Flagstaff, Arizona, with William Pickering and his brother Edward, also an astronomer, to undertake careful observations of Mars. At first, the Pickering brothers thought the enthusiastic amateur intended to tag along on their expedition; they soon realized that Lowell intended for them to tag along on *his*.

By April 1894, Lowell and his crew were constructing an observatory in Flagstaff, and the following month, he was already publicizing his latest project in the press. "This may be put popularly as an investigation into the condition of life on other worlds, including last but not least their habitability by beings like or unlike man," he proclaimed. "There is strong reason to believe that we are on the eve of pretty definite discovery in the matter." He was thinking of Schiaparelli's Martian *canali,* which, Lowell wrote, seemed to be evidence of life on Mars, and not just life, but highly intelligent, technologically advanced life. It was a wonderful conclusion, made even more miraculous by the fact that Lowell had yet to begin looking at Mars.

Once he actually observed the Red Planet, Lowell immediately compared it to the dazzling, colorful Arizona desert landscape surrounding him: "The resemblance of its lambent saffron to the telescopic tints of the Martian globe is strikingly impressive. Far forest and still farther desert are transmuted by distance into mere washes of color, the one robin's-egg blue, the other roseate ochre, and so bathed, both, in the flood of sunshine from out of a cloudless burnished sky that their tints rival those of a fire-opal.... Even in its mottlings the one expanse recalls the other." This was not a bad guess—

Mars does contain many features reminiscent of the Southwestern deserts—but Lowell did not stop at this.

In June, he thought he identified a channel crossing the surface of Mars. "This is the first canal seen here this opposition, and in all likelihood the first seen anywhere," he proclaimed; this would mean he had confirmed Schiaparelli's observations of possible Martian *canali,* or rivers. But only days later, Lowell was overcome by doubt. "With the best will in the world I can certainly see no canals." Lowell returned to Boston, collected himself, and by August he was back in Flagstaff, where he believed he was seeing all manner of marvelous Martian phenomena: seas, and vegetation, and, once again, canals. But there was more. Lowell divined that Mars had once had large standing bodies of water on its surface, water that had long since disappeared, except at the polar caps—another assumption endorsed by modern planetary geology. Then he veered back to fantasy: to preserve the water, an intelligent, technologically advanced Martian civilization had constructed an elaborate system of channels from the melting polar caps to irrigate the dry Martian plains. This wonderful scenario went way beyond Schiapirelli's observations. Lowell perceived grandeur and pathos in the Martians' struggle to survive in the face of drought. "A mind of no mean order would seem to have presided over the system we see, a mind of certainly more comprehensiveness than that which presides over the various departments of our own public works. . . . Quite possibly, such Martian folk"—*Martian folk!*—"are possessed of inventions of which we have not dreamed. . . . Certainly what we see hints at the existence of beings who are in advance of, not behind us, in the journey of life." All this was food, if not for the intellect, then for the imagination.

He immediately popularized his sensational theories in reputable magazines such as the *Atlantic Monthly,* gave lectures, and, in 1895, hardly more than a year after turning his attention to astronomy, published the first of his many books about the Red Planet. In *Mars,* he speculated at length about "a possible inhabitant of Mars" and expressed his specific and chilling ideas concerning its appearance: "Suppose him to be constructed three times as large as a human being in every dimension. If he were on Earth, he would weigh twenty-seven times as much, but on the surface of Mars, since gravity there is only about one third of what it is here, he would weigh but nine times as much. The cross section of his muscles would be nine times as great. Therefore the ratio of his supporting power to the weight he must support would be the same as ours. Consequently, he would be able to stand with as little fatigue as we. Now consider the work he might be able to do. His muscles, having length, breadth, and thickness, would all be twenty-seven times as effective as ours. He would prove twenty-seven times as strong as we, and could accomplish twenty-seven times as much . . . whether in digging canals or in other bodily occupation." This was meant literally.

To demonstrate the existence of artificial Martian *canali,* the underpinning of his theory, he included sketches he had made from his observations in Flagstaff. There was just one problem: Lowell was, in Carl Sagan's words, "one of the worst draftsmen who ever sat down at the telescope, and the kind of Mars that he drew was composed of little polygonal blocks connected by a multitude of straight lines." When Sagan compared Lowell's sketches to images taken by the Mariner 9 spacecraft, he found "virtually no correlation at all." Sagan, who came of age under Lowell's seductive influence, was dismayed. "Lowell always said that the regularity of the canals was an unmistakable sign that they were of intelligent origin," he said. "The only unresolved question was which side of the telescope the intelligence was on." But Sagan could not bring himself to reject completely Lowell's intuitions about canals. "I have the nagging suspicion that some essential feature of the Martian canal problem still remains undiscovered." For most scientists, however, there was no "Martian canal problem" because there were no Martians and no canals.

Lowell left his Flagstaff observatory in the hands of associates and in 1895 returned to Boston, where he reveled in his celebrity caused by the publication of his book; when he tired of Boston, he journeyed to Europe to spread the doctrine of life on Mars. He met with Flammarion and, in Milan, with Schiaparelli himself, whom Lowell addressed as *"Cher maître Martien."* But Schiaparelli, whose observations about Martian *canali* had served as Lowell's inspiration, was a little taken aback by his overenthusiastic American disciple. "He needs more experience, and must rein in his imagination," the Martian master observed.

On his return to his observatory in Arizona, Lowell, having done Mars, turned his attention to Venus. In 1897, exhausted from his travels and pursued by inner furies, he suffered a nervous breakdown and retired for four years during which he managed to drag himself to Bermuda, Maine, Virginia, New York, and the French Riviera. Eventually, he resumed his Martian studies. He published his chef d'oeuvre in 1906, *Mars and Its Canals,* as popular fascination with the subject persisted. Here Lowell made Mars sound bizarrely Earth-like, a little cooler, perhaps, and a little drier, but comparable to "the South of England." The book drew withering criticism from Alfred Russel Wallace, the distinguished, elderly English scientist who had formulated a theory of evolution by natural selection similar to that of Charles Darwin's. Lowell's assumptions, and their popularity, appalled Wallace, who regarded *Mars and Its Canals* as a "challenge, not so much to astronomers, but to the educated world at large." In his retort, *Is Mars Habitable?,* Wallace declared, "Mars not only is not inhabited by intelligent beings as Mr. Lowell postulates, but is absolutely UNINHABITABLE." He persuasively argued that Mars was nothing like the lush planet of Lowell's feverish imagination, but actually frigid and dry.

To arrive at his conclusions, Wallace employed sophisticated concepts similar to those used by planetary geologists today. He measured the planet's albedo—or reflectivity—and concluded that its surface temperature was actually quite cold, about 35° F below zero—too cold to sustain life, he believed, or for water to flow. Any lingering water vapor would escape Mars' thin atmosphere into space, thanks to the planet's low gravity, so there would be no liquid water on the surface of Mars, no water to flow in any canals, and no one to build any canals in the first place. As for the polar caps, whose seasonal melting supposedly provided Mars' inhabitants with liquid water, Wallace demonstrated that they were likely to consist of frozen carbon dioxide. "Any attempt to make that scanty surplus, by means of overflowing canals, travel across the equator into the opposite hemisphere, through such terrible desert regions and exposed to such a cloudless sky as Mr. Lowell describes, would be the work of a body of madmen rather than of intelligent beings," said Wallace.

Although Wallace demolished Lowell's findings, the Brahmin astronomer prospered as the sage of Boston; he packed lecture halls, sold tens of thousands of books a year, auctioned off his latest articles to the highest-paying popular magazines, and continued to insist Mars teemed with life. "Mars at this moment is inhabited, [and its] denizens are of an order whose acquaintance is worth making," he insisted in his 1908 work, *Mars as the*

Abode of Life. "Their presence certainly ousts us from any unique or self-centered position in the solar system, but so did the Copernican system the Ptolemaic, and the world survived this deposing change. So may man."

As the years passed, Lowell's position hardened; where other scientists failed to find Martian canals or vegetation, Lowell saw only confirmation of his beliefs and even compared himself to Darwin. "The theory is my own as are all proofs of it which I have been patiently accumulating and thinking out for thirteen years. As it has been violently combated even as Darwin's was and at last is now on the threshold of general acceptance, it is not right to speak of it as a mere theory in the air at which all were working. It has overcome every objection brought against it." Except, of course, the objections he chose to ignore.

To the end of his days, Lowell believed that intelligent creatures constructed canals on Mars in order to survive, and the impression lingers in the folklore surrounding the Red Planet, evidence of the planet's capacity for bending thought. With the possible exception of the Sun, it exerts the most powerful mystique of any celestial body. Nothing else in the night sky inflames the imagination as Mars does. Even now, NASA reminds the public that Lowell's canals were only a "visual allusion in which dark areas appear connected by lines," in case anyone happens to fall too far under the spell of Lowell's eloquent, befuddled books about Mars.

Lowell was often wrong, but he wasn't *always* wrong. Late in his career, he proposed that the solar system actually consisted of nine planets, rather than eight, as most scientists believed at the time. Lowell was not the only astronomer to suggest that another planet lurked beyond Neptune, but to his many critics this seemed yet another far-fetched Lowellian notion. In this case, though, he happened to get it right. Lowell died in 1916; fourteen years later, the hunt for the elusive "Planet X" led to confirmation of the existence of Pluto, whose scientific symbol consists of his initials: ♇.

When Jim Garvin's meeting ends, he returns in his silver BMW to retrieve me, as arranged. From his terse remarks, I gather his meeting was tedious, grim, and budgetary, and he is relieved to be at large once again. As we drive across the vast expanse of Goddard to our next destination, he points to a herd of deer about a hundred yards from us. We slowly proceed along the road, and a flock of indifferent geese emerges from the brush, bringing traffic to a halt. The rolling hills and abundant trees deceive me into thinking we're in a wilderness.

Suddenly, we pull into another parking lot at Goddard and leave the car behind. We charge up a loading dock, pass a huge white tank containing liquid nitrogen, and enter a gigantic edifice that houses Goddard's twelve-story-tall clean room. Within, behind a giant sheet of plate glass, priests of technology, wearing white gowns and face masks and puffy shower caps, preside over gleaming instruments that will be launched into space. Components of the Hubble telescope were tested here, in the largest clean room in the world. It is one hundred feet wide, a hundred and twenty-five feet long, ninety feet high, and looks like a high-tech garage, a very *clean* garage. Before you walk in, you have to pass through an air shower with your street clothes on. As the air blows over you, you put on your isolation suit, and saunter in.

The technicians here wear goggles; an errant eyelash can ruin an entire spacecraft representing millions of dollars and years of labor. As they work on the spacecraft, they baby it. They float it on a cushion of air, even though it weighs thousands of pounds. When a spacecraft is suspended this way, two or three people can move it from side to side with virtually no friction or heavy lifting. Once they've built the spacecraft with great delicacy and precision, they then try to destroy it. They cool the spacecraft with liquid nitrogen, which brings the materials down to 320° F below zero to simulate orbital conditions, and they keep it chilled for a month. Each part must function perfectly in the extreme cold, or it is useless. The spacecraft must also survive rapid heating when it gets warmed by the Sun, so they bake it. And it has to function in a vacuum; in the absence of air, which acts as a buffer, temperatures fluctuate wildly, so they place the spacecraft in a vacuum and test it. The technicians also spin the spacecraft to simulate twenty times the force of gravity. If the spacecraft survives this brutal treatment, its components qualify as "space grade."

Behind each instrument on Mars Global Surveyor stands the individuals responsible for its design and construction. In the case of the laser altimeter, it's Rob Afzal. He's in his mid-thirties and well-dressed by NASA standards. He's more worldly than many of his counterparts at Goddard. "Laser specialist" is not something his outgoing, ironic demeanor brings to mind. You might guess that he does something financial, or, since he's telegenic, that he works as a fast-talking announcer on CNNfn. In fact, he holds a Ph.D. in physics from Brown, where he wrote his thesis on lasers.

The term laser is an acronym for Light Amplification by Stimulated Emission of Radiation. In its natural state, light, whether from the Sun or from an incandescent bulb, consists of a disorganized muddle of waves; they interfere with each other, spread out in all directions, and dissipate quickly. But the waves in a laser beam are organized; they move in the same direction, and the beam is coherent and concentrated. In this form, it can guide bombs to their targets or be used as a surgical tool. The laser that Rob helped to develop to map Mars is particularly sophisticated. It's powered by other, smaller lasers similar to the ordinary laser pointer in his breast pocket. When they are bunched together, the small lasers form a very high-powered big laser. "If you turned it on and let it run," he says, "it would melt."

"Could it burn a hole in the wall?"

"If you could concentrate the energy, it certainly would."

In his laboratory at Goddard, Rob scrutinizes his newest laser with wry fondness. "This is all pretty expensive stuff. This crystal is about ten thousand dollars. Those pump sources are about seventy thousand for four. And this is just for the non-space-flight version; for the space-flight version, they have to be tested more thoroughly. Even these optics are very special—high precision, the purest of the pure. I mean, this little cube of glass here is fifteen hundred bucks." I peer at the glittering shard. You can't order this from a catalog; it's available only from the military.

Now it's time for a demonstration of laser technology. He turns it on, and I hear a faint buzzing emanate from an indistinct source. It's probably just my imagination, but the air in the room suddenly seems electric. "This is the laser beam on a false-color image going into this camera," he says, pointing to a bright thread of light passing between the laser and a lens. False color, he says, means simply that a color—a ghastly shade of green, in this case— has been assigned to the intensity. "This allows us to actually look at the profile of our beam, as if it were hitting you right in the eye. We measure the quality of the beam, and this is near perfect. Even MOLA isn't as good as this." He holds a sheet of white paper in front of the shimmering green beam. "It's drilling the paper at forty pulses a second." The beam spatters against the paper like quivering drops of greenish blood. It's a very hypnotic sight, an entire *Star Wars* in miniature . . .

Anyway, lasers aren't always red, as should be apparent by now. The one we're looking at, for instance, is actually infrared, which means the wavelength is too long for the eye to see; it is, in effect, redder than red. "But this laser has two colors; one is near infrared, the other is green," Rob explains. "The green beam you can actually see. It looks like a laser beam in science fiction movies, a hot green beam of light."

He tells me this is a diode pumped laser, a design first used on the lost Mars Observer and later on Mars Global Surveyor. The diode pumped laser is an important technological breakthrough—"analogous to radios and stereos going from vacuum tubes to solid state." He says that a diode pumped laser "forces the laser light back and forth, and it resonates like sound in an organ pipe." This technology makes for a laser beam "so concentrated and so bright that we can shine that beam from the orbiting spacecraft over three hundred miles down to the surface of Mars, hit dirt, and some of it reflects back up another three hundred miles." The beam starts off about an inch and a half wide, and by the time it reaches the surface of Mars, it's about 150 yards across. "Then we train the spacecraft's telescope to look at the planet surface hit by the laser beam. The light comes back up, our telescope collects it, and stops our electronic timer. We make a time measurement that's only billionths of a second," Rob says offhandedly.

Theoretically, you could do without the spacecraft and shoot a laser beam directly from Earth to Mars and get a signal back, but it would take a "pretty powerful laser" to do this, and there would be huge sacrifices in precision. Even so, it's not out of the question. Few things are impossible in Rob Afzal's universe; it's simply a matter of overcoming the engineering problems within the limitations imposed by the laws of physics.

Rob readily acknowledges that he received considerable help in the design of the laser aboard Mars Global Surveyor. "Basically, we stood on the shoulders of about twenty-five years of military lasers. We worked with McDonnell Douglas (now part of Boeing) on the technology and took it to the next level where the efficiency was high enough that you could send it to a place like Mars and operate it with only fifteen watts of power and make scientific measurements." But it wasn't easy to reach that level of laser prowess. "We went through an unbelievable amount of pain putting this laser together and making it lightweight. For instance, the box that holds it is made from beryllium. It's dangerous stuff, poisonous to manufacture. Only a couple places in the country manufacture the laser box." Then there were the problems surrounding the delicate Q-switch, which fires the short pulses that come out of the laser. The Q-switch is so sensitive that if it changes temperature by more than a tenth of a degree per minute, it will malfunction, and the laser will fail. "At least, it will fail in space; it's not that finicky on Earth." But there are frequent "temperature hiccups" on a spacecraft, each posing a threat to the Q-switch and the entire laser experiment. Rob and McDonnell Douglas managed to get their Q-switch functioning reliably in the hazardous conditions of space, and now, he says, "We really think that we have a shot at a long duration operation."

He did most of the work on the MGS laser at McDonnell Douglas in St. Louis. For months, he would fly into the St. Louis airport, rent a car, drive across the field, park the car, and go down into the bowels of a McDonnell Douglas building with classified clean rooms, and toil for days on end. If he grew tired, he took catnaps in a gurney in the clean room. When it was time to leave, he would drive back to the other side of the airfield and fly out, only to return days later and repeat the cycle. He didn't get to see much of St. Louis, in fact, none at all.

After his team assembled the laser, they tested it. "We did a 'shake and bake.' We shook the laser altimeter in rocket launch modes, and we put it in a space environment temperature cycle to show that it's going to operate." The tolerances for a laser to operate successfully are extraordinarily slim. If parts move even a millionth of an inch, the laser may not function. Although they contract and expand as the temperature changes (especially in the vacuum of space), Rob says he finds ways to compensate "by matching the materials so they grow and shrink together in the relative positions to change." He compares the process of building a laser to watchmaking. "If watches get hot or cold, they'll change time, so watchmakers balance the expansions and contractions of the parts so they stay together." After successfully passing all the tests, but just before it was to be mated with the spacecraft, the laser failed.

Lasers are likely to be unpredictable, Rob explains, because they don't exist in nature. "You're really forcing nature to behave in a certain way, but it doesn't want to. We call it an unstable equilibrium. It's like balancing a pea on top of an overturned bowl. If you touch the pea, it's going to fall off, as opposed to a pea resting at the bottom of a bowl. If you move the pea resting at the bottom, it comes back to the same place." In other words, lasers tend not to work. And if you put it on top of a rocket, which will shake it, and send it into space, which will bake it, it will have an even greater tendency to break down.

Rob was exhausted, and they were now on a super-crash schedule to repair the laser in time for the launch. "We had to go back out to McDonnell Douglas and take it apart. We found that a microscopic thin film had coated some of the optics, and the beam destroyed the optic surfaces. We had to figure out what was the cause of the contamination, solve that problem, rebuild the whole thing, reclean it, retest it, and get it back to the spacecraft—all in just a few months. That was very, very nerve-wracking." He was behaving as an engineer, he tells me, trying to fix the problem and beat the launch deadline. Scientists, he likes to say, behave as if they have all the time in the world. But engineers . . . well, *engineers,* says Rob, are problem solvers, and they're usually racing the clock. "If you're a scientist, and you find an anomaly, you say, 'Oh, that's interesting, let's take some time to study that.' If you're an engineer, and there's an anomaly, you say, *'H-o-l-y* shit, something's wrong, we have only two weeks to fix it before launch date, the vendors are late, and we have to rebuild.' "The pressure is much greater, and the deadlines are real. He shakes his head slowly to convey that engineers must live within the constraints of the real word, while scientists occasionally seem to inhabit a parallel universe.

"You want to talk about a hard schedule to meet?" he says of the engineering difficulties he faced trying to get the laser altimeter ready in time for launch. "We had a launch window about two weeks long. The rocket *had* to launch on one of those days. There's no slipping that deadline. It's not like, 'Oh, well, we'll come in a week late.' It's not like more money will buy you anything. Once you align with the planets, you're either going, or you're not. Each day had its own window as well." Even when they are aligned, Earth and Mars are rotating, so that daily window lasted just one second; if the launch team missed it, they had to abort and try again the next day.

"You can't negotiate even a little?"

"No, you can't," Rob says. "It's celestial."

Once the laser was rebuilt, Rob delivered the instrument himself, lugging it along on a commercial flight. He secured it in a metal attaché case and bought a seat for it right beside his own. People on the plane started staring at him. Who's the guy with the attaché case riding in the seat next to him? What could be so damned important that it had its own *seat?* The only thing missing was the handcuff securing his wrist to the handle. If he told anyone the metal case belonged to NASA, he would be subjected to endless questions; he'd never get off the plane. And if he told them the case contained a *laser,* others would assume it was a weapon, not a scientific instrument. In general, he tried to say as little as possible, but he broke his own rule when a man sat down in an empty seat next to the laser and placed his cup, filled with hot coffee, right on top of the case. "Look, buddy," Rob said, as politely as he could, *"This is a two-million-dollar laser you put your freaking coffee on, and it's going to Mars, you know what I mean?"* The interloper looked at Rob in shock, hastily removed the coffee cup, and NASA's Mars program was once again safe.

The rocket that would carry the instrument into space was a Delta II, one of the most reliable launch vehicles ever made, as the scientists and engineers kept reminding themselves. Rob went down to the Cape to see the launch. It was a beauty, absolutely flawless. The knowledge that Mars Global Surveyor, carrying his laser altimeter, was safely on its way to the Red Planet came as a tremendous relief. Afterward, he had only one thought: his overwhelming need to catch up on sleep.

When I return to Goddard the following day, I am alone. Without Jim Garvin to drive me in his silver BMW, I walk to my destination, and a very long walk it is. Trudging across the Goddard campus, I think of Claire Parkinson pacing off these steps twice a day, every day. I am Garvin-less at the moment because of a delicate matter of politics. I am here to see a scientist on a rival team, and not just any scientist, but Mario Acuna, the magnetometer expert, the individual who has thrown such fear into the mighty MOLA team with his discovery of Martian magnetic fields. Of all the planetary scientists working at Goddard, none has greater stature than Mario.

When he came from Argentina to Goddard in 1963, the center, then just four years old, reverberated with a pioneer spirit; the public regarded NASA with affectionate optimism, even reverence. "I grew up with the space pro-gram," he says in a gentle voice at variance with his forbidding appearance as we talk in his cramped, overstuffed office. During his career, he has seen the center change from an isolated research facility in rural Maryland to a sprawling industrial park. At the time he joined NASA, funding for space ex-ploration, though modest by current standards, was considered unlimited, and in the mind of the public, NASA was playing a leading role in the fight for the survival of America and the Free World against Communist adver-saries. Innovation was part of daily life and was taken for granted. The Cold War brought about a sense of security and sense of belonging for many, and life at NASA in those days seemed simpler, if not more innocent, than today.

"In the sixties, they would leave you alone," Mario tells me. I hear nos-talgia in his voice, as well as a trace of bitterness. "Politics was not a consid-eration. You had enough time to think. Risk was acceptable; it was a natural by-product of what we do. Even failure was considered a natural by-product of what we do. That was an ideal situation because you could do a lot in a very short period of time. We would launch three spacecraft a year. Later on, it took twelve years to launch just one. This was a very, very dynamic place, alive twenty-four hours a day, lots of young people around, and nobody was afraid of making mistakes. On the contrary, there were sometimes rewards for making mistakes. That was a time when it was an honor to be a civil ser-vant. Somewhere along the line, the civil servant became synonymous with 'lazy' and 'overpaid,' which is simply not true. The Vietnam War didn't help, either. The ailments of American society got blamed on technology, pollu-tion. NASA was obviously headed for a crisis. You could see the pride dis-appear."

I can see why Mario throws his colleagues off balance, especially the MOLA team. He is warm, patient, cordial, and so convinced he is right about everything—the decline of NASA, the magnetic fields he discovered on Mars—that he won't let anyone forget it. And he probably *is* right. He looks around his crowded office, which is packed with the awards and cita-tions he has received, all of them confirmation of his stature. "The country fell in love with management schools, so we followed that, but, in the process, we lost sight of the product. We are *still* more concerned with process than product, although we have recovered somewhat. And then there was the Challenger disaster, which was basically a consequence of what was happening at NASA in the eighties"—the dangerous combination of too much bureaucracy and too little accountability. "That was the low point, and it triggered a total reassessment of NASA, top to bottom. In my opinion, we went overboard and put so many safeguards in place we couldn't do anything. We started trying to do small things that we couldn't do well." He shakes his head in dismay at the idea of NASA trying to do something *small*.

Mario concedes that the "faster-better-cheaper" approach might have shaken up NASA to the point where "intelligent risk is acceptable, but it's not as readily accepted today as it was in the past. Cost is an overwhelming consideration today. We have become much more conservative. NASA has fallen into a bad mindset, where no risk is acceptable, and unfortunately, that means there's an infinite cost associated with the production of very little or nothing." Much of what NASA once accomplished in-house has been delegated to the private sector, but industry won't take the same risks as the bear-any-burden, pay-any-price spirit of the Kennedy era. Often, NASA finds it just can't compete with private enterprise. The other day, Mario was checking on Mars Global Surveyor and discovered that one of the mission's principal programmers had just quit. "The rat fink! Now I have to replace him. He went to work in private industry at three times the salary. NASA is a good training ground for people, but we have a hard time holding on to them."

NASA's parsimony and timidity threatened to extinguish Mario's Martian magnetometer altogether. "When I proposed it, people thought I was crazy. I'm not crazy." To show me what he means, he leads the way from his office to a large but equally shabby laboratory, where the shelves contain thirty years' worth of instruments. This is Mario's workshop; he is a craftsman as well as a scientist, and the tools of his trade are to be found here. He is especially proud of his sensors, an essential component of the magnetometer, sensors he made at this workbench. I pick up one of the instruments; it feels heavy in my hand and resembles a spool of metallic thread. In this case, the thread is fine, gleaming copper wire. Mario flinches slightly as I finger it. The sensor accidentally slips from my fingers to the countertop, where it lands with a thud. He winces violently. "It's very stress sensitive," he says. "We put a great deal of effort into these things. We call them our babies. If someone so much as puts the wrong screw in, it causes enough magnetism to mess it up." As I return the sensor to the shelf and step back from the workbench, he breathes a sigh of relief.

One sensor is up and running in the lab. Mario asks for a five-dollar bill, which I remove from my wallet. "Obviously, there's nothing magnetic on this, no magnets. Let's see what happens when I bring it closer to the sensor." The needles jump; the instrument is measuring the magnetic field on a five-dollar bill. "The magnetic anomalies on Mars that the magnetometer has been able to measure are even smaller than the bill's," he says, pausing to let that sink in. Mario then warns that his stock of sensors is dwindling, and the design, which goes back to 1972, may soon become extinct. "We have abandoned the metallurgy of soft magnetic metals, and when this stock of sensors is gone, it's gone. It will require over a million dollars and a lot of learning to replace it."

Mario also reminds me that taking magnetic measurements is an exceptionally delicate process, especially during aerobraking, which shakes Mars Global Surveyor. "We make our measurements while it's doing all those crazy maneuvers, and we spend a considerable amount of time separating all the motions of the spacecraft from the magnetic field data so we don't get confused. In addition, we have to compensate for the fact that spacecraft themselves generate a magnetic field. Our sensors are located right next to the solar array on Mars Global Surveyor, and because the solar array generates power, it's a source of dense magnetic fields. We worked closely with designers and builders of the solar array to make sure its magnetic field would not cancel out everything else."

Despite all these obstacles, he's getting tremendous returns from the sensors circling Mars. "To the best of our ability to tell right now, Mars does not have a global magnetic field, but it has remanant fields." This is the term for a magnetic field frozen into a rock. "All these remanant fields are trapped in the crust, which means that it had one in the past, but it's no longer active." The fields tell Mario a lot about Mars' geologic history. "The thermal energy from Mars and the composition of the interior of Mars have evolved to the point that the planet can no longer sustain the motions and convection required for a magnetic field. It might still be liquid at the core, but it's certainly not churning fast enough to generate a magnetic field. But it also means that it had one in the past; we can see the effects of that. We need to find out where it came from, how strong it was."

The magnetic anomalies Mario has found on Mars are located in the cratered, southern terrain, but he doesn't yet know how they got there. "We're scratching our heads. There are a number of mechanisms, but there is one requirement: if you get magnetic material hot enough, it loses all its magnetism. So to have a remanant magnetic field, you need a process that cools the material below its Curie point." This is the temperature at which the magnetic material loses its magnetism. If you heated a piece of magnetized iron to its Curie point—over 900° C—there would be no more magnetism. On Mars, iron-rich magma apparently surged from the interior through a fissure to the surface and then cooled off; afterward, the region "remembered" the magnetic field. That's one way the magnetic anomaly could have been created; the impact trait mechanism is another. You disturb the surface, create a crater, and disturb the material underneath. The region cools down, with magnetic anomalies at the point of impact.

"Of course, I may be wrong on both counts," Mario says. "We'll see what happens. One more thing: we have a little piece of Mars here, the meteorite, so we ask ourselves, 'Can we find a magnetic anomaly with the materials in the meteorite?' Turns out we can. But it is not obvious that this meteorite came from the magnetized regions of Mars, so we have a real puzzle here." But, I ask, if Mars once had a magnetic field, why did it lose it? And why did Earth *keep* its magnetic field? "Simple," he tells me. "The amount of thermal energy at the core of each planet. Mars is a smaller planet, with much weaker gravity." And that, more than anything else, helps to explain why Mars lost its water and its atmosphere, and why the simple life forms there, if there were any, died off or went into hiding.

Within days of my leaving Goddard, Jim Garvin urges me to return; it seems I'm already slipping behind the curve. He's been spending his nights sifting through recently acquired data, and he doesn't understand why I am not at Goddard, when he obtains the first image of the Martian north pole, the occasion of a joyous Martian epiphany:

Subject: NEW DATA
Date: Fri, 10 Apr 98 10:15:00 EDT
From: Jim Garvin <jgarvin@nasa.gov>
To: Laurence Begreen <bergreen@NYCnet.net>

Dear Larry,

MOLA passes continue to enthrall and amaze. It's a ten-year dream come true. In 1988, I could only hope that one day we would slice through the unknown aspects of Martian landscapes and plumb the depths of enigmatic craters with laser beams. Now it's a reality. What is so surprising is the wide range of really weird stuff. I have witnessed examples of moated crater-like features and roller-coaster-like fields of dunes and lava plains. The troughs (canyons) in the north polar cap are relatively consistent and display ledges or terraces but little other interior structure and NO stairstep patterns as many had expected.

I am intrigued by more and more impact craters in the near polar regions showing a pedestal appearance, which reminds me of features that are called "table mountains" or *stapi* in Icelandic. These are volcanoes that erupt under ice and core their way up through the ice to form breadloaf-style mountains, often with summit pits.

Spent an almost all-nighter remeasuring another 100 or so impact crater landforms on Mars, as well as dozens of enigmatic depressions. I continue to gain subtle insights by working up close and personally with the magical MOLA data. The consistency of the depths of craters in higher latitudes makes me wonder if there is a 1–2 km thick sedimentary layer in the north that could reflect a huge depositional event. Some would say it suggests a large sea or regional ocean.

The polar dunes are beautiful and attest to amazing wind patterns at the margins of the polar cap. I have been measuring their spacing. Wow! Most are only 20–60 meters in height and 1–2 kilometers apart, much like some dunes in terrestrial deserts.

These new data are just way too much fun and raise so many new questions about the workings of landscapes on Mars.

Jim

In early May, I return to Goddard, wait at the gate, get my credentials, spot the distant herd of deer, and eventually arrive at the faceless building housing MOLA headquarters to find Jim transfixed by another graph. It's similar to the one he showed me in April, but far more detailed, with multicolored tracks for each pass of the laser altimeter. Lately, he has been studying the north polar dune fields of Mars, which he says are analogous to the dunes of the African desert, where they are called ergs—huge, shifting bodies of sand. In particular, a "fresh layer of condensate" sprinkled on top attracts his interest. No one dares call it snow or frost or anything else suggestive of water. It might be frozen ammonia, for all anyone knows. Jim designates it a "frozen, dusty, icy mixture." Whatever its composition, it may contain interesting clues about the composition of the Martian atmosphere and the distribution of volatiles such as water across the planet.

Mars Global Surveyor has also found lots of polar clouds that excite Jim, even though he's a crater guy. "Mars absorbs thermal energy from the Sun into its thin atmosphere every day by using clouds. It also has lots of dust that heats up and cools down as a function of the day. We've taken lots of measurements from a couple of thousand feet above the surface all the way up to forty thousand. Martian clouds show ripple-like effects, similar to our cirrus clouds. This is a very important indicator of the way the atmosphere circulates."

Using the sixty passes MGS made over the north pole, Jim and his colleagues have generated the first relief map of the Martian north pole. This was a tricky operation, since the passes generated 12,000 data points, each about six feet apart. Keeping them all straight, Jim says, "takes a lot of bookkeeping." They entered the data points into a computer, which added color and churned out a picture that looks so realistic it can easily be mistaken for a photograph. To Jim, the size of the polar cap, approximately 600,000 cubic miles, speaks compellingly of the presence of water on Mars. If the cap melted, there would be "enough fresh water on Mars to produce a global ocean uniformly twenty feet deep."

Jim wonders about the distribution of all this Martian water hundreds of millions of years ago and the possibilities of ancient life breeding in it. Again, he seeks answers in analogies to what is known about the rise of life on Earth. "The record of complex animal life on Earth is around five hundred and thirty to six hundred and seventy million years—we don't know exactly. Before that, there's another two billion years of primitive life on Earth. Extremophiles were popping up in hot beds where there were inter-

esting chemicals. If you drive up to those hot springs in Iceland, you see bubbling mud pots and clays sitting in little geothermal pools with pure silica
precipitating out of these little bathtubs. And you have colonial bacterial systems rampant in these systems." The situation might have been the same on
Mars, billions of years ago.

Jim often tries out his latest inspiration on Dave Smith, the MOLA team
leader. As director of the Laboratory for Terrestrial Physics, Dave occupies
one of the more lavish offices at Goddard, although he is more likely to be
dashing from one meeting to the next than sitting behind his desk. Even in
his office, he stands, paces, frets, and exclaims. If you ask him what he does
all day, he will tell you, rapid-fire, "My job is to harness this group as a team
and to get the best science I can out of them, not being too heavy or too
light. If you can't explain what you do, you might as well never have done
it." Dave is not a guy you want to bullshit. "The only thing scientists have is
their reputations, and they're easy to lose. You can write one or two bad papers and lose it. To get paid to do this is a luxury."

Dave got his start in planetary geodesy: the study of the planets' sizes,
shapes, gravity. "We were progressing rapidly with Earth," he says of his early
days, "but not with the other planets. It was an area ripe for major accomplishments. I decided I would take this lab into planetary geodesy in a competitive way, because, in science, you have to compete through proposals to
get funded." He'd long thought that lasers could be extremely useful for
planetary geodesy, and he goes back a long way with MOLA, more than ten
years, and counting. He recalls NASA's deselecting his team's first altimeter,
the radar altimeter, as a "traumatic" experience, but he eventually recovered.
"I had proposals for a replacement experiment from MIT, from JPL, and
from my group here. I ran a review and concluded that what I wanted to do
was build a laser altimeter, which was close to my heart, because I've worked
with lasers all my life. I proposed it."

This time NASA told him to go ahead, but gave him only $10 million,
and insisted the instrument had to be on time, or it would rest comfortably
on a shelf. "It *was* ready on time, but it flew on Mars Observer in 1993 and
was lost along with everything else on the spacecraft." Dave pauses. "You
worry about so many things. You worry about the instruments, about whether

this or that will work, and the one thing that never crossed my mind was that we would lose the entire spacecraft." He shakes his head. "Some scientists were shattered. NASA doesn't lose things! It took us a while to come back. The question became, 'How do you get back to Mars again?' We rebuilt, and we made a comeback." The years it took to accomplish this were often stressful. He says that JPL, which designs and controls many of NASA's robotic missions, "gets a lot of flak, but people there are smart as hell and work their butts off. If you don't have people who are rigorous, you can't fly missions to Mars."

The other thing Dave worries about, besides science, is money. Most scientists prefer not to discuss budget issues with me—they'd much rather discuss their research—but among themselves, they spend a fair amount of time squabbling over funding. Occasionally, they confide they've been "worrying about the budget," as if they've been caught in flagrante delicto. Their worry is fueled by NASA's tendency to shift money away from science projects at any time, and the changes, even the prospect of such changes, take a toll on the scientists and managers who spend years devising and building experiments. There are also budgetary booby traps built into the system. If you don't spend all your allotted money on a project within the year for which it was appropriated, you may not get funded for the full amount the following year. It doesn't matter why the money wasn't spent. If the vendors happened to be late to deliver and so were late to be paid, and you've spent only fifty percent of your budget, NASA may assume that is all you need, and the following year, your budget may be cut in half.

Dave explains these painful realities to me one evening in his office at Goddard. It's well past eight, and he's worked a long day. He offers me a lift in his car, and since it's pouring and cold, I'm delighted to accept. "I'm off to Pasadena tomorrow morning," he says as he gets behind the wheel of his Saturn. "I have to catch a seven A.M. flight from Dulles, which means I have to get up at four, but these early morning flights are good, because I will land by ten-thirty or eleven and go straight to JPL. I have a meeting planned for twelve-thirty." Driving swiftly through the rain, he says he's looking forward to it.

By midsummer, Jim Garvin has received data from about a hundred MOLA passes, and he's flying. He scrutinizes fifty polar impact craters that have never been seen clearly before, tantalizing himself with the possibility that their white condensate may be water ice. He devotes particular attention to two craters. The first is Korolev, named for the Russian rocket engineer. Garvin christened the other "Sasquatch" because of its imposing dimensions: forty-nine kilometers in diameter. Both are filled with something he guardedly calls "white stuff." He spends so much time analyzing and imaging Korolev that I sometimes get the feeling that he's been camping out there, trudging across its snows and measuring features as he did on Surtsey. I wouldn't be surprised if Garvin could tell you more about the crater Korolev, 100 million miles away, than he could about the shopping mall closest to his home.

One morning in July, he declares to me that he's "frantically trying to develop an interpretation of these and dozens of other high-latitude craters to see if we can test the hypothesis that these are feeling the effects of a frozen layer under a thin, dusty surface layer." This is exotic for him, because, he says, "we don't have experience here on Earth with cratering in frozen media, and there are theories that the impact process in ground ice would make very strange landforms. My suspicion is that some craters punctured through the thin upper surface layer, exposing ground ice and allowing it to become involved in the later stages of the cratering process, perhaps inflating the ejecta in the near rim zone and making pedestals—craters that appear like perched mountains with a depression at the summit." In other words, no one's ever seen anything quite like Mars' climate and gravity and water distribution. New concepts are required, and as the exploration of Mars continues, the vocabulary of geology must expand. Jim hints that these shapes, and the evidence of "hydrothermal fluids" they've left behind, might "make a good place to evolve microbes."

I finally comprehend his elliptical thinking regarding the beguiling "white stuff" in the craters. If there is water, even in the form of ice, near a crater, it might have been a breeding ground for extremophile life. Whenever a large object struck Mars, even a very frigid Mars, it would have generated a tremendous amount of heat, melted the ice and surrounding terrain, and kept things warm and amenable to biology, at least for a while. Of course, if a crater were formed by a volcano rather than an impact, so much the better for bringing a source of warmth into contact with ice, melting it, and providing a breeding ground. Jim doesn't say this in so many words; the most he'll do is nod in assent when I describe what I believe he—and many of his colleagues—are thinking. Jim simply lays out the evidence, geologic, chemical, biological, you name it, and allows the conclusion to become inevitable. The longer he and his colleagues study Mars and consider the prospects for life emerging there, the more likely they are to conclude, "How could life have *not* emerged?" But they know that if you go around shouting *"Life on Mars!"* too often, you run the risk of sounding like a latter-day Percival Lowell.

Although Lowell's scientific reputation is tainted, his vision of an intelligent Martian civilization sparked the imagination of science fantasy writers who embellished his ideas. H. G. Wells' 1898 novel, *The War of the Worlds,* picked up where Lowell's speculation left off, but Wells made his Martians far more malign than the astronomer dared imagine. Significantly, Wells did not take the reader to Mars, to explore the wonders of a Martian civilization; instead, he brought the Martians to Earth. The novel derives its power from a calm and credible account of a Martian invasion; in Wells' hands, the event inspires dread from the moment the first Martian appears on Earth. "I think everyone expected to see a man emerge—possibly something a little unlike us terrestrial men, but in all essentials a man," he wrote. "I presently saw something stirring within the shadow: greyish billowy movements, one above another, and then two luminous disks—like eyes. Then something resembling a little grey snake, about the thickness of a walking stick, coiled up out of the writhing middle, and wriggled in the air towards me. . . . A big greyish rounded bulk, the size, perhaps, of a bear, was rising slowly and painfully out of the cylinder. As it bulged up and caught the light, it glistened like wet leather."

Just when the Martian conquest of Earth and the destruction of humanity appear inevitable, Wells delivers a masterstroke of irony—scientific irony, at that. As soon as the Martian invaders land on Earth, and breathe our air, drink our water, and eat our vegetation, they consume bacteria against which they are defenseless. Without warning, the monsters collapse and die in droves. "There are no bacteria on Mars," Wells assumed; thus he spared humanity from annihilation.

Others wrote under Lowell's influence. Edgar Rice Burroughs

(1875–1950) was a former cowboy, miner, cavalry trooper, and tireless seeker after get–rich–quick dreams who turned to pulp fiction to make a living. Best known as the creator of Tarzan, he was also responsible for Captain John Carter, a Civil War veteran who finds new challenges battling demons on the planet Mars in a series of adventures beginning with *A Princess of Mars* (1912). Burroughs spoke of his exotic setting with an unmistakably personal sense of yearning: "I turned my gaze from the landscape to the heavens where the myriad stars formed a gorgeous and fitting canopy for the wonders of the earthly scene. My attention was quickly riveted by a large red star close to the distant horizon. As I gazed upon it I felt a spell of overpowering fascination—it was Mars, the god of war, and for me, the fighting man, it had always held the power of irresistible enchantment. As I gazed at it on that far-gone night it seemed to call across the unthinkable void, to lure me to it, to draw me as the lodestone attracts a particle of iron." The public soon came to share Burroughs' fascination with Mars, and he obliged with a series of death-defying adventures on Mars—Barsoom, as it was called by its inhabitants. No matter how far-fetched the events Burroughs described, the exploits of John Carter provoked millions of readers, including scientists-to-be, into thinking and dreaming about Mars.

After World War II, Ray Bradbury, a much younger science fiction writer in California, began writing whimsical stories about Mars, and in 1950, they were published as *The Martian Chronicles,* which Carl Sagan considered the best science fiction novel about the Red Planet. It is certainly the most romantic of all such portrayals, depicting Mars' inhabitants as serene, intelligent beings. At times, Bradbury sounds a bit like W. B. Yeats on peyote: "They had a house of crystal pillars on the Planet Mars by the edge of an empty sea," he writes of the Martians, "and every morning you could see Mrs. K eating the golden fruits that grew from the crystal walls, or cleaning the house with handfuls of magnetic dust." Bradbury's Martians even have talking laptop computers: "You could see Mr. K himself in his room, reading from a metal book with raised hieroglyphs over which he brushed his hand, as one might play a harp. And from the book, as his fingers stroked, a voice sang, a soft ancient voice, which told tales of when the sea was red steam on the shore and ancient men had carried clouds of metal insects and electric spiders into battle."

Mars remains so distant that it exists on the borderline of fantasy. No matter how whimsical the Mars of Bradbury or Lowell or Burroughs, the scientists who now study the planet grew up under the influence of these visionaries. Some modern scientists, such as Carl Sagan, have freely admitted their debt; others function in a culture conditioned by them. The prospect of life on Mars has gone in and out of scientific favor and popular awareness, but it continues to dazzle, to puzzle, and to tantalize. Mars has become the Moby Dick of scientific investigation, not merely a planet, but a symbol of the ungraspable phantom of life.

Chapter Eight

THE GENESIS QUESTION

When Gerald Soffen heard in the summer of 1996 about possible evidence of life in a meteorite from Mars, he became so excited he had trouble sleeping. Twenty years earlier, he had served as the project scientist for the two Viking missions to Mars, and he remains NASA's gray eminence on the subject of astrobiology. Now, two decades later, the data collected by the Viking missions—the composition of the Martian atmosphere, in particular—were being put to startling new use. No wonder he couldn't get a good night's rest. This was a remarkable turn of events, for in the 1970s, some of that data had been used to demonstrate the absence of life on Mars.

The paradigm was shifting around him, and around the rest of the Mars community. In the recent past, scientists searched for evidence of life on a grand scale, life that was actually trying to contact *us,* evidence based on the premise that extraterrestrial life must be dramatic, intelligent, worthy of Hollywood. Instead, if the meteorite evidence had been interpreted correctly by McKay and his colleagues, extraterrestrial life is *tiny,* between one-tenth and one-hundredth the thickness of a human hair. It is humble; and it is, cosmically speaking, in our backyard.

In *The Structure of Scientific Revolutions,* Thomas Kuhn advanced the idea of paradigm shifts as characteristic of the way science develops. Once acknowledged, paradigms frame issues and direct research until the next paradigm appears and overturns existing assumptions. Darwin established a long-lasting paradigm, as did Einstein, but lesser known paradigms emerge in science all the time. The change is not necessarily orderly or pleasant; there is often considerable acrimony, as scientists struggle to protect or establish their reputations. Even scientists can be resistant to the implications of new and unruly data; the old paradigms die hard. Often, says Kuhn, the paradigm shift is generational, as a new breed of scientist gradually replaces older colleagues. (Kuhn's thesis has lately come under fire, a perverse sort of confirmation of this process.) In astrobiology, the paradigm shift is well under way, and new concepts have supplanted old assumptions. As a result of this revolution, the field has become more rigorous. To many astrobiologists, it now appears natural and inevitable that life starts with the simplest building blocks:

OLD PARADIGM	NEW PARADIGM
Intelligent, exotic aliens	Extremophile, microbial life
Purposeful	Random
Distant	Located on the next planet
Vigorous, possibly predatory	Dormant
A fluke of nature, anomalous	Inevitable outcome of chemistry
Scarce	Widespread
Contains spiritual implications	Confirms accepted scientific theories
Requires an Earth-like environment	Can develop in environments different from those found on Earth
A miracle	A statistic

This paradigm shift is not yet complete. Many scientists belong to the old school, and others insist that life exists only on Earth until proven to exist elsewhere. Nevertheless, the new paradigm is gradually achieving consensus, and scientists now pose a different set of questions. Instead of asking, "Does life exist elsewhere?" they are now inclined to inquire, "How do habitable worlds form, and how do they evolve? How did living systems emerge? How can we recognize others? What is the potential for biological evolution beyond the planet of origin?" All these questions assume that biological activity exists beyond Earth, probably in great quantities, and all we need to do is find it.

Joshua Lederberg, the Nobel Prize laureate, phrased the new paradigm this way: life exists, and Earth is the prime example. This seems too obvious to mention, but it contains profound implications. If the only planet we know well happens to contain an abundance of life, why should we assume it is an uncommon or unique phenomenon? Carl Sagan pointed out that there is nothing remarkable about our corner of the universe, nothing to indicate that it is especially hospitable to life. Since our neighborhood is as or-

THE GENESIS QUESTION   183

dinary as can be, wouldn't it follow that life is similarly common? Not only that, but once the young Earth stabilized, life appears to have taken hold within 100 million years, a geological blink of an eye. "The rapidity with which life arose on the earth may imply that it is a likely process," Sagan wrote. "It is dangerous to extrapolate from a single example, but it would be a truly remarkable circumstance if life arose quickly here while on many other, similar worlds, given comparable time, it did not."

In search of life beyond Earth, Dan Goldin, NASA's administrator, decided in 1996 to form an Astrobiology Institute, a worldwide network of scientists searching for evidence of life on Mars, on other planets, and on asteroids. Known for making decisions quickly and informally, he found himself in an elevator one day with Jerry Soffen; it was a short ride, just eight floors, and during it, Dan asked Jerry if Mars was really the next big thing. Jerry told him yes, it was, and as the elevator door opened, Dan said, "Do it!" And Jerry did it. He took charge of the Astrobiology Institute.

Jerry is slight, wiry, with longish wavy hair, and he wears a goatee, which gives him a philosophical air. He was trained as a biologist at Princeton, and he has been at NASA long enough to see the prospects for discovering life elsewhere in the universe wax and wane. As the acting head of NASA's Astrobiology Institute, he was running the most powerful mechanism yet devised for finding answers to the question of extraterrestrial life. Although the Institute is quite real, it was designed as a *virtual* institute. "The members won't have to meet frequently," he tells me. "We're going to wire them up and give them the Next Generation Internet. They'll be on-line twenty-four hours a day, not just whenever there's a meeting. They'll be sharing data back and forth. It isn't like a science team where you say, 'Well, I need a chemist or a physicist or a biologist.' These people will be working all the time together. You could be in an airport and still participate in the meeting. Everybody will know what's on everybody else's mind. Science is very old-fashioned, very conservative, and it's got to change; this institute could be the beginning."

Jerry administers the Institute from his unprepossessing office in NASA's Headquarters in Washington, D.C. A prominent sign on the wall declares:

ASTROBIOLOGY IS THE STUDY OF
Life in the universe; and the chemistry, physics and adaptations that
influence its origins, evolutions and destiny.

It addresses the question:
Is Life a Cosmic Imperative?

Astrobiology is a Swiss army knife discipline. To deploy the current termi-
nology, it is multidisciplinary; it is integrative. A working astrobiologist functions as a chemist, a biologist, a geologist, an engineer, a mathematician, and
a philosopher. There is little agreement about a hierarchy among these disciplines, yet it has been said that every biologist dreams of being a chemist,
and every chemist dreams of being a physicist, and every physicist dreams of
being a mathematician, and mathematicians dream of being God.

One of the first things Jerry did in his new role was to solicit applications from potential members of the embryonic Astrobiology Institute. The
idea was for scientists representing their institutions to pitch ideas for funding. "To our amazement, over fifty proposals for the initial set of members
came in." The enthusiastic response was really not that surprising. Biology
and planetary geology are hot disciplines, and no other institution can match
NASA's record in planetary exploration. "NASA has had a very strange romance with biology," he says, "but it's not been a very popular romance.
NASA is made up mainly of engineers, not scientists. When NASA was
born, the subject of Mars rarely came up. I entered the game in 1961." At
the time, the prevailing term for extraterrestrial life was "exobiology," coined
by Joshua Lederberg. "The concepts of the 1960s were very simple. You
grew stuff in petri dishes, the way high school students do their experiments.
There were perhaps two dozen of us in the field at the time, and we never
had a chance to get anything on a mission. Finally, we got our life detection
experiments on the Viking missions, and I became the project scientist; I was
at the right place at the right time."

The mission was hailed as a great success, of course, but in terms of the
search for life on Mars, it turned out to be a great disappointment. In those
days, the Southern California desert, which happened to be close to JPL,
served as a rudimentary analogue to the surface of Mars. "Remember, we
were looking for microbial life, and we understood only a little bit of the ge-

ology of Mars. We went to Death Valley, which is kind of funny because it's so hot there, and Mars is cold." He shakes his head, thinking of lessons learned the hard way. When the Viking missions revealed that the surface of Mars bore little chemical resemblance to the Southern California desert, the analogy broke down. "To our shock, there was nothing organic on Mars, let alone what we would call 'life.' " All the more puzzling, because they knew that meteors and comets bearing organic materials had crashed to the surface. But, Jerry tells me, "Mars has a different kind of surface from Earth, with strange peroxides that tear apart the organic molecules and turn them into water and carbon dioxide, among other things. This is true of the entire planet, because planet-wide dust storms redeposit the stuff all the time. Only the polar regions may be exceptions. Eventually, we thought, 'Well, there isn't any life on Mars. It's dead, deader than the moon.' We picked up our marbles and went home. But since the 1970s we have learned more about extremophile life. And then there was the Martian meteorite announcement in 1996 with possible evidence of nanofossils. Talk about a can of worms. It galvanized everybody."

As excited as Jerry gets about the meteorite, ALH 84001, Jerry does not count himself among the believers in life on Mars. He feels there's a fifty-fifty chance that life may have existed on Mars at some time. Just as important, though, "We have changed our paradigm of how we think about life." He notes that microbiologists have focused on medical research to better understand and treat diseases such as cancer, but as for less menacing simple life-forms such as fungi and spores, "Well, we don't know much about them. Maybe *they* hold clues as to what to look for on Mars."

Wanting to know more about the famous meteorite from Mars, Jerry approached the National Institutes of Health; surely the scientists there would be able to tell him more about the nanofossils it contained. He spent several days asking around, without finding anyone who could enlighten him. "They'd heard the word, but they didn't know anything about nanobacteria. I kept pushing. I asked, 'Why don't you know anything about it?' They said, 'Look, it doesn't cause diseases, at least none that we're aware of. Why would we be interested in it? We're paid to fix things and look for things that cause diseases.' "

Jerry also paid careful attention to the arguments aimed at discrediting the claims of the meteorite team about the nanofossils. He thinks the critics could be right; the contents of the meteorite might not be evidence of life on Mars, after all. But he still feels dissatisfied. "Nobody's given me a very good explanation for what the shapes in ALH 84001 actually are. Every time someone tells me they aren't nanofossils, I say, 'Great, if it's not biology, tell me what it is.' You can't just say, 'It's not biology,' and sweep it under the rug. You have got to answer the question, *'What is it?'* Some people say, 'Well, it's an artifact.' An artifact of what? Hell, that's like saying, 'Take two aspirin, and I'll see you in the morning.' It just won't do."

So Jerry keeps an open mind. "There's something about science you learn in graduate school, which is often forgotten. It's that most science is wrong. Today's science will not be tomorrow's science. Most of it will be obsolete." Paradigms don't shift in an orderly, predictable fashion; disorder is more common. "The toughest thing would be to have to answer a senator's question, 'When are you going to find life?' " Once again, no one knows. Nevertheless, Jerry still thinks Mars is the best place in the solar system to find evidence of extraterrestrial life, despite scientific interest in various rivals, especially Europa, the ice-covered moon of Jupiter. Europa looks enticing; it's practically a planet itself, with a hot mantle and, below the ice, liquid water. "I can easily imagine another scientist saying to himself, 'I'll go where the water is. Just give me the tangible,' " Jerry says. "It's a good argument, but we don't have any idea of the history of Europa. At least with Mars we know the history is similar to our planet's. If you had to survive on one or the other, I would go to Mars just because there's an atmosphere of some sort. Europa has no atmosphere. It's a weird place."

I try to imagine life on a distant, airless moon revolving around Jupiter—what would we call its inhabitants, Europans?—and ponder the nature of Europa itself. I ask Jerry whether Europa is a part of Jupiter that broke away or an asteroid captured by Jupiter's gravity.

"That, my friend, is exactly what nobody knows."

Nobody knows: it's a phrase I hear a lot.

Over the years, the controversy concerning the Viking life detection experiments has grown rather than diminished. It has inspired several books, notably *The Search for Life on Mars,* a skillful analysis by Henry S. F. Cooper, Jr.

Yet, after all the discussion, there is little agreement about the ultimate meaning of the experiments, especially among the scientists who designed them and analyzed the data. Even in the planning stages, the experiments engendered controversy among the participating researchers. Carl Sagan urged his colleagues to hunt for any form of life, micro or macro. He wanted cameras to photograph it; he wanted a spotlight to illuminate it in case it came out at night—whatever "it" happened to be. Before Viking's launch, he said, "I keep having a recurring fantasy that we'll wake up some morning and see on the photographs footprints all around Viking that were made during the night, but we'll never get to see the creature that made them, because it is nocturnal." He drove his scientific colleagues to distraction by proposing the spacecraft carry bait to Mars to attract "macrobes"—larger life-forms—which, he theorized, would wander over to the spacecraft and perhaps savor the alien delicacies, their first taste of Earthly cuisine. Sagan lost that argument and many more like it.

The other scientists preferred to concentrate on microbes. It was too cold on Mars for life-forms to wander across the surface, they insisted, especially at night. "If you're a predator," Sagan countered, "what better device would there be than to go around at night with a heat sensor, eating . . . sleeping organisms?" It was difficult to know whether he was being serious when he said things like this or just enjoying himself—apparently both, to hear him explain his approach: "Someone has to propose ideas at the boundaries of the plausible, in order to so annoy the experimentalists or observationalists that they'll be motivated to disprove the idea." Ten years earlier, for example, he popularized the ideas of Iosif S. Shklovskii, a Russian astronomer, who suggested that one of Mars' moons, Phobos, is hollow, and not only hollow, but artificial. Shklovskii advanced a theory that ancient Martians, knowing they were doomed, decided to preserve the fruits of their civilization inside Phobos, as a sort of floating library or museum. This was pseudo-science worthy of Percival Lowell at his most outlandish, as Sagan knew. Think about it anyway, he suggested. Open your mind. Some scientists said Sagan was suffering from an overactive imagination, that he "struggles to create situations where life might exist," in the words of one colleague. Sagan justified his fondness for speculation with a retort that was both serious and flippant: "I think it's because human beings love to be alive, and we have an emotional resonance with something else alive." In short, he explained, "it's fun." Clearly, Sagan was having fun, and just as clearly, many of his colleagues were not.

Out of these debates came a new paradigm of extraterrestrial life, thanks largely to Sagan's irrepressible imagination and his refusal to be bound by categories. While others wondered how to find life, he asked, "What is it?" The old definitions of life don't hold, he said, and the scientific community should redefine it or admit they lacked a working definition of life. Can life metabolize? Is that a useful definition? Well, so does a car. Does it reproduce? So do crystals. In the end, he decided one sufficiently broad test for life might be "thermodynamic disequilibrium," which, in simple terms, means anything that appears to require energy. The problem with this out-of-the-box thinking, of course, was that it included many things not considered life, such as cars—although it could be argued that a car, while not life, is at least indicative of life.

Another scientist involved with the Viking missions, Norman Horowitz of Caltech, took a more rigorous approach toward defining life. Life, he said, was more complicated than nonliving objects, and it was purposeful, dedicated to perpetuating itself or its descendants. Implicit in his argument was the acknowledgment that life was a *system,* as Sagan suggested, but Horowitz narrowed his definition by insisting that the system was highly organized. For him, genes were the key, for only living things possessed a genetic code; only living things evolve when genetic mutation occurs. Horowitz's emphasis on genetics as the defining element of life excluded byproducts (such as a car) that might furnish clues to the existence of life. This was a powerful argument, and Horowitz proved an eloquent exponent: "The genetic attributes of living things—that is, the capacities for self-replication and mutation—underlie the evolution of all the structures and functions that distinguish living objects from inanimate ones. . . . Life is synonymous with the possession of genetic capabilities. Any system with the capacity to mutate freely and to reproduce its mutations must almost inevitably evolve in directions that will ensure its preservation. Given sufficient time, the system will acquire the complexity, variety, and purposefulness that we recognize as 'alive.' " Horowitz expected that life on Mars, if any, would be carbon-based, as on Earth; theoretically, it is not the only possible choice, but it is the most likely. He acknowledged he was playing the law of averages, but, he argued, carbon was so versatile that *extraterrestrial* carbon-based life might be very different from what anyone might expect. "The thing that makes carbon the most likely component of life—its ability to combine in myriad ways with other elements—is precisely what could allow strange and wonderful creatures to develop somewhere else, even if they were based on carbon. Furthermore, carbon is one of the most abundant elements in the universe."

Despite his carefully reasoned set of conditions, Horowitz remained skeptical that any life-form ever existed on Mars. The conditions there, it seemed to him, were too cold and dry to permit biological activity; even creatures that survived in the desert on Earth wouldn't last the Martian nights and winters. "Among all known terrestrial species, only water-vaporizing lichens can even be considered as possible models for Martian life," he said, adding, "If life exists on Mars, it must operate on principles different from terrestrial life." This was not a comforting thought.

Sagan readied an ingenious response: microbes might be able to extract water from Martian rocks. "On Earth, there are kangaroo rats who never drink because they can derive their water from food chemically," he said. "The idea for Mars is not very different. So even if there is no evidence for flowing water on Mars, I'm still sanguine about life." This argument received indirect support from the work of another scientist: Wolf Vishniac, a biologist at the University of Rochester. Vishniac's particular passion was for the Dry Valleys of Antarctica, where, he claimed, simple life-forms such as bacteria must exist, despite a coldness and dryness that almost matched Martian conditions. Time has borne out much of Vishniac's work, but he did not live to see his vindication. First, his life-detection experiment, nicknamed Wolf Trap, was eliminated from Viking because of budget cuts. He was devastated, but remained part of the mission's science team and made field trips to the Dry Valleys to pursue his research. In December 1973, his body was discovered at the base of an ice mountain; he'd apparently slipped and fallen to his death. To Sagan, Vishniac became a martyr to the cause of life on Mars, and after his colleague's death, Sagan redoubled his efforts to find evidence of something—*anything*—biological on the Red Planet.

Throughout all these philosophical and scientific debates, Jerry Soffen struggled to maintain proper perspective; he recognized that they were proposing to work at the limits of science, or even beyond, and the task of finding evidence of life beyond Earth was often disorienting and frustrating.

"This may be one of the most important scientific questions of our time," he said. "It is also one of the most difficult to answer." No matter what their preconceptions, the researchers agreed it was remarkable to find themselves in a position to make the "first test for life" on another planet; only twenty years earlier, the idea would have been considered strictly the stuff of science fiction.

Viking 1 and Viking 2 each carried four identical life-detection experiments. To Jerry, the most useful of the batch was a gas chromatograph-mass spectrometer, known forever after as GCMS. This instrument looked for organic compounds—molecules with carbon—created by biological processes. It could even tell which organic compounds were associated with life, and which weren't. In addition, the spacecraft carried Horowitz's pyrolitic release experiment, used to detect evidence of photosynthesis. There was a gas-exchange experiment, overseen by Vance Oyama of NASA; it was designed to bathe Martian soil in "chicken soup"—actually, a broth of nutrients designed to rouse life from a state of dormancy. Last was the labeled release experiment, which exposed the soil to organic nutrients labeled—or marked—with radioactive isotopes. If organisms lurking in the sample feasted on the nutrients, they would give off gases labeled with radioactivity. Of all the experiments, the labeled release experiment seemed the least likely to produce a positive result. Horowitz said it was "an ideal life-sensing device for an aqueous planet," and Mars, as everyone believed, was desiccated.

Viking 1 was launched on August 20, 1975, and Viking 2 on September 9. The following July, Viking 1 began returning startling, detailed color images. Sagan marveled at the similarity to terrestrial landscapes. "This is not an alien world, I thought. I knew places like it in Colorado and Arizona and Nevada. There were rocks and sand and drifts and a distant eminence, as natural and unselfconscious as any landscape on Earth. Mars was a *place*. I would, of course, have been surprised to see a grizzled prospector emerge from behind a dune leading his mule, but at the same time the idea seemed appropriate. . . . One way or another, I knew, this was a world to which we would return." With these suggestive phrases, Sagan demolished lingering impressions left by the stark Mariner black-and-white television images, which had depicted a Mars as desolate and cratered as the moon.

Shortly afterward, the Viking scientists announced that two experiments yielded positive results. The gas exchange experiment had detected oxygen after bathing some Martian soil in "chicken soup," and even the labeled release experiment showed a positive result. "The odds were overwhelming that nothing would happen," said Gil Levin, who'd designed the experiment, "and when we saw the curve go up, we all flipped." The announcement of life on Mars was greeted with skepticism; the results appeared much too good to be true. Several years earlier, Mariner 7 had apparently detected ammonia and methane in the Martian atmosphere during a flyby, but the results were soon discredited.

Finally, the GCMS, the arbiter of all experiments, registered its findings, which contradicted the positive results of the other life-detection experiments. Apparently, there were no organics on Mars, even though scientists knew that comets and meteorites had transported them to the surface. If there were no organics, as the GCMS data strongly suggested, there could be no life on Mars as the scientists defined it, and the positive results of the other experiments were invalid. Even Sagan was bewildered: "If there is life on Mars, where are the dead bodies? No organic molecules could be found—no building blocks of proteins and nucleic acids, no simple hydrocarbons, nothing of the stuff of life on Earth. . . . Terrestrial soil is loaded with organic remains of once-living organisms; Martian soil has less organic matter than the surface of the Moon."

NASA's scientists had a second chance to detect life several weeks later, when Viking 2 landed on the opposite side of the Red Planet, but the results were eerily similar: oxidation, probably of a chemical nature, but no organics, no "dead bodies." Even Sagan lost faith temporarily. "As the legacy of Percival Lowell reminds us, we can be fooled," he reflected. "Perhaps there is some special inorganic, nonliving catalyst in the [Martian] soil that is able to fix atmospheric gases and convert them into organic molecules." As a disheartened Jerry Soffen pondered the data, he concluded that the prospects for life on Mars were actually very poor. He now knew that the surface was bombarded by ultraviolet rays from the Sun; on Earth, the ozone layer screens out many of those rays, sparing life below, but the thin Martian atmosphere affords little such protection. Striking the Red Planet's surface, the rays tear apart organic molecules, in effect sterilizing the crust. Although some scientists suggested that life on Mars might exist in localized environments, oases protected from the damaging effects of ultraviolet radiation, Jerry realized that Mars was constantly buffeted by fierce windstorms transporting dust across the entire planet; thus, the surface was uniform, with the possible exception of the polar caps. But, at the time, the polar caps were thought to be too cold to sustain life. There would be no chance for life-bearing oases to form.

Two of the other scientists agreed with these negative conclusions about life on Mars. Norman Horowitz wrote that while he believed it was impossible to exclude the possibility of life on Mars, its existence was extremely unlikely; everything they had found could be explained by chemistry. "The field is open to every fantasy," he said. "Centuries of human experience warn us, however, that such an approach is not the way to discover the truth." Late in 1976, the Viking scientists evaluated the contradictory results of their life detection experiments in the pages of *Science* and collectively declared, "No conclusions were reached concerning the existence of life on Mars." Despite discouragement, Sagan continued to argue that Viking should not be considered the final word on the subject. The following year, in an article in the *Journal of Geophysical Research,* he and his coauthors scrutinized the results of the Viking experiments once again and decided that even if life were present on Mars, the data might not reveal it. The paper concluded, "A model of biology can always be invoked which would have evaded detection by our instruments. For example, Martian photophobes could always be poised one scan line away, waiting for reflected light from the nodding camera mirror to disappear." ("Don't you just love it?" Greg Neumann said to me, laughing, when he read this passage aloud. "That must have been Carl.")

NASA ceased to promote the life-detection experiments, which had once been vital to the mission, and the scientists themselves could not reach convergence concerning the significance of their data. The contradictory and inconclusive results were something of an embarrassment. What was the point of spending more than a billion dollars to go to Mars just to prove it has chemistry? Something had gone wrong with these experiments; in the wake of their inconclusive results, it was recalled that in pre-launch tests on Earth they often malfunctioned and failed to detect life in terrestrial soil, and further tests for reliability had been canceled because of budget cuts. Perhaps the four Viking experiments had failed to detect life unambiguously because they had been flawed, not because life on Mars didn't exist. As proponents of this approach were fond of saying, absence of evidence did not mean evidence of absence.

These days, Gil Levin, who believes his labeled release experiment on the Viking mission discovered life on Mars, is president of his own company, Biospherics Incorporated, in Greenbelt, Maryland, not far from the Goddard Space Flight Center. He is the only scientist I've encountered who occupies a lavish office and who keeps me cooling my heels before granting me an audience. When he does appear, I discover that he is a welcoming, exuberant, expressive man, who runs by his own clock. His position in private industry sets him apart from most NASA-supported scientists, who make do with modest, though reliable, salaries. Gil, in contrast, has made the transition from scientist to entrepreneur, and he has also spent more than two decades insisting that he discovered life on Mars, a finding that NASA has refused to endorse.

Leaning back in his chair, his gaze sweeping over his domain, he tells me the long and tangled story of his struggle for vindication. He originally christened his Viking life-detection experiment "Gulliver," which NASA changed to the more technically accurate "labeled release experiment." Still, he preferred Gulliver because "it was a tiny experiment, going to a distant place, but you have to have a bureaucratic name on everything. At any rate, the labeled release experiment did obtain a positive response, and it wouldn't go away. It succeeded in satisfying the criteria that four review panels had agreed constituted detection of life."

When I mention that the findings of the GCMS experiment—*no organic matter*—flatly contradicted the labeled release experiment, Gil has his answer ready: "We took a sample of Mars and anointed it with a little radioactive mixture. If we observed radioactive gas emerging from it, this was taken to be presumptive of life. Obviously, a chemical could evoke the same response, so in its infinite wisdom, the Viking biology team concocted a universal control for this and for the two other life-detection experiments." He means the GCMS experiment, which turned out to be his undoing. "When that experiment was run on Martian soil, we all waited expectantly, because we all thought there must be organic matter on Mars; we only wanted to know what *kind*. But the GCMS experiment came up with zilch and startled everyone.

"Well, this was a problem for everyone, including Jerry. We got two opposing answers. One experiment said there was life, the other said there was no organic matter. Meanwhile, NASA was saying this was potentially the most important experiment in the history of science! If they announced a positive finding, and then later had to retract it, we'd have egg on our face so deep we'd never be able to breathe again." It was up to Jerry Soffen, as the project scientist, to make the call. According to Gil, Jerry simply chose the more conservative course, declaring that the experiments had failed to detect life on Mars, rather than acknowledging the positive results of the labeled release experiment. "I think the job of the science director is to try to *protect* the scientific results rather than *discard* them, but Jerry discarded those labeled release results in favor of the control experiment. I'm friends with Jerry, and I think he's fine in every respect—*except science.*"

In recent years, Gil has increased the intensity of his attacks on NASA's handling of the Viking life-detection experiments. "I now want to say that it is more probable than not that we detected life." When he delivered this pronouncement at a recent scientific gathering, he said, "shrimp went flying through the air. It was horrendous! Actually, we were standing over the shrimp screaming at each other." And when he gives talks these days, he declares he has "definitely detected living microorganisms in the soil of Mars. The only argument left against me is that there's no liquid water on Mars"— a finding he naturally contests.

"Dan Goldin has said the number-one priority of NASA is to detect life on Mars," Gil tells me, "but they won't look for it. So I am amazed at the conundrum here. I have made every possible attempt to be nice and work with these guys. Goldin absolutely refuses to see me. I have called; I have written; I have faxed. I proposed half a dozen experiments in the last ten years to look for life on Mars. They just keep rejecting them. Now NASA wants to go to Europa, one of Jupiter's moons. Why? They say there's probably life on Europa. But they haven't found liquid water. They have found ice, and they say the ice has crevasses in it, and they *think* the crevasses go down to liquid water. They give *that* a higher priority than the Viking data, which detected life!"

Gil becomes so exercised about the scientific wrongs done to him, and to Martian microorganisms, that he starts dumping twenty years' worth of papers, pamphlets, and books in my lap, all of them explaining why his critics are wrong, why the labeled release experiment got it right the first time, and why the other Viking life-detection experiments were flawed. In one paper, from 1978, he suggested that Viking's cameras detected patches of lichen growing on Martian rocks. If true, the finding would be extraordinary, for lichen is a combination of fungi and algae, and algae contain chlorophyll, which would mean there is photosynthesis on Mars. This is a distinctly minority opinion.

The conversation edges toward the twilight zone, but I doubt Gil notices the transition. He offers to drive me to my hotel so he can continue to plead his case. "Poor Percival Lowell," he exclaims as he navigates his white Cadillac along the rain-soaked Maryland highways. "People are still beating up on him. He had excellent analyses, but his *data* were flawed. Now it's the opposite case. We have excellent data from Viking and Pathfinder, but our *analyses* are flawed." Gil also expresses the opinion that Pathfinder, though largely worthless, did return one piece of useful scientific data: the meteorological mast, taking the atmosphere's temperature at various distances above the surface. It showed that the surface of Mars can reach 80° F, much warmer than anyone thought—and much more hospitable to life.

The more Gil talks, the more he sounds like the victim of a conspiracy, but if so, he is a cheerful victim; he wears his status as a badge of honor. He is careful not to call NASA's refusal to endorse his findings a "conspiracy" against him. Sure, he thinks it strange that Dan Goldin ignores him, but that does not make for a conspiracy, exactly, more an example of bureaucratic folly, as Gils sees things, for NASA to suppress evidence—proof!—of life on Mars.

"Why . . . would . . . we . . . do . . . that?" snaps Jerry Soffen, normally a patient man, except on the subject of Gil Levin. "He's paranoid. If we detected life, we'd get all sorts of missions and all sorts of money. We were very close for many years, but I've lost respect for him as a scientist. Scientists are after one thing: the truth. Gil is after glory." I get the feeling that Gil won't be playing a major role in NASA's Astrobiology Institute.

With a sigh, Jerry recounts the checkered history of the Viking life-detection experiments. "Even then, Gil fought with the team. He wanted to do his own thing. 'It's my experiment,' he'd say. 'It's not *your* experiment,' I reminded him, 'it's *NASA*'s experiment.' As soon as Viking ended, Gil came into my office and said, 'Look, Jerry, I detected life on Mars! It's the green color on the rock; it's lichen!' " Jerry subjected the image to spectroscopy, which would extract more information about the chemical composition, and the results were negative. "Gil just didn't like that."

Jim Garvin has also worked with Gil Levin, and I figure that if anyone will give him the benefit of a doubt, I suppose that Jim will. I am soon disabused of that notion. "Gil's very smart, but wacky," Jim says. "He makes no sense. Why would Jerry Soffen, or NASA, *suppress evidence of life on Mars?* For God's sake, that would be among the most important discoveries in the history of mankind! There's no party line or policy at NASA about life on Mars, or any other scientific issue . . ."

"But, Jim, couldn't Levin be onto something with his life-detection experiment?"

"Look, his experiment had a result that *might* be explained by evidence of organic matter on Mars, but that isn't the only possible explanation. There are different approaches to doing science, and maybe the public would like it done one way or another, but the way we do it is to have multiple hypotheses to explain phenomena, and then favor the Occam's razor approach." That means the simplest explanation of a phenomenon is the likeliest explanation. "For example, take the way McKay and others analyzed the Martian meteorite. They considered *all* the possibilities very carefully and eliminated them until they were left with only *one.*"

Jim also dismisses Levin's allegation that NASA destroyed the first image Viking sent back from Mars, and altered the hues of others, possibly to discredit his contention that its cameras photographed patches of lichen growing on Martian rocks. If you correct the color, Gil has argued, the greenish patches of lichen become apparent. I try that one on Jim. "Garbage," he says. But he admits that Gil has raised some interesting questions about the way Mars appears to the human eye, as opposed to a television camera. "The eye sees things differently from the digital cameras we send to Mars. The eye sends an analogue signal to the brain, which interprets it. On Earth, the sky is blue because aerosols in the atmosphere, primarily moisture, screen out other light waves. The Martian atmosphere has only a thousandth of the density of Earth's, and there's almost no moisture in it. If you were standing on the surface of Mars, you would see an ochre sky near the horizon, gradually turning blue the higher you looked. Our cameras see it differently because they analyze each pixel with a spectroscope. They get a lot more information than the human eye, but they don't correct it unless we tell them to. That's why the question of what colors you'd see on Mars with your own eyes is a good one." Jim believes NASA ought to reanalyze Viking's life-detection experiments and even look at images of rocks "with changing spectral properties" to search for biological origins. The issues raised by Gil Levin, though far from evidence of life on Mars, haven't been completely dismissed.

NASA's announcement of the Astrobiology Institute excited Jim Garvin tremendously. He began work on a proposal representing the Goddard Space Flight Center. "We live and die by proposals," he reminds me. He labored on this document for months, mostly at night, enlisting dozens of collaborators, including Claire Parkinson, in a large-scale venture representing Goddard's spectrum of researchers. Giving me a copy, he warns, "If our competitors get wind of specific details of our proposal, that could negatively impact our chances for support in the selection process." He speaks from painful experience. The proposal calls for the establishment of a virtual Center for Dynamic Exoecology (CDEX) and seeks funding in the seven figures over the course of five years. It is an extraordinarily complicated document, but Claire breaks CDEX down to a six-point plan:

- Define and characterize the currently known extremophile environments on Earth. This will include places like the very hot and dark environments in the volcanic vents at the bottom of the ocean, the very cold environments deep in ice sheets, and the very dry environments in deserts and in the Dry Valleys of Antarctica.

- Assess the possibility of finding new or more extremophile habitats on Earth. This will require using remote sensing and integrated Earth system models in a way that's never even been discussed before.

- Examine the limits of extremophile life on Earth in an extraterrestrial context. This will involve laboratory experiments where organisms from extreme environments on Earth are stressed further to see if they can continue to survive.

- Assess where else in the universe, especially within the solar system, environments might exist that could sustain life.

- Attempt to locate sources of extraterrestrial life, using the work described above to help determine exactly where NASA should concentrate its search. We might decide, for example, that looking on the surface of Mars isn't as realistic as probing beneath the surface a few feet and that the best area might be near the polar caps, where there might be water beneath the surface.

- Make further recommendations about the nature of future solar system remote sensing or exploration efforts.

Jim is so invested in this proposal, I realize, that he is bound to take the outcome personally, and I soon find out he is up against formidable competition.

"Pull out that drawer," Jerry Soffen instructs me during my subsequent visit to his office at NASA Headquarters. I'm looking out the window at the moment; it's a lovely spring afternoon in Washington, breezy and cool. We're eating sandwiches at his desk. I reach over and tug on the handle of a large, heavy filing cabinet, bulging with applications to the Institute. "Those are all the proposals. Pick up the top one, the sheets of papers there."

"There must be fifty proposals in there."

"Fifty-three. Just glance down. I mean, you'll see everything. There are museums in there, universities, government laboratories, foreign proposals, anything you could ever want. Celebrities, unknowns. I was bowled over." I suspect the chances of Jim Garvin's proposal being selected are pretty slim. Jerry explains that his advisory committee told him to underpick rather than overpick proposals, which meant he would select only a dozen or so. He assembled a peer-review panel, very much the NASA way of doing things. "The trick was to pick good people who didn't have a conflict of interest. We went through every single proposal word by word and tried to figure out what to do; there was a lot of voting and talk and so forth. The net result was we could not get down to twelve; it was impossible because these are just so good. We turned down Nobel laureates. But at least the ones we did pick have no duplications. They include people looking at the bottoms of the ocean and people who were looking for photosynthetic processes. Some people are looking for life on Mars; some are looking at meteorites. There's everything you could imagine."

Jerry hands me a large, bound volume; it weighs about seven pounds and includes winning proposals on subjects ranging from "The Planetary Context of Biological Evolution" (Harvard University) to "Self-Reproducing Molecular Systems and Darwinian Chemistry" (Scripps Research Institute). I spot a proposal for an "Institute for the Study of Biomarkers" (Johnson Space Center), but there is nothing from Goddard Space Flight Center and James Garvin. I wonder if Jim knows.

"There's one proposal I regret not picking," Jerry continues. "It deals with complexity theory. You know the subject? It is a truly unique thing. Most of our chemistry and physics doesn't deal with it." Murray Gell-Mann, one of the founders of complexity theory, led the proposal team. A winner of the Nobel prize for physics, he describes his vision of complexity in his book, *The Quark and the Jaguar.*

When I ask Jerry how he would describe complexity, he tells me, "It deals with what we'll call simple mathematics, simple physics, and simple chemistry. There's a point where things get so complicated that the simple laws of chemistry and physics no longer apply. I'll give you an example. Try using chemistry and physics to predict the stock market. Does physics operate there?"

"Not exactly."

"Yes, it does. Gravity still holds in the stock market. People are still standing there."

"Well, it doesn't explain the real operation of the stock market."

"That's precisely right. You can't resort to physics and chemistry. You have to invoke a whole new world of thought when you talk about the stock market. And we don't know how it works; we don't even know how it's scaled. At least in physics with conventional space, if you multiply by factors of two, you can predict pretty much along the scale all the way from a microbe to a cosmos. On the other hand, when you start talking about stock markets, you're not sure it's linear. It may not be linear."

I'm having a hard time wrapping my mind around this one, as do many people when they begin to contemplate complexity. That is just the point: complexity is designed to free its adherents from outdated or limited ways of thinking about problems, especially complex systems such as the economy, where there are always surprises and too many variables to consider. Complexity seeks to identify concealed patterns of self-organization in nature, in behavior, in the stock market, wherever complicated interactions predominate. Astrobiology is a natural candidate, because, by its very nature, it requires new ways of thinking.

To demonstrate a simple form of complexity, Jerry brings up an image on his large computer screen and asks what it is. The shapes appear to coalesce into a map of a city, perhaps Washington. The longer I study it, the more apparent it becomes that the image is, in fact, a razor-sharp satellite photograph of Washington. There are the Pentagon, the Mall—unique, unmistakable landmarks—and there, he says, pointing, is NASA Headquarters, where we now stand. How would this new way of looking at these familiar objects have been explained thirty or fifty years ago? How would I or anyone else have been able to conceive of this technology? That's where complexity comes in; it allows for the possibilities of the unknown and undiscovered, the X-factors that we don't yet understand, or haven't formulated, that lie outside the bounds of traditional thinking.

As such, it's a system of thought well suited to astrobiology, Jerry says, "because astrobiology is going to be dealing with things that are very likely beyond what we think of as conventional biology. There is a large volume of literature now by people writing on complexity and terrestrial biology. You can't ignore the subject when you're dealing with biology. Nowadays it's beginning to get into cosmology. It's no longer as simple and universal as we thought." With the aid of complexity, it may be possible to identify principles of self-organization of extraterrestrial life. It may be possible to submit such principles to mathematical analysis. Ultimately, it may be possible to use computers to study, codify, or even to predict the existence of extraterrestrial life. "I really do regret that Murray Gell-Mann's proposal wasn't picked," Jerry said. "We had nobody to review it. Nobody understood it. I read the first two pages, and I couldn't understand it either." And when Gell-Mann tried to explain it, "he lost me by about the fifth sentence. Listening to Murray is like listening to an extreme mystic."

I wonder aloud if complexity is a valid way to approach a subject as multifaceted as astrobiology, but Jerry reminds me of the existence of a large body of evidence supporting complexity and cites Gell-Mann's Santa Fe Institute in New Mexico as confirmation of its growing significance. Still, even Jerry has his doubts about the *usefulness* of complexity. "When the members talk among themselves, it's fascinating to listen to, but you're not sure you understand. It all sounds great until you're asked, 'What did you learn?' "

Anyway, once he selected the initial proposals, Jerry needed to find a director of the Institute. Although he seemed a natural candidate himself, he didn't want the job, at least not in its current form. "Dan Goldin wanted to revive NASA's Ames Research Center in California. Why put it at Ames? Because they have life sciences. They're pretty antique life sciences, but they have some." Jerry thought that the Goddard Space Flight Center, in Greenbelt, Maryland, was a better choice of location for the Institute, but Dan insisted on California, and Jerry didn't want to move out there at this stage of his life. "I agreed to be the architect; I even agreed to be the builder, but not the occupant. I value my freedom too much to be the director of this Institute." He tells me about two highly qualified science managers who turned down the job, one because he found a better job in the East, and another who kept asking, "If this is a *virtual* institute, why do I have to move to California?"

Virtual or not, face-to-face encounters, rather than the Internet, have imparted momentum to the Institute so far. "We scheduled a meeting at Ames, and it was fantastic. About two hundred people showed up. What made it so good was the average age, which was about thirty-two. We had told people, when you come to that meeting, bring your students and your post docs. Bring the next generation. I didn't want to find just old fogies like me. I wanted to see who's going to inherit this thing. Although we had a program, it was not the highlight. The highlight was to see these kids out on the steps drinking Cokes and eating hamburgers and talking about astrobiology. NASA may proclaim things, but the facts of the matter reside in the minds of the thirty-two-year-olds, the serious scientists who want to get in and develop this field."

Despite the Astrobiology Institute's revolutionary potential and its ability to attract younger people, Jerry realizes that NASA continues to give priority to astronauts and politics rather than to science, including astrobiology. "Science is taken more seriously now than it was before," he figures, "but the budget is the same as always. Science has always been about twenty percent of the budget, and it still is. When NASA comes under attack, it can be defended on the basis of astronauts; the public loves them. I wouldn't say the public loves science, but they appreciate it; they think it's basically a good thing. But it's not like everybody wants to be a scientist, or wants their kid to be one. Still, as long as scientists don't do something stupid, people think it's okay to spend some money on them."

Within these financial constraints, NASA has developed new and specific ideas about the best places to find evidence of extraterrestrial life on Mars. They amount to a set of scientific principles articulated in NASA's 1995 report, "An Exobiological Strategy for Mars Exploration." This was the work of many hands; at least eighteen scientists contributed, and Carl Sagan is credited as a consultant, and with it, NASA came close to repudiating the Viking life-detection experiments. "The Viking biology payload was selected and developed with very little knowledge about the possible surface chemical and physical resources and conditions to be encountered," the report said. It also critiques the GCMS experiment, the organic molecule detector whose negative findings invalidated the results of the other tests; the Viking GCMS was simply not sensitive enough to do the job adequately. Most strikingly, the report issues a call to revive the quest for life on Mars after twenty years, and this time, to do it properly:

> Perhaps the most valid critique of the Viking experiments is that they were conducted at the wrong place (and/or possibly time) to detect biology on Mars. All evidence from experiments done at the two landing sites suggests a cold, arid surface environment, apparently suffused with oxidants capable of degrading organic compounds. Future studies must certainly seek sites that are wet (and thus warm) and/or protected from oxidants if extant life is to be detected. Viking results indicated that if biology exists on Mars it does

not imprint an obvious mark on the atmosphere, such as terrestrial life. . . . If there are niches capable of supporting Martian life, it is of paramount importance that they be identified and probed for the presence of living entities.

With the announcement of nanofossils in ALH 84001, nineteen months after this report appeared, its message came to seem prescient. "The idea of searching for evidence of life on Mars may strike some as far-fetched, even fanciful. But there is compelling logic to such a quest, as well as equally compelling excitement," it declared with a Saganesque flourish. "Although a fossil on Mars might seem at first like a proverbial needle in a haystack, experience on Earth tells us that *if we know where to look,* finding evidence of ancient life is not particularly difficult, especially when one considers that such evidence can be relatively widely disseminated in the form of chemical or isotopic signatures. The key is to recognize that the search for life on Mars will involve a logically designed sequence of missions, each of which will focus on defining ever more closely where and how biosignatures may be found."

The chief author of this report, Michael Meyer, looks like a professional athlete in the off-season: tall, bearded, casual, and affable as he offers me a large, meaty hand to shake. His workplace at NASA Headquarters—an enlarged cubicle—is festooned with gremlins and goblins and trolls, mostly of putative Martian origin. Ranging in height from one to three inches, they crouch on shelves next to his books and journals, and peer at us from the window ledge. I remind myself I'm in the office of NASA's astrobiology discipline scientist, not a toy maker's workshop.

Right away Michael starts talking about possible scenarios for life developing in the solar system, with particular emphasis on panspermia, the theory that life spread throughout the solar system from an unknown source. The implications of panspermia expand like ripples. It allows for the possibility that life may have begun elsewhere and come to Earth on a flaming meteorite. "Ten years ago, nobody would've talked about meteorites going from one planet to another," he says. "They figured everything would be vaporized. The energy to escape one gravitational field would be so great that nothing would survive. Well, obviously, that's not true. Something could get accelerated to escape velocity without undergoing a huge compression. Newer models show there's still compression, but it's not huge." Panspermia is still highly controversial. Mention it to astrobiologists, and you get a spectrum of opinions ranging from outright scorn to fervent belief. Michael, for his part, holds with panspermia. "The bottom line is: it's possible that once life started anywhere in the solar system all the other planets would have the opportunity to be seeded."

If seeding ever occurred, Earth, with its relatively strong gravity field, would have captured a large share of life-bearing material; Mars, which is smaller, would have captured much less. Since Mars is the closest planet, he reminds me, it enjoys the highest statistical probability of sending a life-bearing meteor to us. It is much less likely, though not out of the question, for Earth to send a meteor to Mars. The relatively high number of Martian meteorites coming our way could mean the ancestor of life on Earth might be nanobacteria from the Red Planet that came here and evolved. In that case, it has been remarked, we are all transplanted Martians.

Panspermia also allows for other models of how life spreads throughout the solar system and the universe. If a meteorite from Earth happened to transport samples of life to Mars, the presumed fossilized bacteria found in ALH 84001 might have originated here. In that case, we're looking at a specimen of primitive life on Earth. That scenario wouldn't be as exciting as the thought of life independently starting somewhere else, but it would be intriguing, because scientists could study how life adapted to the Martian environment. Keep in mind that meteorites are not necessarily representative samples of the planets or asteroids from which they came. Suppose there were Martian geologists hard at work, prospecting their planet for meteorites from Earth. Eventually, if they looked long enough, they might find a specimen, and it might contain evidence of life on Earth, or it might not. If not, Martian geologists might feel comfortable declaring that the Blue Planet was dead. Michael spends his days puzzling over these quandaries.

"Does stuff ever arrive in this solar system from other systems?"

"Sure, cosmic dust."

That phrase reminds me of a small, transparent box I keep in my desk drawer. The box looks and feels empty, but if you poke your finger inside,

you will touch something brittle and practically invisible. The peculiar substance inside the box is called aerogel, and I came across it during a visit to
a large Lockheed Martin facility near Denver, Colorado. Aerogel is basically
glass and air, mostly air. It goes by the nickname of "mystifying blue smoke,"
and it does resemble a puff of smoke, but it's solid enough to sit in my desk
drawer.

Lockheed Martin builds and operates a number of missions, including
one called Stardust, which was launched on February 6, 1999. Stardust will
swing around Mars and follow a course to the comet known as Wild 2,
which it will encounter on January 2, 2004. At its destination, the spacecraft
will deploy a screen filled with aerogel, which will trap comet dust, and then
it will begin the voyage home, arriving in Utah on January 15, 2006, at about
three o'clock in the morning. NASA predicts a spectacular reentry for the
spacecraft, as its trajectory takes it across the California-Oregon border and
northern Nevada. If you can get to Wendover, on the Utah-Nevada border,
not far from Interstate 80, you will see quite a show on that date. After the
spacecraft makes a soft landing in the Utah desert (which will likely be
snow-covered at that time of year), scientists will remove the aerogel from
the spacecraft and study the particles it has collected. Conventional wisdom
holds that cometary particles consist of one-third ice, one-third organic material, and one-third minerals—rocks, basically. Stardust's particles, gathered
from Wild 2, will be pristine specimens, uncontaminated by Earth's atmosphere, and they may contain clues as to how life began here and on other
planets.

Each year, a tremendous amount of space dust falls to Earth, more than
40,000 tons, Michael tells me, and in the early history of this planet, before
it developed a thick, protective atmosphere, a lot of that cosmic dust reached
the surface. The dust is thought to have seeded our planet with organic
molecules that helped to develop life. If successful, the Stardust mission will
reveal a lot about how that process worked.

Michael cautions that the odds of finding a speck of cosmic dust bearing an actual specimen of life from another solar system to ours are very, very
long. "One of the things that is psychologically amazingly hard to come to
grips with is how empty space is. You're going to spend a billion years in
transit time." In the emptiness called space, there are at least 125 billion
galaxies, and each of those galaxies contains many billions of stars, and, pre-

sumably, planets, yet even if something is out there, it is inconceivably distant from us. "If we were randomly inserted into the Cosmos, the chance that we would find ourselves on or near a planet would be less than one in a billion trillion million (10^{33})," Carl Sagan pointed out. At this scale, Mars seems stunningly near at hand, the planet next door, Iceland without oxygen, a potential home away from home.

Michael's evocation of the emptiness of space unexpectedly takes me back to a time when I was very young and suffered night terrors. They began when I pictured myself as an astronaut floating through space, an infinite blackness punctuated with impossibly distant stars. I was tethered to a spacecraft by a hose providing oxygen, but then the hose would snap and I would drift off into eternal space, until I was no longer alive and my existence forgotten. In the morning, I'd recall the episode with a shudder and go about my life, but it remained lodged in the back of my mind, tucked behind my consciousness. How reassuring that at least one planet looms relatively close.

Michael emphasizes that Mars' proximity to Earth means the Red Planet will eventually be colonized. "There are good reasons to have somebody living on that planet, just for the sake of human survival. And if you start talking about going to Mars and understanding its biological history—if there is one—humans might be able to help solve that, but you need to have a very good baseline *before* you send humans, because once you do, you contaminate the area you're studying. Let's say our robotic spacecraft do discover life on Mars. The question then becomes, 'Do we want to send humans?' "

"Why wouldn't we?"

"Because we start to think about self-protection and accidentally wiping out Martian life. Or coming up with a chimera that nobody's happy about. A chimera is an ancient Greek beast made of different beasties, the head of a lion, the body of a goat, and, I believe, the tail of a serpent. So there are plenty of good reasons to learn about Mars before we send someone."

Like most biologists, Michael assumes all life on Earth arose from a single common ancestor. "That's reflected in the fact that the biochemistry in our bodies, in an amoeba, a cockroach, and a tree are all basically the same," says Meyer's colleague, Carl Pilcher, of NASA's Office of Space Science. "There are too many things about life that seem happenstance yet are uniform across all of life, with only very minor exceptions. For example, virtually all life on Earth uses right-handed sugars in DNA and left-handed amino acids. It could just have easily been the other way around." Handedness refers to the three-dimensional structure of molecules, which permits them to join. Nonliving molecules are ambidextrous, so to speak. Only molecules in living things display handedness, and it is one test for life, at least on Earth. It's possible the earliest living cells on Earth used both left- and right-handed amino acids, but even if that was the case billions of years ago, it's not the way things are now.

Carl invokes an even more startling example of the commonality of life: "Virtually every living organism on Earth uses the same molecule as its main energy-carrying molecule; it's called ATP." Adenosine triphosphate contains two high-energy phosphorus-oxygen bonds. When broken, the bonds release a considerable amount of chemical energy, and they require a considerable amount of energy to establish. Remarkably, all living cells use ATP to store and access energy when needed. "Maybe there is something about that molecule that makes it the gee-whiz, bang-up best, but wouldn't you think that if life had many origins on Earth, it would have found some other molecule that also did the job, and that strain of life might have survived to this day? Apparently, it hasn't. Either some other molecules formed and died out because they couldn't compete with ATP-based life-forms or we've all got a common ancestor, and it just happened to use ATP." If life on Earth arose in this fashion, astrobiologists suggest that extraterrestrial life probably originated the same way. The key to understanding life on other planets, they say, is to look into our own evolutionary past. Only as life develops, and mutation and environment come to play ever larger roles, does it diverge. Nevertheless, the basic building blocks remain the same. Even our own, sophisticated DNA contains primitive elements that go back billions of years; crucial parts of it are ancient and not specific to human beings.

When I ask whether it's possible that there was one kind of life in Africa and another kind of life in Antarctica, Carl quickly reminds me that Africa and Antarctica didn't even exist then. "It's possible life originated in many different places. However, if that was the case, only one of those origins led to a continuing line, and the others all died out. There was really a single common ancestor."

The closest planetary analogue for this paradigm, Carl explains, happens to be Europa, the moon of Jupiter, rather than Mars. "It's ice covered and, we think, may have a global ocean underneath the ice. We look at the surface, and we see that it's very active, jumbled, and young. As recently as a few million years ago, large areas of the surface were molten, and we think there were icebergs floating around. Europa is heated from the inside, so the water is hot, at least at the bottom, and warm throughout. We think that is exactly the kind of environment in which the common ancestor of all life on Earth existed. We think it was a hyperthermophile, a heat-loving bacteria that fed on sulfur. On Earth, they're powered by the internal energy of the planet, which shows that sunlight isn't necessary for life. Energy is necessary, and thermal energy coming out of the Earth will do nicely. It is possible that a similar environment—an energy source and liquid water—may exist on Europa right now. If it does, exploring it would be one of the more remarkable achievements of human civilization. It might revolutionize more things than you would imagine."

"Such as?"

"The pharmaceutical industry is interested in isolated ecosystems, those that have not interacted with the bulk of the Earth's ecosystem, because you can find genetic material and proteins that are new; they have never been exposed to the rest of the biosphere. Pharmaceutical companies are interested in finding out if any of those are of beneficial biological use to treat illnesses. One concern, of course, is that they might be pathogenic and destructive to life on Earth, so we need carefully designed protocols for transporting this material back to Earth, isolating it, and making sure it's safe. But the potential for learning about the nature of life and how it works is immense."

It's time to ask my Genesis Question again. "Does the presence of water on another planet or moon guarantee life forms there?"

"The answer depends on whom you ask. On Earth, any place you find liquid water and energy that can be channeled into a chemical form, you find life, even under a couple of kilometers of basalt. We're discovering that bugs that have been brought up in drilling mud for decades are not surface contaminants, as we once thought, but come from ecosystems well below the surface. These bugs are living off a reaction between hot water that's percolating up through the rock and the iron in the rock. They derive their energy from that reaction."

No matter what form life assumes, scientists have illuminated universal principles of architecture that apply on all levels, from complex biological systems such as humans all the way down to single cells. These principles are known as tensegrity, a term coined by the late Buckminster Fuller, the iconoclastic architect. Among his many innovations, Fuller was known for his geodesic spheres, or "Buckyballs," a particularly elegant example of tensegrity at work. Buckyballs are strong and useful for building because they are biomimetic; that is, they mimic the architecture of living things. Buckyballs are frameworks made up of rigid struts, each of which can bear tension or compression. The struts form triangles, pentagons, or hexagons, all of which give the structure stability. They look a lot like giant atoms. In fact, some common arrangements of carbon are known as "fullerenes"—miniature Buckyballs consisting of sixty carbon atoms arranged in a sphere; the carbon-to-carbon bonds serve as the struts in the sphere.

Donald Ingber, a professor of pathology at Harvard Medical School, has been studying tensegrity for twenty-five years; he considers it the organizing principle of nearly everything in the universe. "An astoundingly wide variety of natural systems, including carbon atoms, water molecules, proteins, viruses, cells, tissues, and even humans and other living creatures, are constructed using a common form of architecture known as tensegrity," he says. The architecture itself is, technically, a "system that stabilizes itself mechanically because of the way in which tensional and compressive forces are distributed and balanced within the structure." Think of tent poles transmitting tensional forces across a network, imparting stability to the entire system, and you have some idea of tensegrity at work. Tensegrity operates on both macro and micro levels, in the way a skeletal system supports itself, or the way components of a protein arrange themselves. In tensegrity, the evolution of a living system is very similar or even identical to the evolution of a nonliving system. "It is no surprise that the basic arrangement of bones and muscles is remarkably similar in *Tyrannosaurus rex* and *Homo sapiens*; that animals, insects, and plants all rely on prestress for the mechanical stability of their bodies; and that geodesic forms, such as hexagons, pentagons, and spirals, predominate in natural systems."

An understanding of the mechanics of tensegrity can enhance our appreciation of how life arose, Ingber tells me. For example, many scientists now think that life emerged in clay rather than in a primordial sea. To Ingber, this is a beautiful illustration of tensegrity. "Clay is itself a porous network of atoms arranged geodesically within octahedral and tetrahedral forms," he says, "but because these octahedra and tetrahdedra are not closely packed, they retain the ability to move and slide relative to one another. This flexibility apparently allows clay to catalyze many chemical reactions, including ones that may have produced the first molecular building blocks of life."

In May 1998, in a talk at NASA Headquarters, Ingber proclaims that eventually the principles of tensegrity will be seen as an expression of even more fundamental laws of nature. "In essence, the creation of life is simply one step in the evolution of the universe," he says. "Through this new understanding of biology, we will create what I call a biomimetic materials revolution—the ability to create new materials that exhibit mechanical features and biochemical processing capabilities previously exhibited only by living materials. This will involve merging computer science, mechanical engineering, and molecular genetics, among other things." Listening to Ingber talking about tensegrity this way is like taking a drug trip minus the drugs: you get the sensation of seeing the underlying principles of the universe and the connectedness of all things, minus the incoherence and flashbacks.

Identifying the architecture of life is not a new idea. In 1917, D'Arcy Wentworth Thompson, a Scottish zoologist, published a study of "biomathematics" in which he deployed mathematical formulas to illustrate how species change over time. Like Ingber, Thompson perceived universal principles at work in the form of life. "Everywhere nature works true to scale, and everything has a proper size accordingly," he observed. "Cell and tissue, shell and bone, leaf and flower are so many portions of matter, and it is in obedience to the laws of physics that their particles have been moved, moulded, and conformed." There is no reason why "everywhere" should not apply to Mars, as well as to Earth.

Thompson's work was limited by the technology of his day. His complicated mathematical relationships, in which so many variables came into play, did not have the benefit of a computer, which would have vastly aided the speed and precision of the calculations. Now, more than eighty years later, NASA has begun to use computers to apply Thompson's principles. The results will eventually be compiled in a massive "Book of Life" cataloging all known microbial life-forms on Earth. Such a guidebook would provide clues to possible biological forms on Mars, on more distant planets, and even on asteroids. Thompson's principles will also be brought to bear in the examination of the nanofossils found in meteorites from Mars. They may even help resolve the debate over whether the enigmatic shapes found in those objects from another world have a biological or chemical origin.

The life-detection experiments aboard Viking, and the more recent search for water on Mars, have all been based on finding the *chemistry* of life. The use of tensegrity, among other mathematical models, may help provide researchers with the *geometry* of life—its characteristic forms, wherever they may be.

Right after Ingber's discussion, I go head to head with Dan Goldin, the NASA administrator. I've been hearing about him ever since I began my journey through NASA ("high energy," "a character," "overall, good for NASA," and "asks a lot of questions" are some of the things most often said about him), I've seen him at a distance, and now here he is, approaching me at a rapid clip, with a shiny black cone in one hand. We're in his office at headquarters, family photographs and models of spacecraft vying for shelf space. Suddenly, he releases the cone, and it thumps on the carpet, spins like an oversized top, and comes to rest on its side. I have no idea what I've just seen.

He takes his seat and looks me over. The testosterone level spikes up. His black cowboy boots gleam against the soft carpet, and he starts talking at a furious rate in a very un-NASA but very New York accent. ". . . it became clear to me in the mid to late eighties that we were going about building spacecraft wrong—not wrong, exactly, but there was a new wave coming," he says. "When NASA built spacecraft, they designed them to fit within the shuttle because it was going to be the only launch vehicle. That was caus-ing us to build just a few spacecraft, and these spacecraft, if you will, were like the last ship out of port. There were so many experiments on one spacecraft that if you lost the spacecraft, you lost the whole system. Mars Ob-server is a case in point. We lost the *whole system*. Those spacecraft cost too much. They took too long to build. Instead of using next-generation tech-nology, which is what NASA's supposed to do to drive the technical base of the country, we went back in time and used old, proven technology, which made the spacecraft bigger and more expensive. Even a ground checkout test became a political event. We had people who watched people who watched people. It was very inefficient. There was no accountability and responsibility . . ."

I still haven't ascertained the exact nature of the black cone, but Dan is on a roll, talking about the future—of NASA, astrobiology, technology, everything.

"When I arrived at NASA, it cost on average $600 million to build a robotic spacecraft, and it took over eight years. We were launching two a year. This year we're going to launch ten. I wanted redundancy in function and number, so we never have to worry about losing a decade's worth of science if we lose another spacecraft. At the present time, we're under two hundred million dollars for a spacecraft, and this year we'll launch ten of them. Eventually, we expect the spacecraft to cost about seventy-five million on average; in fact, we may even get much lower. So that's the concept, that's the genesis . . ."

As he talks on, I recall his nickname, "Captain Chaos." If you've ever worked for a boss who pits two divisions of a company against each other to see who will survive, who changes his mind every week (or every day), who "manages through conflict," then you have some idea what it's like to work at Dan Goldin's NASA. No one expected that someone as charismatic, abrasive, fast-talking, and edgy as Goldin would come to lead NASA. He was born in 1940 and became NASA's administrator in 1992, when the agency was languishing in the wake of the Challenger disaster. Goldin seemed an unlikely choice; he came off as more of a space dreamer than a bureaucrat. He liked to tell the story of how, as a teenager in Far Rockaway, New York, he'd sleep on the beach when the weather was warm and watch airplanes traverse the night sky on their way to Idlewild (now JFK) Airport, and he would contemplate the voyages of ocean explorers. He learned to identify the planets in the night sky and fell under the spell of Mars. After he graduated from City College with a bachelor of science degree in mechanical engineering, he worked for NASA's Lewis Center, in Cleveland. Even then, there was a Mars program, and he was working on it, helping to develop an ion propulsion engine, a technology that has long remained elusive. "Exploration is tough," he once said, "and it does not have continuity. It has *zealots.*" As the race to the moon became NASA's leading priority, the vision of reaching Mars receded, and Dan found himself distracted, ill at ease. TRW, a major aerospace contractor, recruited him, and eventually he accepted a job with them in California. He spent much of his time at TRW in "black" (intelligence-related) areas. Dealing with NASA from the standpoint of private industry, he became disillusioned by the agency's arrogance, inflexibility, and wasteful practices. NASA, he later declared, was "too stale, male, and pale."

Despite his reputation within the industry, Goldin was unknown to the public and to most politicians. As he lobbied energetically for the appointment, he impressed those in Congress who met him. Everything was on the tip of his tongue; he had goals, agendas, vision, boom boom boom. Even better, he sounded like a cost-cutter who would stanch the flow of taxpayer dollars. He talked about "faster-better-cheaper" approaches to space exploration. The Republicans liked his agenda—especially the last item. The clincher was that he was a registered Democrat, which made confirmation much easier. Still, when he received his appointment in March 1992, it appeared he might hold office for only a few months if Bush lost his attempt at reelection, but Goldin survived the change in administrations and continued attacking NASA's cozy way of doing business. At an elaborate luncheon with a highly placed executive from a leading NASA contractor, he exploded. "You've overrun every contract you've ever had with us! You'd never treat a private client the way you treat NASA! It's unacceptable!" Whatever he lacked in polish, Goldin made up for in brass balls.

Despite his strenuous efforts at reform, NASA resisted change; it took years to engineer the early retirement packages and hasten a generational turnover at the agency. Meanwhile, Goldin slowly evolved from an outsider to an insider. He came to be seen as the agency's champion in the face of a hostile Congress and an indifferent public. Throughout these years, Goldin says, he kept faith with Mars, but in a responsible, budget-conscious way. "When I got to NASA in 1992, they showed me a program. They were going to plant seismometers all over the surface of Mars, measure moisture, measure wind velocity. They were going to map the surface. They wanted billions of dollars for that. They were going to have all these big pieces of equipment using old technology. Well, Jet Propulsion Laboratory is now leapfrogging over this and putting whole payloads on a chip." He turns to the mysterious black cone. "In December 1998, we're going to land on Mars.

Two of these cones will be dropped from the spacecraft as it descends through the Martian atmosphere. They will go a few meters into the ground and will see if there's moisture present. And we're now talking millions of dollars, not billions for this mission. It's a completely different way of thinking. People used to measure the vitality of this agency by how big our budget was. That is a very, very bad thing to do."

"Misleading?" I remark.

"It's not misleading—it's wrong, unethical. A budget doesn't measure your output, it measures your input. You can't measure a government agency based upon how much they get. You measure a government agency by what they do for the American people. So we at NASA said, 'We don't want to be judged by how big our budget is.' When we put out press releases we no longer announce how many jobs are involved, because NASA is not about how many people are working on the federal budget; NASA is about how many people will have jobs in the future because of what we do. It's a very different frame of mind. We're a healthier agency because our budget has come down."

Despite a robust, expanding economy and a surplus in the federal budget, NASA's budget shrinks and shrinks—not enough to kill, just enough to hurt: $13.8 billion in 1998, $13.66 billion in 1999, and $13.57 billion in 2000. As NASA constantly reminds the public, the amount comes to less than one percent of the entire federal budget, which now approaches $1.8 trillion. The Department of Defense receives twenty times more than NASA. The Department of Transportation gets around two and a half percent of the federal budget, and Housing and Urban Development, a little less than one and a half percent, nearly twice what NASA will receive in the near future. If the budget were a big pie, NASA's share wouldn't look like a slice—just a thick line between some very generous slices. But when Goldin declares that austerity is actually good for NASA, few within the agency agree with him; the shrinking budget erodes morale and sends a signal that NASA's work is of declining importance to the nation. Critics insist it is time for NASA to relinquish its authority to private industry, but the fact remains that only NASA can coordinate the elements of engineering, science, and industry to pursue the exploration of space. This means launching rockets, tracking spacecraft, maintaining the worldwide Deep Space Network, funding science research, training astronauts, negotiating with foreign governments, among other things. No private company can manage all that, yet. No one else has this vast infrastructure or reservoir of highly skilled, deeply committed employees. In Russia, where the space industry has been privatized, collapse followed. "If somebody could do it privately, I'll applaud," Dan tells me. "The problem that I'm having is, I come from the private sector, and if I went to the board of directors in my corporation and said, 'I need tens of billions of dollars to build all this hardware, and it's going to take a decade, and we're going to find some things, and maybe there will be a commercial payoff in ten years,' I'm not sure I could get financing. Why would a private company go along? Our exploration of the planets can be compared to Jefferson's sending Lewis and Clark to explore the West. That's the kind of thing government does."

Goldin envisions NASA transformed by one scientific revolution after another. "A decade or two from now, NASA will be run, in large part, by biologists because biology will be crucial for developing engineering rules. Software will start to mimic biology. Take a look at the coding system inside a living being, DNA. The system's unbelievable, elegant, and you don't have to have thousands of software engineers encode your replication and your maintenance. The problem we have now is there aren't enough software coders in the world because we use deterministic, hard, numerical computing, and you have to go through every predictive branch point. Computing will become biomimetic, as I call it. You will hopefully put constraints into the software and say, 'Here are the limitations; here's the outcome we want; here are the constraints you must live within. Code yourself, validate yourself, learn, develop, and start operating.' "

In the new NASA, paradigms will fall faster than ever, to hear Goldin talk. "In ten years, we'll be right on the edge of having determined if life is ubiquitous in the universe. In ten years we will have almost completed the first census of our own solar system. By that time, we will have determined if there are Earth-like planets within one hundred light-years of Earth. In ten years, we'll be more or less done with the International Space Station. In ten years we'll be well on our way to either having a permanent base on the moon, a research station on an asteroid, or an expedition to Mars. Ten years from now, NASA will have developed tools that will revolutionize the way Americans live—telemedicine, telepresence, body scans, total immersion, virtual presence, computers that you can't even imagine today. We'll have tools for industry to develop a product in virtual space, and before they commit any significant money, they'll have total confidence in it . . ."

Meanwhile, I'm looking at the black cone on the carpet, and wondering about its role in Mars exploration.

Eventually, I find out more from Jim Garvin during dinner in an immaculate Chinese restaurant near Goddard. Turns out what Dan was showing off that day was a microspacecraft. "I was on the review panel for the science team for them, along with Michael Meyer and David McKay," Jim explains over bowls of steaming rice. "We're going to send two little penetrators to Mars along with it; they're going to smash into the surface, and look for water. They'll conduct experiments with heat flow. And a little camera will pop up and it will look around and take some pictures. And that will be it. We will send them in ballistically. No retro rockets. No parachutes. Nothing. They're just missiles. *Phhhhhhhhhhhhhhhhhhht! Pow!"* His hand dives to the table, rattling the china. Without its protective cone, the minipenetrator strongly resembles the male member; the instrument's science team consists entirely of women. Maybe NASA has a sense of humor, after all.

Days later, Jim learns that his proposal to the Astrobiology Institute hasn't been selected. It's just as I suspected the day I was sitting in Jerry Soffen's office, looking at the submissions. And Jim does take the decision personally, as I feared he would. In fact, he's stunned. Over the next week or two, whenever we talk, he sounds dejected, overwhelmed, taciturn. When he collects himself, he tells me that it's essential for the Goddard Space Flight Center to be represented in the Astrobiology Institute. Next time, he vows, "I assure you we will win or die trying." I believe him.

Part Three

DISCOVERING

MARS

Chapter Nine

ROCKET SCIENCE

As you enter the Lincoln Field Building, the seat of Brown University's Planetary Geosciences Group, you walk across a varnished wooden floor, glance at doors accented with faded gilt lettering, squint in the chiaroscuro left by shades drawn over the windows, and mount a spiral stairway creaking underfoot. Little has changed here in nearly a century. A few computers are in evidence, but they look anachronistic. *Electricity* seems anachronistic in this academic edifice dating from 1903. Then you pass through a door with a sign announcing, "PREPARE FOR SENSORY OVERLOAD" and go down a rabbit hole into a cavern where the ambience abruptly shifts to unreconstructed 1960s Haight-Ashbury headshop. Somehow, the distinguished occupant of this realm, Professor James William Head III of Brown's Department of Geological Sciences, seems completely at home; he wears a button-down shirt, jeans, cowboy boots, and a Mickey Mouse watch, all in contrast to his abundant white hair and thick glasses. Behind him are more than a hundred beer bottles on the wall, each one unique, plus nearly as many beer cans. Professor Head has an announced goal of "tasting one of every type of beer brewed in the entire solar system." There are also wine bottles up there, including one shaped like a lobster. It contains a startlingly bright red liquid; one swallow might make us larger, two swallows, smaller. For today, I will consider it a container of magic Martian elixir.

Once you get past the bottles and cans, you can make out charts and photos from the heyday of the Soviet space program; they're exhibited here as if they were old psychedelic posters from the Fillmore East and West. There are prominent pictures of Lenin in bold, faded red profile, with his demonic, pointed chin. There's a photo of Head with the pope—possibly a skillful fake, since it contains cartoon balloons overflowing with satirical remarks. There's still more stuff hidden behind the beer bottles and Soviet memorabilia. Now I notice the candles in assorted sizes and shapes, the largest of which is a paraffin tribute to the great state of Florida, not to mention countless other curiosities displayed with loving care.

What you can't tell from the decor is that Head is a powerful presence in the planetary community, as it's called; if he were a planet, he would be a Jupiter, or perhaps a Saturn, and other bodies would orbit around him. He has influence over a considerable amount of money available for grants and funding. He has say-so over jobs. And he is one of the better-known members of the MOLA science team. He doesn't wield his power in any obvious way, but it's there, and everyone knows it. Over a lunch in his office of sandwiches and Snapple, he tells me he's actually a very practical scientist, as scientists go. I look around at the furnishings and nod. "I'm an accomplishment junkie. I measure my day by what went out the door. I don't spend a lot of time with my feet up on the table. Sometimes, I get fed up with the pace of things. Dave Smith, in contrast, is an agonizer. I sometimes feel like saying to him, 'Jesus Christ, Dave, when are we going to meet?' " As he talks, I catch a glimpse of a steelier side of Jim Head, but he keeps it at bay.

Jim discovered his interest in geology as a freshman at Washington and Lee University in Virginia. It was an outdoor pursuit, which appealed to him, and when he heard about a summer position as a field assistant in Montana, he immediately applied. He doubted he would get the job, until he discovered he was the only one who expressed an interest in it. Before he went, a professor told him, "You will come back from Montana either loving or hating geology." He came back loving it. A few years later, as a graduate student at Brown, he studied stratigraphy under Tim Mutch, which involved examining lateral layers of the Earth's crust for clues to its formation. Jim compares it to reading a palimpsest: a lot depends on how you decipher the layers and how you decide on the sequence of events and forces accounting for the present formation.

One day, Mutch said, in that folksy, ruminative, Jimmy Stewart manner of his, "You know, there are just no fundamental problems left in terrestrial stratigraphy anymore," and challenged Jim Head to apply Earth stratigraphy to lunar problems. This was the beginning of a revolution in the study of the solar system. "Tim Mutch was really the first person to make the connection between geology and other planets, beginning with the moon in the 1960s," Head says. "Before then, it was mainly astronomers calculating orbits, and frankly they didn't know what to do with a lot of the data. They worried about what to *name* these craters." In the 1960s, the Mariner missions to Mars "blew the field wide open," Head says, because they began to map the surface of the planet. At that point, "the geologists took over planetary exploration. A few people like Carl Sagan bridged the gap, but it became the realm of geologists and geophysicists."

Head worked on the Magellan mission that surveyed Venus in 1990. People tend to forget what a great mission that was; these days, everyone thinks about Mars, Jupiter, and Jupiter's moon, Europa. But Magellan set a new standard in planetary exploration. "For Venus, we have two-hundred-meter resolution for ninety-eight percent of the surface, which is better than we have had for Mars until now, and better than what we have for Earth, because oceans cover two-thirds of the surface of our planet," he says. For all its successes, Magellan proved to be a mixed adventure for Head, who made several bold predictions about what Magellan would find on Venus, few of which turned out to be accurate. Head was by no means the only scientist confounded by Magellan's findings, which showed a planet more like Hell than anyone had imagined. It was hot (900° F), rocky, and hostile to life, even extremophile life.

The experience with Magellan has given him a measure of perspective and patience with the frustrations of the Mars Global Surveyor mission. He finds the occasional delays helpful, because they afford time for fine-tuning. "For example, I know exactly what we're going to do at the Martian south pole now. I had six extra months to think that through, to select segments of the data to study." Like most of the team, he is focused on evidence of water on Mars—ancient standing bodies of water, to be precise—and their link to astrobiology in the form of fossils, or even extant forms of life, such as bacteria. "That's a really important idea, but you don't want to come across as a proponent. We are *testing* these ideas. I can't just say, 'Holy mackerel! There was so much water on the surface of Mars!' Then I become a proponent," he tells me. "The simplest way to describe science is exploration—exploration of the unknown."

Jim has been around long enough to see vast changes in NASA, and despite the promise of "faster-better-cheaper," he's not persuaded that the changes are for the better. When he came aboard, James Webb was NASA's administrator. "He had a vision of partnership among NASA, Industry, and the University that worked very well in the beginning, but the tripartite arrangement is not the same anymore." The scarcity of money has caused more competition and friction, occasionally at the expense of science. "If you can come back with an experiment that costs no more than ten million bucks, that's a go." He considers the amount about half of what is required to do an experiment properly. Still, he's grateful to have any experiment on Mars Global Surveyor. Like everyone else on the science team, he has an indelible memory of where he was in 1993 when he heard Mars Observer was lost: "It was a Sunday, I was at my desk, and I got a call. I thought, 'Oh, *shit . . .'*"

It's taken five years to get from that moment to the early returns of data from Mars Observer's bargain-basement successor, Mars Global Surveyor.

Just after Jim and I finish talking in his office, the MOLA science team convenes here in the Lincoln Field Building for its September 1998 meeting. The members have flown in from all over; there's Matt Golombek from JPL in Pasadena; Dave Smith, Sean Solomon, and Jim Garvin from Washington, D.C.; Maria Zuber from Cambridge; and more than a dozen others from around the country, most with graduate students and post docs in tow. As they wander in to a large shabby classroom right above Jim Head's office, everyone is in a confident mood. They have data, enthusiasm, and a conviction that their experiment has, after all, been worthwhile.

No one can predict how long the good times will last. "We've been waiting for years, but only recently have we had *real data,*" Dave Smith reminds them as they settle in. It may be the only data they get. The Great Galactic Ghoul still lurks in the vicinity of Mars, awaiting an opportunity to spring a nasty surprise on spacecraft (and their makers). Even now, the team learns, there are fears about deploying the high-gain antenna, one of two aboard Mars Global Surveyor. The high-gain antenna transmits data rapidly and continuously to Earth; the smaller low-gain antenna allows a much lower rate of transmission. For that reason, scientists overwhelmingly prefer the high-gain antenna. At the moment, it is stowed against the spacecraft and is supposed to deploy on a long boom in the circular mapping orbit. But it uses the same damper system that malfunctioned when the MGS solar panels were deployed at the beginning of aerobraking. If the high-gain antenna's damper also malfunctions on deployment, the scientists will lose most of their data.

Bruce Banerdt, a team member from JPL, announces that the project is considering going into "Magellan mode," to point its low-gain antenna toward Earth, but that would mean losing forty percent of the data. Having worked on the original Magellan mission to Venus, Jim Head knows from experience the disappointment of reducing data playback. "Once we are in that mode, we'll never get out of it," he laments.

Stumped, the scientists look to an engineer with whom they communicate smoothly; this is Jim Abshire. Jim looks as lean and military and precise as a fighter pilot or astronaut, speaks with a soft Tennessee accent, and has impeccable manners. I can't imagine him ever losing his equilibrium. He's been working on the laser altimeter as long as the scientists, and he reports that it has now been in space for twenty-two months. The altimeter itself has fired 9.2 million shots, more than all the other lasers in space combined. It's been turned off and on 211 times, which no one is happy about. To preserve its useful life, it should be left either off or on until the primary mission commences. He also frets about a solder joint that will fail sooner or later if the temperature keeps fluctuating . . .

Sean Solomon can contain himself no longer. He breaks in and compares waiting for data from the laser altimeter to "Chinese water torture." Everywhere he looks, he sees problems—with the altimeter, with the antenna, and God knows what else in the months ahead. Although they have received a limited amount of data, they haven't begun the primary mission, and the spacecraft seems to be dying by inches. "Can we survive this Magellan mode by keeping the laser on or off all the time?" he asks.

"It's my gut feeling that we would prefer to keep it off," Dave Smith tells him, "as global mapping hasn't started yet." He reminds everyone that NASA will send a pair of spacecraft to Mars late in the year. The Mars '98 Climate Orbiter will circle the Red Planet, testing its atmosphere, while the Mars Polar Lander will settle down at the edge of the southern polar cap. Mars Global Surveyor is supposed to act as a relay for the new arrivals, and Sean has visions of their laser altimeter losing out in all the excitement.

"I think there are some strategic issues here," Sean warns.

"There are no real issues; there are opportunities. I sound like a politician, don't I?" Dave says brightly.

After this discussion, Bruce Banerdt says that tomorrow night the laser altimeter will, after all, be turned on again to fire shots at Phobos, one of Mars' two peculiar moons. The outer moon, Deimos, orbits Mars at a leisurely rate, while Phobos, only 3,700 miles above the surface of the planet, zips around the planet in less than a day. The moons of Mars were discovered and named after the two consorts of Ares, the Greek martial deity, on their discovery in 1877 by Asaph Hall, an astronomer with the United States Naval Observatory in Washington, D.C. Curiously, Jonathan Swift, in *Gulliver's Travels,* published in 1726, 151 years before Hall's discovery, described "two lesser stars, or satellites, which revolve about Mars," and proclaimed that they "are governed by the same law of gravitation that influences the other heavenly bodies." Where Swift acquired his information is not certain, but it proved to be prescient.

Phobos is believed to be a captured asteroid, with a geologic history independent of Mars. As such, it may contain clues to the origin of the solar system. Even by Martian standards, Phobos is a strange place. It is quite small, just eight miles in diameter. If you were standing on the surface of Mars and looked up, Phobos would appear to be about half the size of the Earth's moon. If you traveled to Phobos itself and looked up, you would see Mars filling almost half the sky; it might seem as though you were about to fall to the Martian surface. If you looked down, you would see that the surface of Phobos resembles a heavily gouged and scored piece of dark rock, accented by a large crater, almost one-third of Phobos' overall size. (Asaph Hall christened the crater Stickney, after his wife, Angeline Stickney Hall.) If you walked around Phobos, you would complete a circuit in a day or two. Its gravity is so low that if you jumped vigorously, you would launch yourself into space. Farewell, Phobos.

The Phobos mission will be a quickie, a micromission. The little moon will be in range for only eight minutes, and the scientists will have their data from Mars Global Surveyor within a few days. Their instrument and the spacecraft's cameras will capture more details about Phobos in eight minutes than have been acquired in the entire history of Mars exploration. Phobos will no longer be the mystery it is now.

Having this new data brings its own problems. Who gets it, and when do they get it? "We will release preliminary maps and topographic images to the public rapidly and frequently, but fully validated and corrected data is a more complicated matter," Dave tells his team. "I want to make it possible to get quality data out to you, and for the moment I am talking about *you—* not the public. When you talk about getting data to the public, you are talking about a sequence of events. We are bound to release the validated data to the public on a schedule. We have a period of up to six months to do that, though I hope we can accelerate that schedule. But I'm *not* proposing that every time we have a new data set for you that we release it to the public."

Sean asks, "What are the implications of people who aren't on the team working on data that we know is *not* properly corrected, while we are working with superior data?"

"That means we'd never release any data until the end of the mission," Dave says.

"We're always going to have better data," Maria Zuber says, shrugging her shoulders. "Many of the corrections that matter to us and take months to make are not important for data we release to the public."

"It's a philosophical issue," Sean continues. "It's not proper for us to say, 'I know something you don't know.' "

"We don't want to mislead," says Jim Head.

Maria becomes exasperated with the direction of the discussion. "We're spending all our time arguing about corrections that many *scientists* don't understand, let alone the public."

"But we're the only guys giving out data," Sean reminds her.

"What if we have a statement: 'Caveat emptor?' " Jim Head suggests.

"We are doing orders of magnitude better than anyone else," Maria reminds the MOLA team, but nothing seems to placate Sean.

Jim Garvin, when his turn comes, surveys the room; it holds lots of memories. He spent a significant portion of his Brown career here and slept on the floor, with mice running over his sleeping bag, the night before he graduated and received his Ph.D. He takes a breath and declares, "For once, I will attempt a modicum of brevity. Now, what we've been doing on our summer vacation is going over old Viking images together with the new MOLA data, and many of them belie the craters that we thought were there. They're there, somewhere, but you can't tell where. With the laser altimeter, we have developed a database for one hundred and twenty-three craters in the north polar region." Sean comments, but for once he doesn't criticize; instead, he commends the rigor of Garvin's approach, the close study of craters. As Sean's endorsement of Garvin's work steadily becomes more apparent, the other members of the team cease to heckle the young crater enthusiast.

At this point, Mike Malin is supposed to present his most recent images of Mars, but once again, he's absent. His conciliatory gesture, dispatching an assistant to take his place, has the opposite of its intended effect, reminding everyone else on the team that Malin has missed another meeting.

"We want to know if you've convinced Mike Malin to give us any images," asks Matt Golombek, the project scientist for Pathfinder.

"That's a whole separate issue," the assistant replies, enigmatically.

Matt has prepared for this response and brandishes a recent copy of *Space News,* an industry trade journal. He reads from an article chastising the slow pace at which Mars Global Surveyor has been releasing data. "MARS RE-SEARCHERS COMPLAIN OF DATA LAG FROM MGS" says the headline, but the article mainly concerns Mike Malin's pictures: "Several Mars researchers contacted by *Space News* said there is a sense Malin is hoarding new Mars data, and that the large and data-hungry Mars research community is suffering as a result." As he reads through the article, he slows when he gets to Malin's rejoinder, which he injects with sarcasm: *"I'm hurt and offended by all the criticism."*

Everyone erupts with laughter when Matt reads these words. Golombek doesn't worry about mocking Malin in his absence and doesn't seem to care if word of his theatrics gets back to the camera genius. With Malin's assistant looking on, how can it fail to do so? This must be the point of the exercise: to send a message back to Malin that people are beginning to take notice.

Maria Zuber—diligent, circumspect Maria—mutters, "This is such a load of crap." It's not just Malin's no-show that she's ticked off about, although that's part of it. Mainly, she's very unhappy about having to devote time to mapping the Cydonia region, the location of the so-called Face on Mars. By her estimate, the MOLA team's recent investigation of this optical illusion has already consumed a month, and she hasn't told anyone of her involvement with this aspect of Mars research because she doesn't want to be associated with it. A tenacious little community of conspiracy buffs considers this ordinary rock formation to be evidence—heck, it's *proof!*—of intelligent life on the Red Planet, the handiwork of an ancient, lost civilization, and any new data from NASA concerning the face is sure to attract their unwelcome attention.

Just before the meeting at Brown, Mars Global Surveyor passed over the Face, and Malin's camera acquired crisp new images of the region. To the relief of NASA scientists, but to the disappointment of cultists, they showed that the Face was not, after all, a remarkable likeness of Teddy Kennedy or a giant statue of some sort or the Martian Stonehenge; it's just a rock formation and eroded ridges. You would think the new images would end speculation about the Face, but the findings have only encouraged the Face fanatics, who insist that NASA is *hiding* the real photos of Cydonia or has sent a secret mission to cover up the Face. A rogue website explains that MGS is actually a covert mission designed to return false images of the Face back to Earth, while secretly recording the truth: that the Face was actually fashioned by intelligent beings.

The folly isn't confined to the Internet. Around the time of the Brown meeting, I came across a copy of *The Mars Mystery,* by Graham Hancock, an English journalist intrigued by the Face controversy. My eyes quickly fell on the following passage: "Like other big state bureaucracies, NASA has lied and will lie again. We think the evidence suggests that it has lied about Cydonia ever since the Face on Mars was discovered." Hancock goes to fairly extreme lengths to justify his allegations. For instance, in 1985, Carl Sagan wrote an article for *Parade* explaining that the Face was no more than an optical illusion. Sagan's dismissal was telling, because he was unusually partial to madcap theories concerning intelligent life on Mars, such as Shklovskii's floating library concealed within Phobos; yet in this instance, he used his reputation to discredit this particular piece of scientific nonsense. In his book, Hancock responds that Sagan was privy to the both the original and "doctored" images of the Face and deliberately misled the public. Of course, Sagan is no longer alive to refute the accusation.

After all this, I was not surprised to read Hancock's theory about why the Face exists. "Cydonia is indeed some sort of signal—not a radio broadcast intended for an entire universe, but a specific directional beacon transmitting a message that was intended *exclusively for mankind.*" Decoded, it's a pretty dire message, and it goes like this: there's a dangerous asteroid aimed directly at Earth, and the Cydonia region points to it, and if we're smart enough, we'll destroy the asteroid before it turns Earth into a barren desert like Mars.

At times, science and superstition proceed in lockstep. After her years of hard work on Mars Global Surveyor, Maria fears she will be best remembered for her work on the Face on Mars.

At the end of twelve hours of nonstop meetings at the Lincoln Field Building, Jim Garvin bounds down the staircase. It's great to be in the sweet-smelling air again. He strides past a small quadrangle, where he pauses to watch students playing softball by twilight. He is suddenly overcome by a sense of déjà vu. That was him, playing ball in the same quadrangle, almost twenty years ago. "Nothing has changed," he says to Jim Head. "The only difference is that now, when I make a mistake, I have an excuse ready. The hair on my toes is turning gray. I'm over forty . . ."

"It's hard to believe," says Head solemnly. "Garvin at forty."

That night, during dinner at the Brown Faculty Club, Garvin leafs through a fragile relic: the notes he took as a graduate student at Brown, in another lifetime. Jim Head and Tim Mutch were his professors then, and Jim Head has kept Garvin's notes all these years. Garvin looks at them in astonishment; they're so primitive, like home movies of his childhood. There are the first programs he ever wrote for entering planetary data, his lecture notes. He finds it difficult to pull away from his young self, speaking to him from these brittle, discolored pages, but he must return to Washington in time to take his son, Zack, to the doctor first thing in the morning to have stitches removed.

Over the course of the next few days, the Mars Global Surveyor performs its micromission: surveying Phobos. The laser altimeter, when it is turned on, works exactly as planned, and NASA's fears about the high-gain antenna dissipate. MGS won't have to go into Magellan mode, and the mapping of Mars proceeds, despite the odds. At the Goddard Space Flight Center, at MIT, and at Caltech, the scientists revel in their newly acquired data.

Three months later, on December 11, 1998, I drive toward Launch Complex 17 at Cape Canaveral Air Force Station in Florida, where the next mission to Mars stands in readiness atop a sleek white and teal Delta II. Adorned with colorful decals, the rocket trails cables and hisses like a snake about to strike. It's a gorgeous, intimidating thing; I can't take my eyes off it.

Little has changed at the Cape since the days when Alan Shepard and John Glenn began their Mercury missions here in the early 1960s. The facility has a down-at-the-heels, what-the-future-used-to-look-like feel. There have been a few adjustments, of course. The rockets are more powerful these days, and they are launched not at the push of a button but with the click of a mouse. Otherwise, it's the same old place I saw on television as a kid when I watched rockets heading for space. In the car with me now is Omar Baez, a NASA mission manager. Young, affable, generously proportioned, Omar is the son of Cuban emigrants. When we pull up to the launch pad, a couple of guards stare at us blankly. There is no press in attendance, no sightseers with fanny packs and flip-up sunglasses; this is strictly business. About a dozen people work purposefully at the site; they are cheerful, but they never take their eyes off the rocket.

A cluster of smaller engines surrounds the base of the rocket. "Those are the GEMs," Omar tells me. "They're the graphite epoxy motors. They're solid rocket motors, sort of like bottle rockets. You turn them on; you can't control them; you can't turn them off or throttle them. They're just solids. All four of them will light at takeoff. When the rocket passes through the middle atmosphere, it will jettison two of them. Three seconds later, it will jettison another two. They fall into the ocean. This is only a minute and some seconds after liftoff. Then the rocket flies on a single main engine for about three minutes. It also has two vernier engines; they're very light, they control the roll motion. Then you get separation on that, and you'll fly on your second stage, which works on oxidizer and hydrazine fuels. And we have a third stage, which is a solid rocket motor that boosts the spacecraft into its final trajectory. The spacecraft has its own propellant module on board to get it into its correct orbit. Over the coast of Africa, or just after it, the Mauritius Islands, I believe, the third stage will light, and it goes into its Mars trajectory. The whole flight's lasted about eleven minutes at that point." Less than fifteen minutes after launch, the spacecraft will be beyond Earth orbit, on its way to Mars.

The payload atop the rocket is the Mars Climate Orbiter, the next in NASA's series of "faster-better-cheaper" Mars missions. It's small, as spacecraft go, about the size of a refrigerator, and it weighs 1,387 pounds. In about ten months, on September 24, 1999, after completing a roundabout, 416-million-mile trajectory, MCO will reach the Red Planet, where it will join Mars Global Surveyor in making the rounds. MCO will deploy just two experiments, a pressure modulator infrared radiometer (PMIRR) and a Mars color imager (MARCI), a new generation of camera from the workshop of Mike Malin. The PMIRR will measure the Martian atmosphere, which consists mostly of carbon dioxide and small amounts of nitrogen and oxygen, and track water vapor. This atmosphere sustains clouds of frozen water, carbon dioxide, and dust stirred up from the surface, but it's too thin and too cold to produce rain.

Omar recollects that the previous Delta II that was launched from this site blew up and scattered flaming debris far and wide. There's the bunker that was destroyed, he tells me. Here's the parking lot where cars were destroyed; we're standing in it. The explosion burned a six-foot crater in the parking lot. No one was injured, but the episode could have been a disaster. Security has tightened since then, and no one remains anywhere near the rocket during launch, not even in a bunker. The countdown proceeds smoothly, but the skies are overcast, and there is always a chance that the launch will be scrubbed. The weather can change, or the software can malfunction, or rockets can blow up—all without warning. But the Delta II has an excellent track record, I remind myself. "It's the workhorse of the fleet," Omar says in agreement.

Several months earlier, on the day that Mars Climate Orbiter underwent its final testing, Dan McCleese, the chief scientist for Mars exploration at Jet Propulsion Laboratory, was sitting in his office, which was enlivened with images and models of Mars. Dan has a reassuring yet authoritarian manner enhanced by perfect hair, perfect teeth, and a perfect smile; he looks like he holds a statewide elected office. But at the moment, he was worried. Very worried. He had a great deal riding on the outcome of this mission. One of the instruments, the pressure modulator infrared radiometer, was his, and he was watching over it with extra care. The PMIRR was a replica of an experiment aboard Mars Observer. Dan was devastated by the loss of that mission, as were all the other scientists involved with it, including Jim Garvin.

Too many billion-dollar flops, and these scientist-managers wind up in a little office down some narrow hallway that nobody can find.

Risk was something he'd learned to live with. "Failure is as close as the next launch," he told me with a sweep of his hand, "because the way I look at it, this program is not about science, it is not about manifest destiny, it's about exploration. It's about attempts to do things beyond what we think we can confidently do. We will find out what we don't know, but more than that, we will find the answers to questions we haven't even asked because we don't know enough. That element of discovery is what made Pathfinder a success, and it's what I want to keep breathing into this program."

Dan said that his objective is to observe the state of the weather and climate on Mars today, through experiments similar to those in Earth orbit. "We make vertical profiles of atmospheric temperature that are similar to the balloons that are launched from the Earth and the remote sensing that goes on from satellites. By remote sensing, I don't mean just pictures of clouds; I mean maps of the atmospheric temperature, atmospheric water, the dust content in the atmosphere, which behaves very much like the ozone does in ours in that it heats the atmosphere." He leaned back in his chair, becoming expansive. "One of the great conundrums of the Mars climate is the way in which the poles of Mars interact with the atmosphere. About twenty percent of the total mass in the atmosphere freezes out at the poles each winter."

Winter on Mars is not quite the same as a terrestrial winter. For one thing, it lasts twice as long, as does the Martian year. It's unclear whether it actually snows on Mars. At times, water ice and carbon dioxide come out of its scant atmosphere, but the effect might look more like fog accumulating, or something never seen on Earth. At any rate, this seasonal mass and energy exchange influences the Martian climate and may also echo changes throughout the Red Planet's history. "By understanding the modern climate we can begin to work back and try to understand the ancient climate," Dan says. Like many other scientists, he believes the ancient climate might have been warm enough for the sustained presence of liquid water on the surface: the key for life.

BBBRRRHHH . . . AAAAAAAAAAAAA . . . GGGHHH!

"That blast is telling us to get out of here." Omar says, dashing to his car. The warning siren keeps up its litany: *BBBRRRHHH . . . AAAAAAAAAAAAA . . . GGGHHH!* The workers promptly disperse. I follow Omar, and as we tear out of the parking lot, I glance back at the rocket shimmering in the rear window—ominous, mournful, poised to begin a journey to another planet or to blow itself up. Omar drives quickly to the Mission Director's Center, and when we arrive, slams on the brakes.

From the outside, the building looks like a weathered tin can. My ears still ring from the siren as we enter. An announcement reverberates through the halls: *"Weather forecasters at this time are still indicating about a thirty percent chance of violation of the launch constraints. Cloud decks and rain are being monitored in the general area, but we still remain optimistic as the clouds appear to be thin and should be okay for launch as they continue to move off the coast. . . . Weather balloons have been released and are reporting back at this time that the upper-level winds are also looking to be well within constraints. Winds on the ground are looking to be well within constraints. So winds will not be an issue. Again, the only concern that we would have, given any discussion on the weather, would be the cloud coverage and the thickness of the clouds that the Delta vehicle will have to travel through. We're at T minus twenty-seven minutes."*

While all this is going on, I take a seat in the small VIP area with Noel Hinners, a former center director of Goddard who is now with Lockheed Martin, the aerospace contractor for Mars Climate Orbiter. Dr. Hinners, as everyone calls him, has seen space exploration from the perspective of both industry and government. "The interesting thing," he says, "is that when you're in the government, you think you understand and know how a contractor operates. But, in reality, you don't. When you're on the industry side, a lot of people think they know how the government operates. But they don't, either. The government has no concept of business." Hinners acknowledges that ever since Dan Goldin arrived at NASA, the agency has become somewhat more businesslike, but not entirely. "Within government, the incentive to do things at a reasonable cost is tough to come by. . . ."

I want to continue this conversation, but launch time is approaching. Omar tugs on my shoulder, and we racewalk out of the building, pile into his car, and zoom to a hilltop with a clear view of the launch pad. We are as close as anyone can get to the rocket. A distant loudspeaker snarls:

"We're at T minus four minutes . . . T minus two minutes thirty seconds and counting . . . T minus ninety seconds and counting . . . The Mars Climate Orbiter

*has been fired up and remains on through the duration of the launch to Mars . . .
NASA is reporting that the spacecraft is go for launch . . .*

"Ready to go for launch!" Omar shouts at me. We brace ourselves.

"T minus sixty seconds and counting. . . . The liquid oxygen tank room is reported at 100 percent. . . . T minus fifty seconds . . . T minus forty seconds . . . T minus fifteen seconds and counting . . . twelve . . . eleven . . . ten . . . nine . . . eight . . . seven . . . six . . . five . . . four . . . three . . . two . . . one . . . We have ignition, and we have liftoff.

The teal Delta II disgorges a swelling column of smoke lit from within; the yellowish-pink luminosity grows until it challenges the dull Florida day, and still the rocket remains as earthbound as a building. Within seconds, the sound of the engine hits us like a detonation, ripping and piercing its way through the sodden air, but the rocket remains immobile. Finally, it appears to shudder, although the impression may be an optical illusion, the result of the energy shaking up the atmosphere, and it lifts off the pad slowly and reluctantly, seeming at first to stagger, then scaling the pillar of incandescent smoke. Now it slants, seeking a target, but at the same time unwilling to leave the Earth behind, ascending without much conviction, until it suddenly summons strength, and you see in an instant what it was designed to do. The rocket soars into the sky: power made visible. I am assaulted by the loudest, lowest roar I've ever heard, and then the sound redoubles in strength, until it feels like rodents tearing apart my eardrums from the inside. I blink, and the rocket is overhead, distant, almost out of sight. If it deviates from its course, technicians will detonate the machine, but it holds its bearings in all three dimensions, canting slightly, pulling away from us, heading higher over the Atlantic, returning lightning to a thunderhead. It swiftly pierces the cloud cover and disappears, leaving behind a jagged, drifting white trail. A bullet has been fired at the planet Mars.

Eight seconds have passed since liftoff, and more than $100 million have been expended. My eyes return to the launch pad, still wreathed in fine blue smoke, proof that this spectacle was not an illusion. To observe a launch is to become a believer. You may not remain a believer, but if you do, that belief can be traced to the moment the rocket conquered the sky. You believe, if only for a moment, that anything is possible, and under the influence of the noise and excitement, you may believe in magic, but there is no magic here, just a bold display of Newtonian physics.

We expect to see these boosters burn out and jettison at sixty-four seconds after launch, coming up in ten seconds . . . and we have burnout on all solids . . . at this time, we have jettison of the solids . . .

. . . A beautiful view in the second stage of the Delta vehicle as it continues to rise, now one minute and thirty-nine seconds after liftoff . . . Vehicle now at an altitude of twenty-one point five nautical miles, downrange distance forty-five nautical miles, velocity over three thousand miles per hour . . . It continues to look good . . . We are now at an altitude of thirty-four nautical miles, downrange distance ninety nautical miles . . . Velocity is over forty-eight hundred miles per hour . . . Flight continues to go nicely . . .

With the rocket out of sight, we observe the rest of the launch sequence on a monitor set up on the hilltop; it displays a real-time picture captured by a television camera strapped to the rocket as it ascends. The launchcam, as it's called, points straight down, revealing the Florida landscape as it recedes; the individual trees merge into a landscape, the roads become geometric patterns, and the rocket continues its ascent . . .

The Delta first stage begins to tumble back into the atmosphere . . . second stage engine continues to burn well . . . velocity over thirteen thousand four hundred miles per hour . . . second stage cutoff, eleven minutes, twenty seconds after launch now . . .

The cloud cover below the camera swiftly breaks up, the curvature of the Earth becomes visible, the Atlantic Ocean turns from a hazy gray to a dark, rich blue, as the sky darkens, and suddenly, the view is unmistakably that of space. The rocket levels off, and climbs at a slower rate. By now, it's in orbit, one hundred and seventeen miles above the surface of the Earth.

It will coast for about thirty minutes at about 15,000 miles an hour, until the second stage is restarted and burns briefly. At this point, the third stage, bearing the payload, begins to revolve on a turntable, propelled by small guidance rockets. Spinning rapidly, the third state separates, ignites, and lifts itself out of Earth orbit during an eighty–eight-second burn. To slow the spinning, the third stage extends weights on flexible lines, as if the spacecraft were an ice skater thrusting out her arms to slow herself. When it has stopped revolving, or nearly so, Mars Climate Orbiter separates from the third stage of the rocket and follows its trajectory toward Mars.

About twenty observers are standing on the hilltop, bewildered, staring at the trees and an empty launch pad. Almost everyone here is from NASA, except for a gaggle of launch groupies intent on ridiculing the countdown announcements. They've been hoping for another rocket explosion, a massive fireball, confirmation of NASA's incompetence, but they have been disappointed; today's launch has been flawless, and by now the spacecraft is several thousand miles away from their boisterous maledictions.

I turn to observe a man standing near me, packing his camera equipment. He's bearded and wears a T-shirt commemorating another planetary mission. He looks affable enough, so I introduce myself. "Mike Malin," he says in reply, giving me his business card. He picks up his cases and walks rapidly down the hill. The man's a genius, I remind myself.

There are twenty-three days until the next Mars launch, the Mars Polar Lander, on January 3, 1999. Like Mars Climate Orbiter and Mars Global Surveyor, it will commence its journey from Launch Complex 17A at Cape Canaveral Air Station. It will land on Mars, bringing an artificial presence to the edge of the southern polar cap. An entire robotic fleet will be circling Mars, and the data return promises to be spectacular. There's never been a more exciting time to be a planetary geologist than at this moment.

Naturally, Jim Garvin's wired by the time of the next MOLA science team meeting, in La Jolla, California, in late February 1999. He's not the only one who's excited. Life in the Mars community is good these days; the Mars Global Surveyor is relatively healthy, and two more spacecraft, the Mars Climate Orbiter and Mars Polar Orbiter, are streaking toward the Red Planet at 17,000 miles an hour for a rendezvous later in the year. And there are at least six more missions in the pipeline. Even Sean Solomon is smiling. Having three Mars missions under way is unprecedented, but, given the dolorous history of missions to the Red Planet, I'm not about to mention this remarkable streak aloud, any more than you'd pat a pitcher on the back and shout, "Hey! Do you realize you're working on a perfect game?" as he heads out to the mound in the ninth inning.

The MOLA gathering takes place at Scripps Institution of Oceanography, perched on a rocky cliff above the Pacific, a few miles north of San Diego. The MOLA team chose to meet at Scripps because it's in Mike Malin's hometown, and Malin was naturally invited, but once again, he's a no-show. Beyond the picture windows, surfers clad in slick black wet suits bob on the ocean swells, awaiting a decent wave. When a wave comes ashore, you can hear the tympanic boom of the Pacific all the way up in the hall where the meeting takes place, much to Dave Smith's displeasure. There's business that needs attending to, and the Pacific, with its surfers and soporific flickering sunlight, amounts to a distraction rather than a backdrop. There's a certain California funkiness to the place, redolent of redwoods and salt-impregnated breezes and hot-tub escapades, and its laid-back, artsy feel makes many of the scientists who work back East envious. One complains, "It looks like they actually spent some money here to make people comfortable," and he shakes his head ruefully, thinking of his own practically windowless cinder-block cave at Goddard, where he spends the best part of his days and years.

In a few days, the Mars Global Surveyor mission, in progress for nearly two and a half years, will finally commence global mapping of the Red Planet. (Until now, it has covered limited areas.) At the same time, the spacecraft's instruments will look for indications of water, volcanoes, and ultimately, of *life*—well, Arden Albee, the senior scientist of the entire Mars program and self-proclaimed skeptic on the subject of life on Mars, is in attendance, so it's best not to use the "L" word. At the moment, though, Arden proves to be affable, witty, self-deprecating, and full of gruff, sardonic comments on the state of the universe and NASA, with which he's been associated since 1959, two years after Sputnik circled the globe. It's hard to think of NASA entering advanced middle age, but it is. Arden tells me he plans to stay with the agency in some capacity or other until the Mars missions come to an end, which means forever, in effect. In the days before Mars became a prime target for NASA, Arden worked on lunar samples, and then, he says, "Somehow, I became chief scientist at Jet Propulsion Laboratory for six years." Arden is also dean of graduate studies and professor of geology and planetary science at Caltech, yet he'd rather talk about his appointment as chair of the House Committee of the Caltech Athenaeum, its faculty club, which gives him control over its wine cellar, valued at $250,000. "But as I told the wine steward, my favorite brand starts with a 'C'—*Chivas.*"

Arden is a tribal elder among these scientists; to the extent that they defer to anyone, they defer to him. Lifting his chin, which is covered with a distinguished bristle of silvery whiskers, he recalls, in his wonderfully resonant, sagacious voice, the first Announcement of Opportunity for Mars Observer eighteen years ago. When NASA decides to go ahead with a mission, it puts out an A/O, which veterans consider a fishing expedition whereby the agency endeavors to determine how little they will have to pay for the engineering and scientific components. Since each mission tends to be unique, and the bidding is blind, it's hard to arrive at a fair market price, and contractors become wary.

These days, Arden is intrigued by the prospect of real scientific cooperation between the Americans and the Russians in Mars research. The idea was unthinkable during the Cold War, but he says the possibility first emerged during the Reagan-Gorbachev summit in 1986 in Iceland. Historians regard this encounter as a standoff, at best, an occasion for Reagan to lecture Gorbachev on the error of Soviet ways. But Arden says that at the time, NASA suggested the two superpowers try sharing people and data, rather than hardware and technology, for a change. So the Americans started to learn a lot about the Russian Mars program, which had once been far more ambitious and vital than NASA's. The ambitious Russian Mars '96 mission, he tells me, "was originally supposed to be Mars '90 and kept slipping. There was a lot of French involvement; it was one of the first truly international missions." To hear Arden describe it, Mars '96 would have been the coolest Mars mission ever. In addition to forty science experiments, two landers, and two penetrators, Mars '96 carried a balloon designed to float wherever the Martian winds carried it, taking measurements as it went. As the sun heated the surface and the atmosphere during the day, it would rise and travel, and as the temperature dropped, it would sink to the surface for the night and resume its windborne peregrinations the following day. It was not to be. The booster rocket never escaped Earth orbit, and the spacecraft limped around Earth three times before disintegrating over Chile, which was upset by the thought of a nuclear-powered contraption falling out of the sky onto its snowcapped mountains, if that's where it came down. No one knows for sure.

Although Mars '96 failed miserably, Arden tells me that it has directly inspired Mars Express, a new European rival to NASA's "better-faster-cheaper" missions that will use spare parts from Mars '96 and include some scientists who worked on it. The point he wants to make is that there are always lessons, sometimes very big lessons, to be learned from disasters such as Mars Observer and Mars '96. From one perspective, Mars Observer was an abysmal failure; from another, it led the way to a new generation of Mars missions. Ultimately, what matters most is a nation's willingness to engage in the risky business of planetary exploration, and at the moment, the United States, for all its budget-cutting and angst-ridden scientists, is the only nation mounting a serious, sustained effort to explore Mars. The reasons are largely economic, of course; few countries have the money or the infrastructure to sustain a serious planetary program. And there are intangible cultural forces at work; they are powerful but difficult to quantify.

Dan Goldin and others at NASA repeatedly cite Lewis and Clark's exploration of the West as a precedent for charting and cataloging the unknown, and it seems more than possible that explorers will one day post their encounters with the Red Planet on the Internet for all to read, just as Lewis and Clark offered Americans their first glimpse of the West, its terrain, its flora and fauna, and its inhabitants. It's possible that in some more distant time, people will look back on those early journals of Martian discovery with a sense of nostalgia for the days when the Red Planet was terra incognita—for the days when it was not even known for certain whether life flourished there—in other words, for the present moment, when Mars, apart from a few tiny spacecraft that dot its surface, is still virgin territory.

Arden enjoys administrating, which is unusual among his colleagues. As the occasionally jaded, world-weary comments he drops about scientists suggest, he knows you can't manage them in the conventional sense—especially not the members of the MOLA team. At the moment, his role is to lean over their shoulders, get a sense of what they're up to, and pass the information around. Arden has opinions, but he doesn't care whether you agree with him, as long as you speak up. He considers Mario Acuna's magnetometer team uncontrollable, thanks in part to the machinations of the team's leader, who has a way of retaining new and important information until the last possible moment. This was the case last year, with their discovery of magnetic anomalies on Mars, and it seems to be the case now. "They have some superhot discoveries they are keeping to themselves," Arden says. Even *he* doesn't know what they are, and he won't find out until they hold a press conference to announce their findings. "There's nothing I can do about it," he growls.

Finally, Dave Smith claps his hands and calls the MOLA meeting to order. His bracing, brusque vowels give the meeting the air of an RAF squadron briefing before the Dawn Patrol takes to the skies. "This is the last meeting before we get turned on. We came to the end of aerobraking a few weeks ago. We are very close to a circular orbit. JPL have not messed with the orbit, and they even canceled a trim maneuver because they didn't need it. When you're in a four-hundred-and-five-kilometer orbit, circling Mars every two hours, there's not a lot of time to think. Getting the spacecraft into orbit was a remarkable achievement. They got us into the right local time as well. Right now it looks great, just great."

Actually, maneuvering Mars Global Surveyor into its mapping orbit was a lot riskier than David allows. By international treaty, any spacecraft actually landing on Mars must be carefully sterilized to prevent contamination (Pathfinder being one example). An *orbiter*, such as MGS, need not undergo the same process, but if it drifts into a low altitude where there's a chance of its crashing into the surface, JPL will immediately execute a burn and bump it into a much higher "quarantine orbit." A spacecraft unlucky enough to get stuck in a quarantine orbit has probably used up a great deal of irreplaceable fuel to get there and is too high for detailed remote sensing. If MGS hadn't completed its year-long aerobraking maneuvers in exactly the right way, it would have wound up in orbital exile and lost most of its scientific value. That's what Dave was trying to touch on, but the scientists, in general, have about as much interest in the engineering and navigational skills that placed MGS into a precise orbit around a planet more than a hundred million miles away as they do in where they've parked their cars in the lot next door. All they really care about are their experiments and their data. "Now we come back to what needs to be done . . ." Dave says.

"Dave! Dave!" Sean says, hijacking the agenda. "Are there voices at JPL still arguing *not* to deploy the high-gain antenna?" Dave assures Sean that NASA intends to rely on the high-gain antenna, and they won't have to resort to Magellan mode.

"You guys made a tactical error," Sean shoots back. "Don't call it 'Magellan mode.' Magellan was a successful mission. Call it the 'permanent failure mode.' "

"Call it 'Mars Observer mode,' " someone else suggests.

"He's got the right idea," Sean says.

Trying to get the meeting back on course, Dave reviews the status of the mission's science. Everything will change within days, once mapping commences, but this summarizes the accomplishments of Mars Global Surveyor during its unanticipated period of elliptical orbit mapping:

EXPERIMENT	DATA	HIGHLIGHTS
Mars Orbiter Camera	2,140 images acquired	Ubiquitous dunes, craters, layers
Thermal Emission Spectrometer	11,000,000 spectra	Hematite deposits
Mars Orbiter Laser Altimeter	206 profiles	N. plains; N. polar cap
Magnetometer	1,000 periapses	Crustal magnetic anomalies

The big question concerns those 2,140 images acquired by Malin's Mars Orbiter Camera. The team wants to know when they'll be released, and Dave tries to explain, in technical terms, that they are forthcoming. Everyone mutters, "The check is in the mail." (Malin will eventually redeem himself months later, when he makes more than 25,000 Mars images publicly available on the Internet.)

It's Maria Zuber's turn to report on the status of Mars Global Surveyor, and all she can think about at this moment is the flood of mapping data about to overwhelm her impatient, high-strung team. For the past month, she's done something quite unusual for her; she has taken time away from work to stay home with her husband and children, because she knows she'll be working "impossible hours" once millions of bytes of data arrive. She planned to write some research papers, but, she told me, "I decided to bag it because while I knew it was perfectly respectable work, I realized it wasn't going to be nearly as important in the scheme of things as what was coming down the pike. Something inside of me knew that it was finally going to happen. When you work toward something for a decade, the journey becomes the focus and you think less about the destination, but I have been thinking about the destination all the time now." And she wants to make sure the rest of the MOLA team will be prepared for the onslaught. "We'll be taking data continuously, so anyone who wants to show up at Goddard is welcome." The invitation is best not refused.

"How much data do you get on the first day?" Sean inquires.

"Eighteen hours," Dave advises.

For the next twenty minutes or so, they try to figure out precisely when the data will start coming in, converting Goddard time to JPL time to "Zulu" (Greenwich Mean Time), always coming up with different answers. How many rockets scientists, I ask myself, does it take to tell the time? They turn to Greg Neumann, their data expert, for an answer. "When will we get the data?" he responds. "Well, it depends on what you pay. I expect that all of our software will break." He isn't smiling.

"That's the spirit, Greg," Maria says.

Greg sounds worried as he explains the ramifications of a 115- to 117-millisecond error he's detected in the rate of transmission, which might mean as much as a 1.3-meter error in measuring the height of the surface of Mars. He can't determine the cause of the problem, and he assumes the worst, that it affects the entire spacecraft. Greg detests imprecision of any type. Dave tries to reassure him that a 1.3-meter error is good enough for government work. "When we started out fifteen years ago, we said we'd be good to within fifteen meters, and people said we couldn't do that. And now we're down to one point three meters."

Maria emphasizes that the map they are about to create will be used to select a site for the Mars Polar Lander when it touches down at the edge of the Southern ice cap in December. You can feel the anxiety level rise at the prospect.

Sean Solomon, twitching excitedly, decides it's time to go after Maria, perhaps because it's the least popular thing he could do.

"Maria, what do we owe, and whom do we owe it to, and when do we owe it?" he asks, his tone deceptively pleasant.

"We . . . owe . . . a . . . map," she tentatively replies.

Sean leans forward. "To *whom?* That's the question."

"To the Mars Surveyor Project."

"When?"

"One month after we collect the first month's data. And we owe *Science* a paper, for another thing," she says.

"You guys are pulling teeth here," Sean says. "I shouldn't have to be asking all these questions."

The contemplative silence of scolded children descends over the meeting. Sean wants closure, but Maria, along with most everyone else in the room, knows that there's rarely anything like closure in a planetary mission, except when it's imposed by a disaster or a budget cut. Their job, their science, is a *process:* an endless thread of learning and debunking. Sighing like a teacher confronted with stubborn students, Maria patiently summarizes the mission's obligations. "Is everyone happy with that?"

"No! I am *not!* When do we owe the paper to *Science*?" Sean repeats.

"One month after the map," she tells him. "I know many people, including you, would like to work on a different schedule."

"You're redefining 'better-faster-cheaper!' "

Maria has an inspiration. "Let's say we have the whole ball of wax ready for the Fifth Mars Conference in July." This conference, the first of its kind in a decade, will bring together scientists from around the world, and the MOLA team, with their new data, will play a leading role. Professional acclaim is always a pleasing prospect.

"That makes more sense," Sean says.

She's won—at least for now. Arden Albee calls out from the back of the room. "I have told *Science* we will have a special issue for them in July." A special issue! This tops the Fifth Mars Conference, and the scientists murmur approvingly.

Dave rushes to Arden's support. "And we'll give them sufficient advance notice of the great stuff we've got, because we will start mapping on Saturday or Sunday."

Sean wants more. "Tell 'em if they want it to give us the cover."

"There've been a lot of e-mails back and forth about the cover," Maria says.

"You've got to know what the competition is," Jim Head advises. "We were hoping for the cover with our paper on Mars' north polar cap, but they gave it to a tethered cockroach instead."

"That's higher up the food chain than a *worm!*" Sean says.

Dave Smith interjects, "We've been getting questions from NASA like, 'What are you going to see?' "

"Yeah, they've been asking, 'What are you going to *discover?*' " Maria adds, mocking the bureaucratic absurdity of such questions.

"The Face on Mars!" Sean shouts. "Ask Greg about it!" The team looks at Greg, who remains mute, impassive. The Face, as everyone knows, is unworthy of discussion.

Sean yields the floor to Jim Garvin, whose voice is unnaturally soft and raw. He explains he's caught the flu from his children. When his miniature laser pointer misfires, Head and Solomon can't resist making fun. "Unauthorized child use," he pleads, ignoring the distractions. More enthralled with his Martian craters than ever, Jim displays his new understanding of the features he's been studying with data from both the camera and the laser altimeter. He believes he's seeing evidence of relatively young volcanoes on Mars, about ten million years old, or less; this observation defies the conventional wisdom about volcanic activity on the Red Planet, which holds that Martian volcanoes are ancient—billions of years old. The MOLA data has given Jim the necessary ammunition to distinguish between recent impact craters and genuine volcanic craters formed by internal upheavals.

"If I can show there are little volcanoes formed by oozing, like the ones we saw in Iceland, without an explosion, then I can say how old they are," he says. "I can say how much heat they generated, and I can ask, 'Do I have a set of conditions that would be hospitable to life, if they were to exist on Earth?' Maybe, *maybe,* I have a favorable location for the development of life. Keep in mind that you need a lot of these opportunities, a zillion of them, and hope that a few of them take. It's like throwing a bunch of seeds on infertile ground. Only a few are going to find the right amount of water and sunlight to sprout."

Despite the growing evidence of recent volcanism on Mars, Garvin wins few converts at La Jolla. Even Dave Smith, his team leader, remains skeptical. Undeterred, Jim continues to spin his theories. One crater in particular, near the northern polar cap, has inflamed his intellect. "The MOLA track tells me this crater is not a billion years old. Now, how old is it, exactly? Let's say it's a hundred million—pretty old, but not *that* old. Well, Jesus, Mary, and Joseph! Now we have a view of a crater that's been buried and exhumed! It still looks fresh and tells us that the polar cap had to increase in volume by a factor of two to accommodate that degree of burial."

The waxing and waning of the polar cap provokes Garvin to wonder how climate cycles might have affected the development of life on the Red Planet. "One of the issues in astrobiology is to address the viability of life in cold storage," he explains. "For example, on Earth you can freeze-dry organisms; there are some that have been freeze-dried for as long as twenty-seven million years. Most biologists now argue that you can't go much longer, but what if that number is wrong? What if it's a *billion* years? Could we have on Mars truly dormant—and I mean suspended animation—microbes that are waiting for those rare times every hundred thousand or million years when it's clement? And then they bloom! They do their thing, they survive, they take energy from the Sun, and when the climate turns cold, they go to sleep again. If they live that way, they're never going to get big, like Redwoods or elephants. They're not going to be mammals, because mammals require too much energy. They're going to be these hardy little things that know how to wait."

Garvin presents his latest findings to the team members with mathematical precision, free of interpretation. Sean tries to shoot holes in the data, but Jim, prepared for questions, ably defends his conclusions, which point to, among other things, the existence of much more water—in the form of ice—on Mars than was previously believed.

"Is there such a thing as a *normal* crater?" Sean asks. "I suggest you start with those." Garvin returns to his crater analyses, wondering aloud how the craters got filled, and then emptied, and what the process reveals about the geologic history of Mars. As he plows on, Sean occasionally breaks in with more questions, but each time, Garvin recovers his equilibrium and proceeds. At the end of the presentation, Sean suddenly compliments him and says this will make for a terrific presentation at the Lunar and Planetary Sciences Conference next month in Houston. Jim wearily takes his seat.

Matt Golombek, who is next to address the MOLA team, knows how to work a crowd of Mars scientists; he knows just which buttons to push. He's full of ironic, mostly harmless digs that keep everyone entertained before he gets down to business on the specifics of site selection for the Mars Polar Lander. A celebrity in the Mars community ever since the Pathfinder mission, he's here today to deliver the bad news about the site where the next spacecraft to Mars will likely land. "It's not about, 'Gee, what a nice place to go, and wouldn't it be fun to go there?' It's about *safety.*"

Sean rises to defend the scientists' honor. "Will the scientific criteria for selecting a site be secondary to safety?"

"If you don't like it, talk to the Mars Program Office." Matt lets his words sink in, as the scientists bid farewell to visions of data concerning Martian mountains, volcanoes, ice caps, and canyons. Matt lobbies for flat and boring and safe. "If the lander's leg hits a rock and does this"—he leans sharply to the side—"that's an engineering problem they haven't even addressed." He explains that according to NASA standards, there must be less than a one percent chance of landing on a rock. Sean caustically remarks that such a strict, unrealistic standard would force them to distort the odds.

Matt has more bad news: the rover on the 2001 mission will be the spare from the Pathfinder mission. Using four-year-old technology on the new mission sounds like taking "better-faster-cheaper" to an illogical extreme, and the scientists protest. Why are they flying an old *spare?* What if something's wrong with it?

"No, no," Matt says, "there's nothing wrong with it. It's just the spare. It's as good as the original rover that went." The scientists realize it is hopeless to influence NASA on the question of the rover, and Matt resumes trying to sell them on the virtues of a safe landing site. He explains there are actually hundreds of sites that are reasonably acceptable "at first blush," but they conceal hazards. "You could set down in the bottom of Valles Marineris if you wish," he concedes. This is the giant Martian canyon that dwarfs even the Grand Canyon, and landing within its walls would generate geological excitement. "You could get a room with view." But the scientists don't want a room with a view, they want a steep cliff, a volcano, an impact crater, an ice cap, an enigmatic landform—anything but flat and boring. As the day ends, the scientists turn disgruntled, even petulant at the prospect of sending a spare rover to a nondescript site on Mars. As they file out of the conference room and into the California sunshine, they are already making plans to lobby for a more interesting landing area.

That night, Rob Afzal and I get to talking over Dos Equis and chips at a Mexican restaurant in La Jolla. We're sitting at a long table, ten or a dozen people from the team lined up on either side, with Dave Smith presiding at the other end. After the waitress slaps down large platters of burritos and fajitas and enchiladas, Robs begin holding forth on NASA. The agency, he says, is a highly political place—"not that I think that's a bad thing. NASA, you have to understand, was originally a publicity front for military reconnaissance missions. Years ago, NASA was sending monkeys into space to distract the public from the real purpose of the mission, which was intelligence gathering, spy stuff. The Apollo missions to the moon were absolutely political as well as technological marvels. There was no secret we were trying to get to the moon before the Russians, and the implicit message to the world was, 'If we can get to the moon, you'd better beware.' These days, even the space shuttle carries classified payloads from time to time."

"Well, how much at NASA is actually secret?"

"Everything at NASA is open," Rob says, "but parts are restricted, for instance the communications satellites, what frequency the space shuttle uses to communicate, so people can't screw things up." Listening to Rob, I realize I don't want to become *too* disillusioned about NASA, just disillusioned enough to see it clearly. Even now, after having seen lots of NASA, I prefer to think that there are more Garvins and Parkinsons and Soffens in the agency than time-serving bureaucrats and that its quasi-military culture hasn't compromised its dedication to pure science. A little idealism can be a good thing, sometimes.

My son, a generally pragmatic high-school student, thinks I am naive. He is one of many who hold an opposing view of NASA, convinced the agency maintains a hidden trove of captured alien spacecraft, which automatically gives it access to superior technology, the catch being that it must be kept secret. He finds it astonishing that I traveled all the way to Lockheed Martin in Denver, where many NASA planetary spacecraft are built, and failed to inspect the alien ships, which he believes are kept there. I explained that security was tight, you couldn't just wander around as you pleased. Exactly, he said; I'd just proved his point. He finds ample evidence to support his views on cable television, where documentaries purport to explain how much more NASA knows about extraterrestrials than it lets on to the American taxpayer. The evidence, as presented on television, is plentiful and seductive, while I have mostly logic and common sense to fall back on when I try to persuade him that NASA is up to no such thing.

During a tour of Goddard one day, I mentioned my son's opinion to the guide and insisted on seeing the captured alien spacecraft. "I would show them to you," he said, "but then I would have to kill you."

The next morning in La Jolla, when the science meeting turns intensely mathematical, I duck out onto the porch overlooking the Pacific and the surfers, where I bump into Dave Smith. Apparently, things have gotten too mathematical for him as well. He's in an exuberant mood, rubbing his palms together at the prospect of the impending data avalanche. "Once the data starts flowing, they're going to get more stuff in a few days than they've gotten in the last year and a half. If the mapping lasts only thirty days, and it's scheduled to last a lot longer, we will get a *fantastic* amount of data"— enough to revolutionize our thinking about Mars and the prospects for life on Mars, which, by the way, Pathfinder, for all its success, did not. I notice that Dave and I are feeling more comfortable with each other; it's taken a year to reach this point. He has been wary about allowing me into the scientific inner sanctum; scientists have a hard time talking to each other, let alone to someone outside the calling.

He reminisces about how he became interested in planetary science in the 1980s, at a time when he began to feel that the fundamental questions about Earth had been answered, and all that remained were nuances and details. Mars, in comparison, was largely unexplored, an entire planet, relatively close by, about which fundamental questions remained. He decided to shift his focus to Mars. "I wrote the original proposal to explore Mars back in nineteen-eighty-five and expected it would all be over by nineteen-ninety. There were delays, Mars Observer wasn't even launched until nineteen-ninety-two, and then it was lost the following year. When I was putting together this team, Arden Albee said to me, 'My God, they're all prima donnas! How are you going to get them to work together?' Yet it seems to me we have become a team. Now, it's not like the other teams. It's older, bigger, brasher, livelier, more contentious, yet more cohesive."

We return to the conference room, where the booming Pacific waves can be heard and felt. Tossing a piece of chalk in the air, Dave entertains his MOLA science team with the prospect of exotic and unlikely locations for future meetings. Does France sound tempting? Iceland, anyone? France keeps getting postponed, like a far-off dream.

"Whatever happened to the Baltimore airport meeting?"

"How about Montana? I can get you a magnificent dude ranch for a week."

"Nude ranch?"

"Dude."

They can dream all they want; they know the next meeting will probably take place in a severely functional NASA center.

Rob Afzal wanders out onto the terrace overlooking Scripps Pier and the Pacific Ocean, shimmering in the late-afternoon sun. As he watches the exhausted surfers emerge from the water, lugging their boards as they trudge across the beach, he smiles broadly and exhales. "Ah, the end of another hard day at the office."

Days later, as the data finally begin to flow, Maria becomes giddy at her station back at Goddard. "When we did the turn-on to initiate continuous mapping, I was sure the instrument would work fine and that the data would be spectacular. We had watched our instrument perform for a year and a half at Mars in an orbit it wasn't designed to operate in. Now that it was going to operate as designed, I honestly expected clear sailing, tempered with the expectation of intermittent crises that are typical of robotic spacecraft operations. Nonetheless, there was no lack of adrenaline in the operations room when the first data came down and it was every bit as good as we'd hoped. No matter how much you know that you designed it right, it is hard to come to grips with the fact that your experiment is at Mars and doing what you tell it to do. When the first download showed a pass that crossed near the center of the south polar cap, we popped the champagne. Now we're getting a million measurements a day." It's an odd feeling to have so much new data about Mars, knowing, at least for a time, new things about the Red Planet that no one else knows. The moment of discovery, for Maria, is not exhilarating, except, perhaps, in retrospect. She wants to understand it all and, diligent scientist that she is, make sure it's correct. As a result, she feels temporarily overwhelmed, as though drowning in a sea of data. She neglects to return phone calls and to respond to e-mails. Nothing much matters but Mars. Occasionally, someone at work will remark to her that she looks stressed, but, she says, "I am having the time of my life."

Immersed in the raw data, Jim Garvin finally has before him the challenge of a lifetime. "To me, it's a crater-lover's candy store. We should cross hundreds of craters in a few days and most of the biggies. We will traverse south polar features including craters on the ice. . . . We will all be foaming at the mouth." He's looking for evidence of recent volcanic activity near the polar cap, with its implications for the "L" word—life. Days later, he trembles with exhilaration. "I am awash in beautiful impact craters," he says, reverently. This sacred moment is evanescent; this is *his* data, and for now he understands it better than anyone else. As the weeks pass, his urge to shout about it becomes overpowering.

Subject: FIRST LIGHT OVER THE SOUTHERN HEMISPHERE OF MARS
Date: Tue, 2 Mar 99 11:21:34 EST
From: Jim Garvin <jgarvin@nasa.gov>
To: Laurence Bergreen <bergreen@NYCnet.net>

After a 6-year wait, the very first circum-martian topographic profiles of the Red Planet appeared on March 1, 1999, after a playback from the intrepid MGS. MOLA acquired approx. 10 complete orbits covering virtually all of Mars from pole to pole. I was staggered when I saw our first global views of Mars. While first impressions can be deceiving, I found the topsy-turvy relief of the region around the "wedding-cake-like" south polar cap to be as enigmatic as ever! In only 20 hours, MOLA returned nearly one million unique, high-quality measurements of new landscapes on Mars, as well as a flock of interesting cloud features near the s. polar cap. Of course, my first love are the omnipresent impact craters, which are abundant, although somewhat battered, in the Southern Hemisphere. I observed beautiful craters near the south polar cap that appear to have some degree of ice fill (as we have described near the north cap) and others which appear to have escaped such processes. What this all means for the history of water on Mars is still uncertain. It is my hope that the craters will provide clues to the structure of the upper few miles of the Martian crust, where water or ice is likely stored in different kinds of reservoirs, either as isolated pods of ice/water or as a distributed deposit of water in the pores of materials.

We are all frantically trying to work with 30 megabyte files of MOLA profiles and get back to the business of unraveling the mysteries of Mars. I must admit the wait has been worth it. When we started on the odyssey that became MOLA way back in 1985, I always hoped that we would be traversing Martian landscapes as we are now. To many of us, MOLA embodies planetary mapping as it will be done in the new millennium, and it's fabulous to be there. To paraphrase my favorite rock band (U2): with MOLA, I have found what I am looking for!

Jim

After a month's intensive study of the new data, Jim becomes still more excited by the prospect of the first detailed charting of Mars, the proliferating signs of water, the implications for life, and the growing rationale for sending people to the Red Planet.

Subject: STATUS REPORT
Date: Thu, 01 Apr 1999 08:17:57-0500
From: Jim Garvin <jgarvin@nasa.gov>
To: Laurence Bergreen <bergreen@NYCnet.net>

We have seen a NEW MARS, thanks to MOLA's global topography after only 22 days of pre-mapping. Now we are paying off with data that will REWRITE THE TEXTBOOKS. I believe we have topographic evidence that OCEANS could really have existed on Mars and that the south polar cap is far larger (by a factor of 3 or 4) than previously thought. Thus the Mars water cycle (and hence habitability) may be larger and more interesting than anyone may have dared to dream.

Jim

Wherever he looks, Garvin finds evidence of ancient water on Mars. "I have now measured the total relief of Olympus Mons, by far the largest volcano in the solar system," he says. "It's over twenty-two kilometers in relative relief, and consists of more than three million cubic kilometers of rock, and that's enough to coat the entire surface of Mars with a layer of lava twenty meters thick. If the three million cubic kilometers of rock produced water by degassing at very conservative rates, then it would result in over three thousand cubic kilometers of water, and that would be enough to fill the Great Lakes." He would like to tell me a great deal more about the water on Mars and its implications for life, but, he says with apologies, "I am immersed in figuring out the south polar cap cratering story right now."

At the end of April, Mario Acuna, the occasionally secretive magnetometer expert, is finally heard from. He is predictably and unbearably elated. "We are really excited about this totally unexpected and serendipitous discovery," he says, in language that makes his scientific rivals' blood run cold. His previous discovery of localized magnetic fields on Mars was only the start of a far more complex and suggestive story concerning the geological history of the planet. On the basis of the data returned by Mars Global Surveyor after more than a thousand orbits, Mario and his team realized that seemingly random traces of magnetism detected by the magnetometer formed a pattern of stripes on the surface of Mars, similar to bar codes. They are areas of alternating magnetic polarization, positive, negative, positive, negative. Mario visualizes the magnetic stripes this way: "Imagine a thin coat of dried paint on a balloon, where the paint is the crust of Mars. If we inflate the balloon, cracks can develop in the paint. The edges of the cracks will automatically have opposite polarities, because nature does not allow there to be a positive pole without a negative counterpart."

The most dramatic implication of this discovery is that Mars once had a genuine magnetic field, very much like Earth's, and in that important respect, Mars is more Earth-like than was believed before Mario and his magnetometer came along. It also means that Mars once had a molten core, as the Earth does now. That is only the beginning. The striping detected on Mars also exists on Earth, notably on the ocean floors, where it was caused by the movements of the Earth's crust. Detecting those patterns on Earth led to the discovery of plate tectonics. The existence of similar magnetic stripes on Mars strongly implies that the Red Planet once had plate tectonics, as well.

Until Mario's discovery, the prevailing belief was that Mars had no plate tectonics. Their presence drastically changes scientists' understanding of Mars' past, for plate tectonics are associated with volcanic activity, as the plates pull apart, consume, and regenerate the crust out of the planet's molten interior. On Earth, the process involves water. So the existence of plate tectonics on Mars offers new indirect evidence of volcanism *and* water—not just a little water, but oceans full. The combination of water and energy (from volcanism) is highly favorable to the germination of life; these are the conditions under which astrobiologists believe that life began on Earth: under water, near volcanic vents. It suddenly seems likely that similar conditions once existed on Mars, now that there is strong evidence that it had plate tectonics, oceans, and volcanoes. Although confirmation of life on Mars is still missing, the indirect evidence in favor of it mounts month by month.

There are some important differences, though, between plate tectonics on Earth and on Mars. Earth still has its plate tectonics, which roil its surface, while Mars' apparently faded away billions of years ago. Why? Once again, no one knows. But Mars' present condition suggests Earth's distant future. Eventually, our storehouse of internal energy will dissipate, and our magnetic field will fade, just as the Red Planet's has. Without the protection offered by a magnetic field, the Sun's radiation will bombard the Earth's surface mercilessly, stripping away our water supply and atmosphere. Without oceans, without atmosphere, an inhospitable Earth will circle the Sun for an eternity. In Mars' dire situation, we can see Earth's ultimate destiny.

Later that Spring, when I assume Garvin has sequestered himself at Goddard, measuring Martian craters night after night, he startles me with the following bulletin:

Subject: NAVASSA

Date: Thu, 06 May 1999 07:48:13 -0400

From: Jim Garvin <jgarvin@nasa.gov>

To: Laurence Bergreen <bergreen@NYCnet.net>

Larry:

I survived.

I am covered with poisonwood welts and near dehydration. Our Coast Guard flew me 360 miles round trip to get to this amazing LOST ISLAND.

More soon.

Jim

Survived *what*? Where has he gone now? Trying to keep track of Garvin is like playing a global version of *Where's Waldo?* It turns out he's just returned from another field trip in search of the common geological heritage of Mars and Earth. This time, he's visited the island of Navassa, which, he tells me, is a "little-known U.S. possession that lies in the Windward Passage between Jamaica (to the West) and Haiti (to the East). It shares Surtsey's teardrop shape, but the similarity ends there. Navassa is a terraced limestone island forty miles from the nearest land, and it is a festering hotbed for various life systems and their interactions with Mother Nature, human visitors, and much else."

He sketches in the island's peculiar history. "Apparently, Columbus discovered Navassa, and several of his men canoed to the island in 1504. It is not a friendly place: there are no beaches, forty- to fifty-foot sheer limestone cliffs, and many species of trees that will cause one great discomfort." The most common is the poisonwood tree, whose leaves cause painful blisters, similar to those caused by poison sumac, but worse. The United States claimed the island of Navassa in 1857 under the Guano Act, which permitted taking possession of an island for the purposes of mining guano, which might be considered the fertilizer equivalent of oil. "During the phosphate fertilizer heyday of the late nineteenth century," Jim continues, "Navassa's rich deposits of avian 'red phosphate' were mined by a Baltimore firm. The phosphate must be pried out of a honeycomb-like grid of ten-foot-thick limestone pinnacles. An insurrection led to some killings on the island because of the inhuman conditions faced by the strip miners. Mining ceased, and the only other human involvement with Navassa was the establishment of a lighthouse. By the end of World War II, the island was once again uninhabited." Sounds like a perfect spot for spring break. Once Garvin heard that the island harbors many species found nowhere else, he naturally had to visit. He devised a remote sensing control experiment, which sounded a bit like a pretext, and linked up with an expedition to the island led by the Discovery Channel.

"I left Goddard Space Flight Center on Sunday, May 2, for Great Inagua, Bahamas, on commercial flights," he later wrote in a report to his superiors and for Secretary of the Interior Bruce Babbitt. "Early on Monday the third of May, the U.S. Coast Guard graciously transported me 180 miles over ocean to land on Navassa island. The USCG also flew a 360-degree circle around the island so that I could take some oblique aerial photographs. We must thank them, as the cost of flying one person (me) to the island in their 18-million-dollar long-range helicopters is over a thousand dollars an hour."

The teardrop island came into view, the pilots left him in a clearing, and bade him farewell. Within five minutes of landing, he began his experiments, or, as he puts it, "I embarked on my adventure in remote sensing and field checking a biodiverse little island." Eventually, the Discovery Channel

crew joined him, and they hacked through the poisonwood trees, Garvin enjoying himself immensely, thriving on exhaustion and hardship. He ne-
glected to eat, drink, or sleep; it didn't matter. Once again, he'd come to a
place where few have ever been, or will go, and achieved a state of mind that
perhaps no one else can fully share.

That night, he spent hours planning the field measurements he would
take the following day. At first light, he and another member of the expedition headed out across the terrain, trying to avoid the poisonwood trees.
Along the way, he discovered what he believes to be the first fungus ever
found on the island. "We moved slowly, discovering various biological features as we trekked the limestone pinnacles of the island. The darned poisonwood tree appears to have colonized these regions. It is extremely hard
to walk in these areas. We traversed the contacts between the mined-out surface and natural surface, where the litter of leaves and dried vegetation
makes walking equally difficult. The canopy is just tall enough to make GPS
navigation spotty. We got lost several times, and I had to climb a tree once
to sight the lighthouse and get bearings. Field studies of landscapes and
landcover in the presence of an experienced team is enthralling, and I felt a
bit like an explorer moving about Mars."

The temperature reached 95°; Garvin and his companions wilted. They
staggered across an open field to a rendezvous point, where a small helicopter
plucked them out of their plight and deposited them into base camp on
Navassa. "I had my measurements and some new theories about the landscapes on this island but was dehydrated beyond belief." After recovering partially, Jim resumed taking measurements; to his dismay, he found an enfeebled
feral dog wandering disconsolately inside a large sinkhole. He assumed the
dog, possibly left over from the days when the island was inhabited, would
eventually die of dehydration, for fresh water on the island is scarce. Later in
the day, Garvin, drawing on his last reserves of strength, trekked back to the
lighthouse, took more measurements, which brought him into contact with
the ubiquitous poisonwood trees, and by six-thirty P.M. returned to base
camp, exhausted. "The surface was rougher than the most spiny lavas of
Surtsey or Iceland. Solution-weathered limestone is akin to a bed-of-thorns
texture. My field boots were ruined in one day."

A short-range helicopter picked him up to convey him to the Discovery Channel's waiting boat. He was so tired he fell asleep on the ride despite

the vibration and the engine's racket. The pilot decided to have a little fun with the groggy NASA scientist and engaged in a few aerial stunts to wake him up. When Garvin came to, he was not laughing. On Wednesday morning, the long-range Coast Guard helicopter transported him to Great Inagua, where he spent a couple of hours debriefing local Coast Guard pilots; he then caught a plane to Nassau and another plane to Washington. He arrived home by eleven that night, drained, and itching furiously from the poisonwood. "Go to a doctor," his wife ordered. He slept a few hours and early the following morning sent me his initial bulletin.

"Navassa never ceases to amaze me," he insisted as he was recovering from dehydration. "My experience there was so enthrallingly positive, it makes the joys of scientific exploration worth the pain and effort. This is much like exploring Mars! Now I long to return to Navassa to continue field and remote sensing studies of this 'lost island' and its mysterious recent history. Isn't Mother Nature wonderful?"

So Garvin experienced his scientific epiphany, confirmation that he inhabits a tumultuous universe to which most people remain indifferent. Like the Time Traveller in H. G. Wells' *The Time Machine,* he has returned from another fantastic journey, somewhat the worse for wear, but eager to rejoin his family, friends, and scientific colleagues. "The Time Traveller was one of those men who are too clever to be believed: you never felt that you saw all round him; you always suspected some reserve, some ingenuity in ambush, behind his lucid frankness," Wells wrote of his imaginary voyager. I am convinced that Garvin is a kindred spirit who has just returned from another dimension in time and space. It's easier for him to reach Mars than for most people; in fact, he's made the journey several times.

Chapter Ten

GHOSTS AND GHOULS

"The landscape of Mars may be the most exhaustively described piece of real estate that no one has ever seen."

—L. D. MEAGHER

On May 27, 1999, Dave Smith and Maria Zuber occupy the stage of the main auditorium of NASA Headquarters in Washington, squinting beneath the glare of the lights, under the scrutiny of the press, to usher Mars into the known universe, to bring the future suddenly closer, and they are not entirely happy about it. They would rather be anywhere else at the moment than in this auditorium, but this is the burden of success. Even Jim Garvin, who warms to an appreciative crowd, is nowhere in evidence; other duties claim his attention, and he is delighted to leave the responsibility of explaining the latest findings of the Mars Orbiter Laser Altimeter to Dave and Maria. Usually, these things are done via a paper in a scientific journal, which causes ripples to spread through the scientific community, but science, especially planetary science, is different these days; the scientists need to get out in front of the story, to justify their funding, to explain what it all means to taxpayers who have unwittingly underwritten their research.

Dave Smith takes a deep breath and summons his esprit de corps. "We've been operating continuously for about two months now. We're going to show you some results from about thirty million measurements." And then it's Maria's turn. Usually implacable, she anxiously grips the arms of her chair and begins to talk about the early findings. "One of the incredible aspects of the planet is the difference in elevation between the high southern hemisphere and the low northern hemisphere." This means that on Mars the geological and topographical action can be found in the southern hemisphere. That's where all the big mountain ranges, canyons, and volcanoes are—Olympus Mons, Valles Marineris, and the Hellas Crater. That's where those large magnetic stripes Mario just discovered are. The northern hemisphere, in contrast, is flat. In fact, it qualifies as the flattest place in the entire solar system. There's no magnetism in these latitudes, nor are there large volcanoes or mountains.

As Maria likes to say, it's all downhill from the southern hemisphere. "You can see there's a uniform slope of about point zero three six degrees," she tells the audience in Washington. "So if you landed on the south pole of Mars and started walking north, there's nowhere to go but down." Maria and the rest of the MOLA science team can tell a lot from this slope about the patterns of water drainage on Mars in the past. And the most recent data from the laser altimeter taken during March and April has prompted the team to more than double their estimate of the "water inventory" on Mars at present. Where did all the extra water come from? The south polar cap, which most everyone believed consisted primarily of carbon dioxide, now appears to consist mainly of water, a lot of water. "If we add the volume of the two caps together," Maria says by way of illustration, "we get a volume of present-day water ice on Mars to be about one and a half times the ice in the Greenland ice sheet today." Even this tremendous amount of water is, at most, "only forty percent of the minimum amount of water that people once believed flowed over the surface of Mars."

Maria also brings news of the Hellas impact basin, perhaps the most provocative feature in the entire southern hemisphere. In the old Viking era pictures of Mars, Hellas Planitia, to give the feature its formal name, seems unremarkable, a mostly circular depression in the southern highlands. Using those dated images, you would scarcely suspect you were looking at the largest hole in the entire solar system. The brand-new topographical maps, assembled from MOLA's measurements, reveal Hellas as the result of a cataclysm of monstrous proportions. It is six miles deep and 1,300 miles across. It is surrounded by excavated material rising more than a mile above the surrounding terrain and extending 2,500 miles in every direction from the basin's center. That's enough material to bury the continental United States under a blanket two miles deep. If a similar asteroid hit North America, there would be no more United States. Continents would tremble, oceans would boil, and the atmosphere would partially dissipate. It's possible that the asteroid that created Hellas permanently altered the Red Planet's destiny, draining its oceans, robbing it of its atmosphere, and denuding it of life. But that is just one of many possible geologic scenarios; it will take Garvin and his colleagues a long time, and many millions of measurements, to reconstruct the history of Hellas.

Maria clears her throat. "The last thing I'd like to show is a close-up of the landing site of the Mars Polar Lander," she says. This will be the first spacecraft to touch down on the surface of Mars since Pathfinder in July 1997. At the moment, the public is oblivious to its existence; only JPL tracks its journey to the fourth planet from the Sun. Unlike Pathfinder or any previous American mission, MPL will not land in a monotonous, rock-strewn plain, far from features of geological interest and biological potential; it will land, if all goes according to plan, at the edge of the southern polar cap, amid alternating layers of dust and ice littered with occasional craters. But it will need the help of Mars Global Surveyor to make a safe landing in this exceedingly tricky terrain. "We've been able to get a high-resolution elevation map of the lander site region," she tells the audience. "We are collecting nearly a million measurements of elevation on Mars every day, so the map gets better as I'm speaking."

This report from another world is successful, as press conferences go, especially if you measure success in terms of publicity. The next day, MOLA and its gorgeous polychrome maps appear on the front page of the *New York Times,* the *Washington Post,* and other major outlets. It's as though they are publishing a glimpse—a very detailed, scientific glimpse—of the future. Mars is no longer terra incognita; it is at hand, and if we can see it this clearly, we can go there to explore. This is the first time much of the American public has ever heard of Mars Global Surveyor, and its existence comes as something of a shock. Fact is, throughout its entire existence, Mars Global Surveyor, and its key components such as the laser altimeter and Mike Malin's camera, have been overshadowed by Pathfinder, although it can be argued that in the record of Mars exploration, MGS is infinitely more significant, more revealing of Mars, than Pathfinder ever was. Now, with the publicity resulting from this press conference, MOLA begins to win recognition beyond the scientific community. Of course, most of the public has no idea what Dave and Maria are talking about when they discuss Hellas Crater and the possibility of ancient standing water on Mars, to say nothing of that uniform slope in the southern hemisphere. The MOLA science team, and NASA, aren't packaging their discoveries for public consumption; they are putting out a ton of raw observations, suggesting a few interpretations, and leaving it at that. It will take a while for the implications to sink in. No wonder the public is a bit baffled by all the data. But the public isn't the intended audience for this material. The intended audience is other scientists—the one big endlessly quarrelling family known as the Mars community—and they, more than any others, will pounce on the data, and feed off it for years to come. The process will start in a few weeks, when Dave, Maria, and the rest of the team take their MOLA roadshow to a succession of giant science conferences, beginning with the spring meeting of the American Geophysical Union in Boston.

In preparation for the spring AGU, Jim Garvin studies the latest MOLA measurements with his characteristic fervor. He's working with data that's only a few weeks old, and, in some cases, even less. A few days after the Washington press conference, he sends an account of his revelations that seems to have bypassed his rational faculties and emerged directly from his unconscious. This is the nearest I have seen him, or anyone else, come to scientific rapture:

Subject: MARTIAN EPIPHANY
Date: Mon, 24 May 1999 08:33
From: Jim Garvin <jgarvin@nasa.gov>
To: Laurence Bergreen <bergreen@NYCnet.net>

Larry,

I have experienced another Martian epiphany. Mike Malin, the creator of the Mars Orbiter Camera (MOC), and I have started what I think will be the most fruitful collaborative science work of my career. We are collaborating on MOC-MOLA analyses of impact craters on Mars. Given their abundance, formation process (instant landforms), and ubiquity, they can serve many purposes as one tries to unravel what has happened and what is happening now on Mars. I received an installment of over 100 MOC images of Martian craters early last week from Malin. This treasure trove of aesthetically intriguing images, many at spy satellite resolution, will facilitate new science that combines landscape morphology with MOLA topography. MOLA has vertical precision that is less than a meter, while in the MOC images, one can resolve things 6-10 feet in size from SPACE.

When I first loaded the CD's containing the images, I did not expect to see the new Mars that greeted my eyes. It literally brought tears to my eyes—just ask Cindy, my wife. Several of the exquisite images of the north polar region stunned me to total silence! (This is a rarity for me.) In some cases, I observed surfaces with a pebbly texture on which were inscribed dark scratches, seemingly by Mother Nature's Martians. What I think I am seeing, adjacent to amazing ice-blanketed impact features in this region, is a dust-veneered surface of ICE, perhaps like a frozen pond, with a light mantle of powdery snow. In these cases, the surface of Mars resembles that of Europa, one of Jupiter's moons.

In one other case, I nearly fell off my chair in my basement, next to my dog, a 120-pound Bouvier des Flandres happily sleeping amid my Martian discoveries. My eyes were greeted by a region surrounding a tremendous rim of a relatively small impact crater (say a few miles across) that at first glance reminded me of late-winter dirty snow and ice. I am sure you recall those last patches of ice where a pinnacle-like surface develops as the ice finally melts away. The area outside this crater, at two-meter resolution (100 times better than the Viking-based global mosaics we usually use), suggests impact melting of an ice layer that could have been a major component of the local target at the time of impact.

Many of these MOC images are aesthetically interesting as well as information-rich. This is one of the points that Tim Mutch used to make about the Viking Lander views of Mars from the mid-Seventies: one can appreciate the beauty of the landscapes without being overtaken by the scientific content. This brings Mars closer to home for people, even if they are not scientists.

How this links to the question of Martian life is fuzzy, but a glance at many of the superlative images from MOC, acquired while MOLA was pinging the surface, will facilitate improved analyses of the role of water in the high-energy hypervelocity impact cratering game. Since impact cratering is the process that most likely delivered Martian meteorites to Earth, then human-scale images of craters may be more informative and life-related than older, larger ones. I think I can see frozen water surfaces in some tiny craters (say, 100 meters across) as well, nearest the north polar cap. Wonderful stuff, and much too engrossing.

Jim

All this is on Garvin's mind when he arrives in Boston on June 1, 1999, for the spring AGU. The AGU claims more than 30,000 members, and thousands of geologists fill the Hines Convention Center and various annexes. You could never imagine that so many geologists have ever lived, let alone would congregate in the same place. But when you understand that geology now claims the entire solar system (including countless asteroids) as its turf, you begin to realize why there are so many geologists assembled here.

When the MOLA science team convenes in a private room at the Hines Center to sort through their findings, they immediately face withering financial problems. Bruce Banerdt, from JPL, reminds everyone that it takes $1,700,000 annually to operate the spacecraft, and several million more to sustain all the science teams. In all, the Mars Global Surveyor mission consumes nearly ten million a year, and at the moment it appears there won't be any more money for the mission after 2001, even if the spacecraft is still healthy and returning data. If unfunded, the spacecraft will simply be ignored as it maintains its Martian vigil. There is little the scientists can do about NASA's funding plans, with the exception of Dave Smith, who can squeeze a little more out of NASA when necessary, but even his leverage is limited.

At the moment, though, the team has cause for celebration; they have made the cover of *Science* (again!) with their latest false-color topographical maps of Mars. In the issue of May 28, 1999, the images reveal, for the first time, Martian topography in accurate three-dimensional detail, and more than that, they provide significant clues as to what lies beneath the surface of Mars. This time, *Science* has cooperated with the MOLA team, but copies are in short supply. Greg Neumann flashes a cover, with the dramatic hemispherical maps, electrifying Garvin. "Where did you get *that?* Can I have one?"

"It's my only copy. I won't get any more for weeks."

Meanwhile, Sean Solomon bristles over the possible inaccuracy or incompleteness of the published material. He frets about outsiders using unrefined but released MOLA data without authorization. "I'm not worried about credit, I'm worried about *bad science.*"

"How do you feel about the release of this data set?" asks Dave Smith.

"It will have to be referenced to the geoid, or it's not usable data," Sean says. The geoid is a reference point, a critical baseline, an equipotential surface. It can be used for determining how high a mountain is, for example. On Earth, the geoid is sea level; on Mars, a putative surface level. "If you're going to post the topography, you'd better post the geoid," he admonishes. "You've got to be responsible here. It's got to be trackable."

Glowering, Dave asks, "How are they going to be making mistakes by using the topography? I don't think you could make a mistake."

"Three-quarters of the people using it will say, 'How do I subtract the geoid?' " Maria interjects.

"But you have to define the geoid," another voice adds.

"Imagine somebody using this data set five years from now," Sean says.

"Oh, we'll have new data," Dave answers.

"But it's out of our control!" Sean insists. "Just give it a label, preferably one that no one will understand."

"Greg, what can you tell us about data acquisition?" Dave asks.

"It's coming faster than we can handle. We have a month of precision orbits not applied to date."

Maria Zuber sees something beyond the visual in these data. She sees gravity. She may know more about gravity on Mars than anyone else on Earth. With her radio tracking data and her new topographical maps, her understanding will expand dramatically. But it's all happening so fast. Minutes before her presentation here, she's still preparing her slides. "Arden Albee said I would be showing these slides this morning," she tells me as she frantically works. "The only problem is, when he said that, I hadn't gotten the graphs and materials from Dave. I was about to say to Arden, 'Slides? What slides?' I just got them! All of this should sing together like a symphony, but it doesn't, yet."

"Why measure gravity in the first place?" I ask.

"To understand changes in Mars," Maria says. "Gravity equals mass. Gravity equals changes in Mars. If you subtract the attraction of topography gravity, you can tell how density varies inside of Mars. You can tell what's beneath the surface, the underlying landscape, if you will. You can tell if there's enough strength under the surface to support volcanoes, for instance. And you can tell if mass has moved around, like the polar caps coming and going. So it's a very important piece of the puzzle, and it's been missing until lately."

For Maria, measuring gravity on Mars is very simple, in principle. You just measure the Doppler shift of the signal that communicates with the spacecraft. "Gravity is an acceleration," she remarks. "It accelerates the spacecraft, for example, so all you have to do is measure those fluctuations to tell if there is more or less mass beneath the surface of the planet." However, getting good, trustworthy measurements is fiendishly difficult because you have to correct for an unbelievable number of effects. Maria lists a few off the top of her head. "You have to compensate for fuel sloshing around in the spacecraft, for general relativity, for radiation pressure from the Sun, for drag on the spacecraft from the planet's atmosphere, and so many other things. Correcting the data is a career in itself." But not her career. "There are people at Goddard and JPL who do that," she tells me, "and who are very good at it. I trust them to get it right. I'm more of an end user."

Minutes later, Jim Garvin, wearing his Snoopy tie, armed with his laser pointer, takes center stage in a large hall, standing before several hundred scientists. He reduces his latest findings to one dramatic announcement: "We have ghost craters."

The instant he utters the words *ghost craters* into the microphone, heads lift from laptops and notebooks, cell phones snap shut, and exclamations of shock and dismay reverberate throughout the hall. *Ghost craters on Mars?* That's like taking an X ray of a valuable painting and discovering another work of art hidden beneath the surface. Without pausing to draw breath, Garvin draws dizzying laser circles on a ghostly image detected by MOLA. "This particular ghost crater is fifty kilometers across with one hundred meters of total relief. It's *undetectable* in the Viking imaging. One of the craters that we've sampled extensively shows this characteristic pattern of a central infilling of material far in excess of any structural uplift we'd anticipate; in fact, at some levels, it has the topography that is above the pre-impact surface that we infer for this crater. This turns out to be the norm for the larger craters at the high latitudes. This is really rather revolutionary."

Garvin runs through his calculations, showing why the crater holds more material than anticipated and seems to defy gravity and logic. "We've measured a lot of craters," he says, "and I continue to measure them from my basement at home, a wonderful thing to do. Keeps me busy and off the street. We have a first global assessment. We're starting to break that into major units, clusters. One could imagine a long period of cyclic erosion and sublimation to form the topography." Sublimation occurs when water passes directly from a frozen state to gas, without becoming liquid. "I think MOLA has taken us on a roller-coaster ride through the craters of Mars, which I'm very excited to share with you. And, we're progressing . . ." His allotted time—fifteen minutes—suddenly runs out, and he sighs in frustration. Another Garvin minute has run its course. "I'll just stop right there. Thank you." Jim Head hands him a two-word note. "Good job!"

Later, Garvin tells me what he had planned to say about ghost craters at the conference, if he'd had the time. "We suspect they are heavily infilled craters whose expression has been muted by millions of years of dust deposition, at a pace that is higher than expected for the global Mars." The pace is important, because higher sedimentation rates suggest that the ghost craters may contain evidence of an ancient, lost ocean on Mars. In this scenario, the ghost craters were once covered with liquid—the scientists' risk-free euphemism for water—and when the water disappeared, they were nearly invisible under the sedimentation, and they continued to recede still further under windborne deposits. If this is the case, Garvin's ghost craters provide geological evidence for a very different, very ancient Mars, an ocean planet, and one perhaps teeming with microorganisms. "Craters," he explains, "leave behind an amount of residual heat. Let me give you an analogy. Suppose I shot a bullet into the wall over there. The bullet would go into the wall, of course. Now, if you touch a bullet hole right after it has formed, it would be hot. When you hit a planetary surface with a speeding bullet in the form of an asteroid, it's also hot. And the impact can sometimes trigger melting; it can melt frozen ice, and produce, say, a little pond that would exist in a liquid state before it refreezes and then sublimates." While the pond remains in a brief liquid state, something could grow in it, something could replicate and mutate, something could *live*. And when the pond freezes, that growth would be preserved in the ice as surely as a fly in amber. So the location and structure of *craters* may offer clues to the location of *life* on Mars. Jim has again been careful not to draw these conclusions for skeptical scientists, but to lay out the evidence and allow them to form their own conclusions. An artist with his data, he shows but rarely tells.

Several weeks later, and several thousand miles away, in Pasadena, California, another artist is at work: Everett Gibson of the Mars meteorite team. He's in attendance at the Fifth International Conference on Mars, which takes place at Caltech. At the moment, he's planted himself at the edge of a large lily pond on the Caltech campus, buttonholing scientists as they come and go. More than three hundred of them have assembled here, in July 1999, for the gathering, the first of its kind in a decade. As in so many conferences, the real work gets done in the hallways, over dinner, or, in Everett's case, by the lily pond, rather than in the plenary sessions.

He's a blunt, personable man with a slab of gray hair coming down one side of his face, and a significant belly, which he carries well. He's wearing a black knit polo shirt with the emblem "MARS METEORITE RESEARCH TEAM," in case anyone has to think twice about who he is. We're standing eyeball to eyeball, and he's jabbing the air in front of my chest. "When I'm retired," he tells me, "I'm going to publish my own book." His voice, imbued with a West Texas accent, carries across the Caltech campus, causing heads to turn.

"What about?"

"All the bullshit that's gone on behind the scenes with ALH 84001! I can tell you the names of eight books that are filled with lies and distortion of our information." His critics, he says, are sloppy, malicious, and just plain envious. His blue eyes bore into me, and he looks like he's ready to punch me out if I spread any more misinformation about his work. I carefully explain that I think he's on to something, and even if his team hasn't *conclusively* demonstrated the existence of Martian life-forms in ALH 84001, no one has come up with a better answer. "It seems to me," I tell him, "you've found something remarkable, whatever it may be." Once he hears this, he calms down a bit. I ask for some examples of distortions. Everett says the media has made up its mind that most scientists no longer believe ALH 84001 contains evidence of life on Mars, and they go back to the same four or five so-called experts to reinforce that point. This tactic infuriates him. He tells me almost word for word what David McKay, the team's leader, told me: the members of the meteorite team are more convinced *now* than they were at the time of their original announcement in 1996 that they have found evidence of ancient life on Mars.

"There's a near-conspiracy to discredit our findings," Everett says, glancing at the lilies as they gently rise and fall. "We were so cautious and deliberate. You know, before we published anything, we spent *two years* trying to convince each other there was no life on Mars, and we could not! But we knew we were going to be challenged, and we were pretty proud of ourselves, because when we were, we had a lot of answers in the bag. You can't just announce an idea and drop it. You have to defend it. You have people saying we just did this to help NASA—*hey, we didn't do that!* We didn't try to help NASA's budget. That's completely false! Only my immediate supervisor had any idea of what I was doing with the meteorite, and no one at NASA Headquarters had the slightest idea until our article in *Science* was in print. Then it got involved with politics, and well, that's the system we work in."

He reviews the four lines of argument for life in ALH 84001, folding his fingers into his massive fist as he goes, then ticks off his team's refutation of the criticisms leveled at the arguments. No, the nanofossils are not too small. No, the PAHs are not caused by contamination. And no, they are not relying on a contaminated part of the meteorite for evidence. "Now, we don't want to get into a pissing contest with our critics, but it's hard to avoid. People distort the truth to serve their own agendas, and that disturbs me, because it goes against the way I was raised and what I believe in. It's a scandal, the way the media has ignored our refutations, but we have more in the pipeline."

"Such as?"

"I won't say any more until I make my presentation on Friday, and this is just Wednesday." On Friday, I gather, there will be more life on Mars than ever before.

He turns and kneels beside the pond, then gives me a conspiratorial wave. "Lookee here," he whispers. I crouch down beside him, straining to see whatever he's pointing at. "There's one of the little critters." I lower my head, until I can see under the leaves, and there, about six inches away, a little frog rests on a leaf, its protruding eyeballs staring back at us. He swipes at the frog, which leaps from his outstretched, wiggling fingers to a more distant lily pad.

"Scientific advances take a long time to be accepted," he says, turning back to me. "Plate tectonics took forty-five years to be accepted by the scientific community. Our data have *not* been refuted. For all the challenges that have been raised, there's equal doubt cast on them. It's almost like a court trial that we're going through here. We are like detectives who have gathered the evidence and have presented it to a grand jury, and then they say, 'Yes, it's worth pursuing.' Now, it has to be prosecuted. We have to show that life is an evolutionary process. It should still be there, on Mars. The Nakhla meteorite has structures suggestive of life. It's one point three billion years old, and it contains structures formed seven hundred million years ago. And now we're looking at Shergotty"—the same meteorite Jim Garvin examined several years earlier for impact metamorphism—"which is one hundred and sixty-five million years old, and it's also got these structures in there. We're saying there may be biology going on. Once life as we know it gets a foothold, it's extremely difficult to extinguish it. It survives in pressures and temperatures that one or two decades ago were considered lethal."

By now several other scientists have begun hunting for frogs in the lily pond. They reach over and lift lily pads and scoot around to the side of the pond, trying to ambush the little creatures, who dart here and there with amazing speed. The hunt finally ends when David McKay comes over to confer with Everett.

"There's a CNN crew that wants to talk to you," Everett tells McKay.

"Aw, to hell with them," McKay says softly. "You talk to them, if you want to. I don't want to talk to them. I've got better things to do." McKay moves out of sight before CNN, or anyone else, can engage him.

Then Kathie Thomas-Keprta saunters over. Her fine blonde hair flutters in the breeze, her gold-rimmed glasses glint in the sunshine. She looks burdened with jet lag and deadlines. With Everett Gibson and David McKay, she is still searching for fresh evidence of life in Martian meteorites. Like Everett, she's avoiding many of the presentations going on within Ramo Auditorium; unlike Everett, she has scant interest in working the crowd or informally defending her work. The Mars Meteorite Research Team will be the biggest draw at the conference, and while they revel in the attention, they are determined to underplay their influence among planetary geologists and astrobiologists.

"You look exhausted," Everett tells her. She *is* flustered; the poise that I noticed when we spoke months ago at the Johnson Space Center has temporarily deserted her.

"Well, Ev, you should say I look lovely. Are you going to the conference dinner tonight?"

"No, I'm not, and neither are you. David, you, and I are meeting to prepare for tomorrow's presentation in my hotel room, and I'm bringing the
food, so tell me what you want for dinner."

"Fried onion rings, french fries, fried anything." And she saunters off before anyone else engages her in conversation.

Arden Albee and Matt Golombek chair the Fifth Mars Conference, but
it's really Arden's show, to the extent that anyone can tell scientists what to
do. Three hundred of them have come to Caltech to participate in five days
during which the sum of knowledge about the Red Planet will be revealed,
reviewed, hailed, debated, analyzed, and eventually subjected to revision and
invalidation. As I register for the conference, I receive a CD-ROM containing several hundred Mars papers to be given here. They include "Mor-
phometry of Circum-Chryse Outflow Channels: Preliminary Results and
Implications," "Small Valleys Networks on Mars: The Glacial Meltwater
Channel of Devon Island, Nunavut Territory, Arctic Canada, as Possible
Analogs," "Polygons in Southern Utopia Planitia: Initial Results from the
Mars Orbiter Laser Altimeter," "The Role of Water/Ice in the Resurfacing
History of Hellas Basin," "Subsurface Volatile Reservoirs: Clues from Mar-
tian Crater Impact Morphologies," "Global Geometric Properties of Mar-
tian Impact Craters: A Preliminary Assessment Using Mars Orbiter Laser
Altimeter (MOLA)," "Implications of MOLA Global Roughness, Statistics,
and Topography," "Nature and Origin of Martian Surface Materials,"
"Where Is the Geochemical Evidence for a Warm Wet Mars?" "The Dust
Cycle: Sinks and Sources, Cause and Effect," and "Driving Stresses in Mars
Polar Ice Caps and Conditions for Ice Flow," among others.

The Fourth International Conference took place a decade ago, when the
notion of life on Mars was considered borderline folly. Ten years ago, the
concept of extremophile life was just beginning to emerge, and the implications for astrobiology were unclear. Ten years ago, Jim Garvin was hoping
for his shot with Mars Observer, and Matt Golombek was another young
planetary geologist working on one phantom mission after another. Ten
years ago, Mars was not generally considered essential to understanding the
solar system, the universe, or the origins of life, except among a few zealots
such as Carl Sagan, who stubbornly held out hope that the Red Planet
might contain evidence of life. Ten years ago, Pathfinder didn't exist, and
NASA had no firm plans to send a rover to Mars. Now, a Mexican scientist

tells me, "In the scientific community, everyone has their eyes on Mars, and it's going to be that way for the next ten or twenty years." Although this seems like a good time to hold a big Mars conference, a summing up of everything scientists have learned lately, Maria Zuber insists it's not *exactly* the right moment. The MOLA data are too new, only a couple of months old in many cases, a couple of weeks in others, a couple of days in a few instances, and no one has had time to make sense of it all. Only ten percent of the scientists in attendance have the new data; the rest know about it only by report or rumor, and that's not enough, in her estimation, for the conference to work as well as it should. "So much of what we're saying now about Mars, based on Mars Global Surveyor, is heresy, at least compared to the view provided by Viking. This morning, as we were making up slides for our talk, I turned to Jim Head and said, 'Are you ready to *horrify* everyone?' "

Meanwhile, I await the arrival of Jim Garvin, who's back in Washington, happily lost in his craters and other MOLA data. "MOLA will fire its hundred millionth shot," he informed me just before I came to Caltech. "I recall a certain senior and extremely bright official telling me in private in 1989 that MOLA would never get to Mars and never fire a shot. To him and the rest of the naysayers, I say, 'Oh, ye of little faith.' We now have better topographic information about Mars than for several continents on Earth." That was the last I heard from him, until he arrives late for the conference, exhausted, his eyes rimmed with red after an all-night flight from Washington, D.C., to Los Angeles.

As we walk to Ramo Auditorium, he tells me about his agonizing good-bye to his daughter. "Danica is two now, old enough to know I'm leaving, and she was crying, her lip was trembling, and she was saying, 'Daddy, don't go!' and I was telling her I would be home in just thirty hours, but it didn't matter to her how long I would be away." Jim tells me that his wife, Cindy, is also perplexed by his frequent absences and his deep involvement with his work. "A few years ago, I was describing to Cindy my theories about astrobiology, and how life started on Earth with material from space seeding the planet with the necessary molecules, and it was all so beautiful that my eyes were filling with tears as I was describing it to her. She stopped me, and said, 'Jim, who cares? It all happened *four billion years ago.*' "

Garvin strides down aisle and bounds onto the stage at Ramo Auditorium to address the assembled Mars scientists about the latest crater data from

Mars. "I have just come from NASA Headquarters," he announces in a hoarse voice. He has a certain swagger and immediately gets everyone's attention; he speaks with authority. "What I am going to talk about is work in progress. I'd like all of us to relax and contemplate these data, from MOLA and MOC—they're beautiful, and they're confusing." As he walks through the initial readings from MOLA, his laser pointer lighting the way, he adds, "We've got a thousand simple craters measured. We're starting to sort them by units. These are all topographically fresh. We barely touched those degraded craters. We do have a tomorrow." He would like to stay longer (the conference has three more days to run) to answer questions and win converts to the cause of remote sensing, but he must return to Washington. After only a few hours in California, he flies home, trying to keep his promise to return to Danica in thirty hours.

At Caltech, the Fifth International Conference on Mars proceeds, a happy self-contained society of Marsists. Many of the participants have come to feel a sense of kinship with one another, and they have dropped their guard. The possibility of life on Mars has become the main topic of speculation, and as we picnic on the grass, the conversation turns cheerfully unguarded and conspiratorial. "I left a secure, chaired professorship to do this insane thing, and it turned out to be a barrel of fun . . ." I turn to see who is talking. It's Ken Nealson, discussing his new job as head of an astrobiology unit at JPL. In his shorts and sport shirt, he has the relaxed demeanor of someone vacationing rather than matching wits at a science conference. For years, he was ensconced at the University of Wisconsin, esteemed as an astrobiologist, but not directly involved with the meteorite team at Johnson. This status conferred a certain objectivity on him, and he rather enjoyed being on the outside, looking in. This is his first Mars conference, and it's apparent that he relishes challenging scientific assumptions and orthodoxies. While many of the scientists here congratulate themselves on their discoveries, Ken, smiling his avuncular smile and scratching his head, says, "I've mainly been impressed by how much we *don't* know about Mars. It *amazes* me."

He has thought more broadly about the subject of extraterrestrial life than anyone since Carl Sagan, which becomes apparent as he talks. "There are many traps in thinking that you can fall into when you are looking for evidence of biology. The main temptation is to look for similarities to life here on Earth. But if it resembles life on Earth, it probably *is* life on Earth. And it probably is *not* life from Mars. If it has DNA and RNA, it is from Earth. Life on Mars is going to be different. If we have a four-billion-year history of life here, as scientists think we do, we have no idea what the first billion years of life on Earth were like, because nothing has survived." But those vanished specimens probably resembled life on Mars, if it ever existed.

The best way to get a definitive answer to the question of extraterrestrial life, Ken believes, is to bring back a sample from Mars. With current technology, it's probably the *only* way. "Ultimately, we would like to have samples from many places in our solar system and beyond, but Europa is a 10-year round-trip and a journey to Saturn's moon Titan and back would take 20 years. Realistically, Mars will have to be our next laboratory," he explained in a recent article in *Engineering & Science*. Despite his skepticism about our current state of understanding, he believes we will eventually find life beyond Earth. "Given the number of other solar systems already known to exist, and the emerging numbers of planets around far-away stars, it seems unlikely that life will *not* be found elsewhere. Development of the proper strategy, and definition of those conditions that do and do not support life will be key to the ultimate discovery of extraterrestrial life. With the proper strategy and approach, the question seems to be not one of whether there is life, but when we will find it."

Ken often discusses extraterrestrial life in short phrases whose implications expand as you ponder them. "What are the limits of life?" he asks. "It must be measurable. Once you establish the limits of life, then everything is easy. . . . What this meteorite [ALH 84001] has really taught us is that we have a lot to learn about how to distinguish life from nonlife." He urges scientists caught up in the search for extraterrestrial life to shift perspective; their approaches may be limited by our knowledge of life on Earth. "If, from space, you have been looking for complex life on Earth, you would have thought it dead until the last few hundred million years; and if you were looking for signs of intelligent life, you wouldn't have found any until 70 years ago when the radio was invented."

Unlike Ken Nealson's self-effacing approach to the issue of life beyond Earth, Everett Gibson likes to show off. He has a taste for spectacles, crowds, and confrontations. Every year during his vacation, for instance, he stages an air show outside of Houston. He loves flying, and so does his wife, who is a pilot; when they fly together, he serves as navigator. Everett belongs to an historical aircraft organization informally known as the Confederate Air Force, which claims 8,000 members. What the Confederate Air Force does, basically, is keep World War II–vintage aircraft flying. He has rebuilt a B-17 himself and flown more than a hundred missions in it. Keeping these old crates flying is a hell of a lot of fun for Everett, but he sees a higher purpose in his pastime. "I tell kids, 'The B-17 and B-24 are the aircraft that kept you from speaking German.' "

Everett has been around Johnson Space Center since the place was in its infancy, thirty years ago. "JSC was built by the Army Corps of Engineers ahead of schedule and under budget," he says, to which I am tempted to add, "And it shows." You can see that he's intensely proud to be associated with the place. He arrived at an exciting time. Harold Urey, the Nobel laureate, was his first office-mate, and the first moon rocks were coming in for study. Still, the exhilaration of those days barely registered on his psychic seismometer compared to the worldwide controversy when his meteorite team announced possible evidence of nanofossils in ALH 84001. He's still getting over that shock. "I've learned more about human nature in the last three years than I did in the previous fifty-six," he says of the experience, and for a moment, pain clouds his face. He's going to learn more about human nature today, when he and his team announce their latest findings to the Fifth Mars Conference. They know they'll be facing a basically hostile auditorium and that most of the scientists will have come prepared to debate and tear apart their arguments.

David McKay, the team's nominal leader, speaks first. He takes it very slowly, as a Texas drawl creeps into his voice, a drawl that was barely in evidence when I talked with him months before. Perhaps his plain-folks approach will make his newest findings more palatable. The show begins. He displays a series of slides of nanobacteria in three Martian meteorites—the familiar ALH 84001, the less familiar Nahkla, and then, surprisingly, Shergotty, the relatively new (165-million-year-old) Martian meteorite. All at once, I realize that Everett inadvertently revealed the substance of McKay's talk: the team believes they've found Martian life in a *third* meteorite, Shergotty. The audience is silent, staring at the slides, and a fair portion of the scientists are astonished, both by the revelation of nanofossils (if that's what they are) in Shergotty and by their uncanny resemblance to terrestrial nanobacteria. McKay has prepared a set of side-by-side Earth-Mars comparisons, and the differences between the two, if there are any, would not be apparent to anyone except, perhaps, a specialist trained to notice morphological subtleties.

McKay deftly underplays the implications of these mirror images. "I'm not trying to convince you that these *are* nanofossils. We don't know for sure what they are, but we're saying it's a reasonable interpretation. All I'm doing is pointing out the morphological similarity that allows for fossils as a reasonable interpretation," he says, emphasizing the word "reasonable." He possesses a sure touch when addressing an auditorium humming with professional malice, having mastered the knack of simultaneously underplaying and overplaying his findings. Most scientists, when facing a crowd as intimidating as this one, will stick to their presentations. You can do that when you're discussing dust storm cycles on Mars, but when you're discussing a subject as emotionally charged as extraterrestrial life, almost everyone feels an urgent need to express an opinion.

McKay directs attention to side-by-side comparisons of Nakhla nanobacteria fossils and bacteria grown in the laboratory from Columbia River basalt. This is the stuff that came to the rescue of the life-on-Mars team. When their critics insisted the formations in the ALH 84001 were too small to qualify as life because nothing that tiny could be considered alive (at least on Earth), these basalt-loving organisms proved them wrong. Now, McKay's slides show an amazing resemblance between the bacteria he's grown in the lab and those in the Nakhla meteorite. To clinch his point, he explains that nanobacteria aren't some exotic item found only on Mars or below the Columbia River. It took him only a few months, he says, for the nanobacteria to fossilize at room temperature down at Johnson. His drawl, the way he slouches in his jeans, heck, everything about him suggests that his discovery is no big deal: you add some water to rocks, and you get life. Happens here; happens on Mars; it could happen anywhere.

But this proof may not be as convincing as it looks. For two years now, McKay's critics have been making the point that we're not looking at extraterrestrial life in the meteorite; we're looking at Earth life that crept into the meteorites and made itself at home. If you examined some pristine Martian rock, they argue, the nanobacteria simply wouldn't be there. Of course the "Martian" nanobacteria resemble terrestrial nanobacteria—they *are* terrestrial nanobacteria. McKay lowers his voice when addressing this issue to indicate that he takes it very seriously. "We think every one of the Mars meteorites has some contamination," he admits. "You have to accept that and work out ways to tell contamination apart from what is indigenous."

Everett Gibson, in his talk, takes McKay's latest findings a step further. If Martian nanofossils exist in three meteorites of varying ages, he argues, "It's likely that life *still exists* on Mars, and has been there for the last four billion years." He figures that life on Mars settled in "niches," probably in water below the surface, where it survives to the present. This assumption underlies his current thinking about life on Mars: "If life was present seven hundred million years ago, in Nakhla, and also one hundred and sixty-five million years ago, in Shergotty, then probably it is present today because nothing significant has happened to the planet in the past one hundred and sixty-five million years." This chain of reasoning contains two "ifs" and one "probably," which prevent it from being conclusive, but it represents the meteorite team's best guess about the history of life on Mars.

Next, Kathie Thomas-Keprta tells the audience about her latest morphological analysis of bacteria. She has been studying them in three dimensions, taking pains to distinguish nanofossils from crystals. The similarities in size and shape between terrestrial bacteria and Martian nanofossils make another powerful argument for life on Mars. What's surprising about her presentation is her demeanor. She appears edgy and talks so rapidly it's nearly impossible to understand her, yet her analysis is precise and rigorous, extremely difficult to challenge.

But she will be challenged, as will everyone else on the meteorite team.

The main rebuttalist this morning is Allan Treiman of the Lunar and Planetary Institute, who has made a career out of evaluating the conflicting claims of the meteorite team and its critics. He is the best known arbitrator on the subject, and as far as I know, he doesn't stand to gain if one side or the other wins. Treiman carefully balances one argument against another, and then he decides which side has come out ahead. His talk is entitled "Biomarkers in ALH 84001???" and as the proliferating question marks suggest,

Treiman is a worrier. In recent months, he has seen approximately equal strengths and weaknesses in the arguments of both sides, but now, if I'm hearing him correctly, he puts the skeptics in the lead. He's unhappy with the way the meteorite team has advanced four separate but related arguments in favor of life on Mars; he considers that approach close to sleight of hand. In the formal paper accompanying his talk, he writes: "Can scientific arguments be treated like sticks—four weak sticks tied together equal one strong stick?" Three out of the four lines of evidence put forth by the life on Mars team look pretty weak, he says. Worse, he thinks McKay is playing fast and loose with the evidence, manipulating and withholding findings that might help his critics. McKay's hypothesis, he writes, is now just "one hypothesis among many, all vying to explain this wonderful meteorite and its enigmatic clues to Mars' distant past."

McKay stands, and with as much dignity as he can muster, he says, "We do disagree with a number of your statements."

Pointing to the images of "Martian" and terrestrial bacteria on the screen, Treiman says he thinks they merely show clays and perhaps "a little bit of terrestrial weathering." He waits for that devastating assessment to sink in. "I wish you had waited until you had better evidence," he says, zinging it to McKay.

Gibson springs to his feet. "Hey, our papers are peer-reviewed," he reminds Treiman. "We can't do more than we're already doing. We all must seek the truth, and not seek errors in each other's analysis." The line, designed to elicit applause, is greeted with silence. By now, each side sounds like Groucho Marx insistently asking, "Who are you going to believe—me, or your own eyes?"

Later, Ken Nealson remains his gentle skeptical self about the meteorite team's approach. "Although I greatly enjoyed this morning, the fact that these things look so much like Earthly organisms makes me tend to think they *are* Earthly organisms." Ken is doing his job here; he is supposed to play the part of the loyal opposition to the meteorite team; that's one of the reasons why NASA invited him into the Astrobiology Institute. Yet there are questions that neither he nor any of the other skeptics have answered conclusively at this conference. Although Ken likes to emphasize that life on Mars is likely to be very different from life on Earth, a growing volume of evidence suggests that Martian geology resembles terrestrial geology and that Martian chemistry resembles terrestrial chemistry. All of this suggests that life on Mars is similar to terrestrial life, not strikingly different. If life on the two neighboring planets turns out to bear a family resemblance, it may be that life on Mars came from Earth via the meteorite express. Or it might suggest that life on both planets came from somewhere else, a common ancestor, some form of panspermia, seeding both planets with organic molecules necessary to stimulate biology.

Near the end of the conference, Bruce Jakosky takes the stage to deliver his thoughts about the state of the search for life on Mars. Unlike so many others here, he dresses stylishly, with a carefully cropped beard. The creases in his trousers are sharp, and his shoes are shined. Jakosky adopts a humanistic perspective, and suddenly the scope of the conference widens to include a new range of philosophical issues. He is unafraid to lecture his colleagues about what they ought to be doing. "By exploring the world, we are exploring ourselves, and what it means to be human," he says. "We're not doing science for the results; we're doing it for the process. . . . Research isn't over until it's explained to the public. . . . It's frustrating to think that you pursue a scientific career for thirty or forty years, and two years after you retire, everything you've done is obsolete . . ." For once, the audience murmurs in agreement.

Things happen rapidly in science, and scientists often change their minds about fundamental matters. Science is, among other things, an endless process of revision. Scientific evidence is really a palimpsest, preserving a record of earlier ideas, many of them erased or obscured. Scientists forever try to free themselves from the obvious errors of the past, only to confront eternal questions concerning Mars, the solar system, and the nature of life. Many of the ideas enthusiastically promoted here at the Fifth Mars Conference will be obsolete by the time of the Sixth Mars Conference. Looking back to the present moment, the scientists will likely wonder, *"What was I thinking?"*

In the late hours of September 22, 1999, Mars Climate Orbiter approaches its destination at about 12,300 miles per hour. In the 285 days since its perfect launch from Cape Canaveral, MCO has enjoyed a flawless mission across a trajectory of 416 million miles. Confidence in MCO at NASA and JPL runs high.

This spacecraft will be the first weather satellite to circle another planet. It will measure the Martian climate in more detail than ever before, and, as it circles the Red Planet, it will search for signs of water. MCO will also serve as a relay station for crucial data from its companion spacecraft, the Mars Polar Lander, which is scheduled to land near the south pole of Mars in a few weeks, on December 3, 1999. NASA will then have a fleet of three spacecraft—the Global Surveyor, the Climate Orbiter, and the Polar Lander—simultaneously investigating aspects of Mars and interacting with each other.

On September 7, Mars Climate Orbiter returns a fuzzy image of the Red Planet; it isn't much to look at, but the image demonstrates that the spacecraft's camera, known as the Mars color imager, is functioning. When the spacecraft gets closer to the Red Planet, the camera, supplied by Mike Malin, will capture images with a spectrum ranging from infrared all the way to ultraviolet, a range greater than the human eye can see. It will detect ozone clouds and water vapor; it will read the invisible ink of the Martian climate and decode signs of water and possible life. Everyone agrees it will be a great thing.

All that remains is for MCO to enter into orbit around Mars. Two years earlier, the Jet Propulsion Laboratory accomplished the task with the successful orbital insertion of Mars Global Surveyor, an engineering feat that dispelled doubts hanging over Mars missions since the loss of Mars Observer in 1993 at just this point in the mission. The Mars Orbital Insertion (MOI) is scheduled to begin at 1:50 A.M. Pacific daylight time, on September 23. In the JPL control room, adrenaline-stoked mission controllers anticipate another demonstration of what they do best: interplanetary navigation. Sam Thurman, manager of flight operations for JPL, issues a confident statement: "The curtain goes up on this year's Mars missions with the orbit insertion of Mars Climate Orbiter. If all goes well, the happily-ever-after part of the play will be the successful mission of the Mars Polar Lander that begins in December." There is no reason to imagine things will not go well.

To commence insertion, the Mars Climate Orbiter will fire its main engine for sixteen minutes and twenty-three seconds to slow its velocity to 9,840 miles per hour. Martian gravity will gradually capture the spacecraft. The spacecraft will disappear behind the Red Planet, a period known as occultation, for about twenty-one minutes. When MCO reemerges, JPL will reestablish contact with the spacecraft. All of these events will require an extra eleven minutes for confirmation that they have occurred as planned—the time it will take for the radio signals to travel 122 million miles from Mars to Earth. The reacquisition of the spacecraft's signal is scheduled for 2:26 A.M. Pacific daylight time. Once JPL confirms that the MOI has been successful, that they have a mission, the exploration of Mars will proceed as planned. MCO will enter a slow aerobraking orbit around Mars, until it reaches a uniform height of 262 miles above the surface.

At JPL, engineers controlling the mission follow the script to the second. The burn occurs precisely as planned. The spacecraft disappears behind Mars, hidden for long moments from Earth. The engineers wait to confirm they have a mission. The mood in the control room becomes quiet, expectant, as it always is at times like these. Sam Thurman sits in front of his desk, staring at the monitors before him.

The deadline approaches . . . arrives . . . and . . . passes.

The engineers attempt to reestablish contact with the spacecraft. Mars Climate Orbiter is out there, somewhere, they just have to find it and lock onto it. These things don't just disappear—except for Mars Observer, but that was a catastrophe, and this is almost certainly a momentary hiccup. It was a good burn—they've confirmed that. The spacecraft's computer might be down or it might be disoriented. Glitches occur frequently in planetary missions, and there are well-established recovery procedures.

"We're in a state where we're not quite sure what's happening," says one project manager. "At this point, we're still confident we're going to see the spacecraft signal in the next few hours."

Another manager adds, "We do have an indication that due to some navigation problems, the spacecraft was lower than intended at Mars. As a result, we are cautiously optimistic."

Three hours later, Richard Zurek, the project scientist, admits, "The longer we don't hear from it, the worse off we are."

The morning passes without any sign of Mars Climate Orbiter. Nothing changes throughout the long afternoon. By day's end, the weary crew at JPL acknowledges that the Great Galactic Ghoul has claimed another spacecraft.

Chapter Eleven

HUMAN ERROR

"I have not failed. I've just found 10,000 ways that won't work."

—THOMAS EDISON

Subject: MARS CLIMATE ORBITER
Date: Sat, 25 Sep 1999 13:21:00 -0400
From: James B Garvin <jgarvin@nasa.gov>
To: Laurence Bergreen <bergreen@NYCnet.net>

Larry

The loss of Mars Climate Orbiter is SEVERE, but we have resiliency. The Mars Polar Lander can send back its treasure trove of data through our Mars Global Surveyor. Such adaptability is a hallmark of our robotic exploration program. I believe, in my gut, that the science will be recovered—the good folk at NASA HQ and at JPL who manage and implement our Mars Surveyor Program will want to get it back and it may be possible to add back the two sensors and the novel camera to future orbiters (maybe Mars '03 or Mars '05). Further, we can use MGS, if we can keep it alive, to do some of the missing science—an extended mission for Mars Orbital Camera would allow recovery of some of the lost science at much less money. I think MOLA could help too, but keeping our laser transmitter running beyond its design point of 687 days (continuously for 500 million shots) is UNKNOWN territory.

What this all suggests is that NASA must resolve to implement the first piece of Mars infrastructure for relaying information to Earth independently from our science/robotic satellites. If we had a Mars orbital telecommunications satellite, we could always get more data back to Earth from Mars. We have developed new technologies that may allow a fully optical telecommunications link to Mars at TV bandwidth (or, in English, we could have coverage of stuff happening at MARS at the same volume and quality as the TV we see when we watch the Olympics anywhere on Earth). One of my goals is to see an optical communications link to MARS from Earth for this purpose. It would facilitate getting more data back of interest to the common person, and permit better coverage and better science.

Doing amazing things in space is HARD! Mars is 100 million miles away, and threading the needle with a spacecraft by trying to insert it into orbit is RISKY business. Mars has always been a harsh mistress and between the Russians and us, we have lost a bunch of missions. Now we mourn loss, and we recover by devising clever solutions amid the frightening uncertainties of the political system that wreaks havoc on NASA's budget every year.

Jim

NASA tracks down the reason for the loss of Mars Climate Orbiter within days of the event, and it turns out to be embarrassingly simple. Engineers at Lockheed Martin calculated the amount of thrust created by the spacecraft's small engines in pounds, the traditional English unit. Engineers at JPL took these numbers to refer to the metric unit known as a newton. The difference is noteworthy. A *pound* of thrust accelerates one pound of mass one foot per second. A *newton* accelerates a kilogram of mass one meter per second. Although a pound of thrust is nearly four-and-a-half times greater than a newton, no one had detected the discrepancy. When the thrusters aboard MCO fired, the difference between the anticipated and the actual correction sent MCO into orbit around Mars at an altitude of thirty-seven miles, instead of sixty to eighty miles. Coming after a 400-million-mile mission, the miscalculation seemed relatively minor, but it was big enough to place Mars Climate Orbiter into a dangerously low altitude where it burned or broke up in the Martian atmosphere.

The Great Galactic Ghoul lurking around Mars finally has a name and a diagnosis: human error.

The mix-up in measuring standards doesn't come as a complete surprise to me. I've found that NASA scientists and engineers, Garvin included, habitually alternate between yards and meters, miles and kilometers, at least in conversation. NASA has added the metric system to its collective mindset, but it has not replaced the old ways. Recent NASA documents that I've seen use meters and feet side by side, on the same page—and these concern prospective spacecraft for *human* missions. My next thought after hearing about this disastrous mix-up is, "Where is Sean Solomon when you really need him?" I could see him asking the tough questions, serving as a bracing reality check. "Are those measurements in newtons or pounds? . . . You mean you don't know the difference? . . . What *is* the difference? . . . Why are you using *both* systems? . . ." and so on. I could see the poor soul who confused his pounds and his newtons, thereby wasting nearly $200 million contributed by the American taxpayer, not to mention all the years American scientists and engineers invested in the project, dissolving into a puddle during Sean Solomon's inquisition, and the mission being saved as a result. But there was no inquisition this time, and no last-minute rescue. Mars Climate Orbiter does not belong in Sean's bailiwick.

Although highly embarrassing, the problem is readily fixable. There was nothing inherently wrong with the spacecraft or the launch vehicle or even the computers. The bad news is that the elementary nature of the goof—and that's exactly what it was, not an error in judgment, nor a miscalculation, nor the intervention of an unexpected event, just an obvious blunder that any engineering student would have caught—caused NASA to look incompetent, even laughable before the American public and before Congress. If the goof had occurred on a Russian mission, everyone would have shrugged their shoulders and said, "Well, what do you expect?" But it has happened to an American spacecraft, and there is no way to hide the mistake. In a frantic attempt to save face, improve morale, and all the rest of it, Dan Goldin appoints no fewer than three panels to look into the problem—and this move comes *after* the cause of the disaster has been discovered.

Six weeks after the loss of MCO, on November 4, 1999, the mighty MOLA science team returns to the ancient Lincoln Field Building at Brown University. More than a year has passed since their previous convocation here, and during that time NASA's Mars program has taken on a different cast; the euphoria is now tempered by anxiety. Going in, I expect big changes to have swept over the MOLA team since I last saw them. I expect that the ebullience of the La Jolla meeting will have vanished. I am wrong. In recent weeks, NASA managers have effectively convinced the public, Congress, and themselves that the loss of the spacecraft did not mean the loss of science; in fact, they have done such an effective job of explaining away the loss of MCO that you have to wonder why they sent it in the first place.

As always, the meeting opens with Bruce Banerdt reporting on the health of the spacecraft, but what he's really doing is imparting the scuttlebutt from Jet Propulsion Laboratory, where he works. "JPL reacted to the loss of Mars Climate Orbiter by reorganizing." The comment incites snickering; Bruce keeps a straight face throughout. "Basically, it was a process error," he says, delivering the JPL line, without appearing to realize it *is* a line, nothing more. "No one's been fired . . ."

"Yet," Maria interjects. "I've heard there's a press conference coming next week, and word on the street is that they will make unfavorable comments about JPL management. There was a desire to not make the report public before the MPL landing, but there is a lot of pressure to get this out. It seems to me that someone will eventually be removed."

"Meanwhile," Bruce says, "Mars Polar Lander"—which will touch down on the south pole of Mars—"has been having its own problems, which are keeping people busy." He immediately has the attention of everyone in the room. "When MCO went away, people at JPL started looking more carefully at Mars Polar Lander, and it made people uneasy. It turns out that the way the landing sequence was set up, the engines were blowing out solid chunks of hydrazine, for example, so they're scrambling to rework that."

Greg Neumann, as solemn as ever, displays a map of the south polar region of Mars where MPL will touch down. "As you can see, the landing region is a little elevated over the surrounding area. I think there are dunes."

"I've been looking at that," says Maria, "and they may not be dunes. In any case, it's pretty bad." She turns to Mike Carr, a geologist who has spent his career studying Mars, and, specifically, water on Mars. "What do *you* think they are?"

"I don't know."

"What's the general feeling out there just now?" she asks Matt Golombek, who peers gloomily at the laptop before him. His long pause and drawn breath hint that Mars Polar Lander is heading for oblivion; there's none of the unshakable, I'll-stake-my-career-on-it confidence that he brought to Pathfinder. He finally answers Maria's question by contrasting the uncertainties surrounding Mars Polar Lander with the success he experienced guiding Pathfinder to its landing site in the much smoother, more reliable Chryse Planitia. "I had the most robust lander ever designed," he reminds the team. "You could have rolled that thing down the side of a mountain. Honestly, I just don't know about MPL. No one knows." He shrugs in dejection.

"My sense is that it's like landing on hard snow," Maria says, as she raises her arms, then plunges them toward her lap and exhales, miming the spacecraft sinking into the crust.

"That is why it is called 'robotic exploration,' " Jim Head says. No lives are at stake, only reputations. His comment does not placate the troops. The members of the science team normally complain about safe but boring landings sites; now that they face the prospect of sending a lander to a truly interesting but hazardous site at the edge of the southern polar cap, they become anxious as they consider how rocky, icy, and slippery the site is. To add to their worries, Matt invokes the "Malin Rule" to convey the complexities of trying to select a landing site using available data, even data from the high resolution Mars Orbiter Camera: "Anything that looks really smooth in Viking images turns out to be rough in MOC, and anything that looks rocky in Viking turns out to be basically smooth in the MOC images." As Matt and everyone else in the room realize, fifty percent of Mars missions end in failure. The current series of "faster-better-cheaper" missions is two for three, still ahead of the average. The failure of Mars Polar Lander would bring the series in line with the statistical norm; its success would, therefore, be all the more remarkable.

A sense of insularity takes hold of the team. I don't hear much talk this time about their eagerness to share data across team lines, only an emphasis on maintaining a buffer between the functioning MOLA and glitches afflicting the other instruments on the spacecraft. But even MOLA has been showing signs of age. When I was down at MOLA headquarters at Goddard a few weeks earlier, Greg Neumann told me, "Our laser has been doing hiccups lately. There are sudden drops in energy and then it doesn't recover to its previous level. Engineers have no explanation; it doesn't fit their scenarios. Curiously, it's returning better data than ever before, which makes life interesting. Nobody's ever run a laser this long in space, so we don't know what to expect."

Now, at Brown, Greg ominously reports that a heater designed to warm the laser has "gone past the optimal point" and might affect the instrument's performance, "so be prepared" for what might be the last of the great MOLA data dumps. He points to a large multicolored mural of Mars topography fastened to the blackboard behind him. There is even more detail in these maps than in the images he showed me early in the summer; they are the products of laser altimetry's extraordinary accuracy and Greg's obsession with correcting the data returned by the instrument. "I don't just do this for my own amusement. It's to make sure the corrections are, in fact, correct. We've now generated data for one hundred and ninety-five days of precision orbits . . ." he says, and embarks on a fugue of worry. "I shudder to think what people can do with this data. . . . We're running out of disk space . . . *don't send data, folks!*"

"We are going to release data on a three-month cycle," Dave Smith interjects.

"When NASA cut our budget twenty percent, we told them 'no intermediate products,'" says Maria. She realizes, though, that it is almost impossible to draw lines in the sand when distributing data.

Sean Solomon, sitting next to Maria, comes down hard on this point. "If we keep putting out intermediate data products, they will have ill-described elements, and people won't know that."

"If we don't put them out," Dave reminds him, "we put out nothing!" Suddenly angry, he adds, "We simply can't do any more than we are already doing."

"I'm not saying do *more,*" Sean replies, with infuriating precision. "I'm saying do it *once* and do it *right.*"

Sean looks significantly at Dave, and Dave glares back, his blue-gray eyes turning to marbles. "Greg has worked long and hard to correct errors, explain his methodology, and add the appropriate caveats," he says. "We simply can't do any more." Dave looks more exasperated than I've ever seen him.

The meeting's host, Jim Head, intrudes. "Can you parallel process for a moment, gentlemen, while I take a head count for lunch?" The timing of the interruption is not accidental, and once the distraction has ended, the conversation moves on, but not for long.

"We've had nothing but plaudits for our data," Dave reminds Sean, who remains infuriatingly calm, although his face is flushed.

"We've gone from *no one* believing our topography to *everyone* believing it and thinking our data sets are perfect," Maria adds. "They believe whatever we say now, even when we say nothing. That's the problem." She elicits the laughter of recognition from the rest of the team. When it comes to the topography of Mars, there is no challenging the MOLA team these days. Sean Solomon folds his arms across his chest and falls silent, temporarily held at bay.

I've often wondered whether the concept of family (or even a dysfunctional family) adequately describes the MOLA team and its dynamics, given the odd combination of loyalty and conflict they display whenever they convene, but I am not convinced the comparison does them justice: it's too sentimental, too static. The analogy overlooks the essentially intellectual and strategic nature of their enterprise; as I look at them arrayed around a large square table, I realize the controlling metaphor for their enterprise is chess, played out on a cosmic scale. In this scheme, Dave and Maria represent the king and queen, while Garvin, Head, Solomon, Neumann, and other team members are various knights and bishops, and their retinue of post-docs and graduate students naturally play the role of pawns. Their adversaries are many—the limits of human knowledge and engineering, the immutable laws of physics, all those factors that add up to the Great Galactic Ghoul and account for the fifty percent failure rate of Mars missions—but the prize is irresistible: the opportunity to explore Mars, to expand the known universe, to seek our place in it. Like chess, the pursuit of Mars is a game, a very serious and time-consuming game. But if you can tolerate the pain, the disappointment, and the uncertainty, it's the contest of a lifetime.

The next day, when the team convenes, they learn that their proposed special session at the next Lunar Planetary and Science Conference has been rejected. This is a nasty surprise. Rejected! This is hardly the treatment they have come to expect. The team laughs in disbelief and immediately makes plans to reverse the decision. "How dare they!" Maria says. "This is the best science those guys have seen in decades, since the lunar samples, for God's sake." Which of course was almost thirty years ago.

Listening to Maria's outburst, I am struck by how far this team has come in only two years. They've gone from a group of hopefuls desperately waiting their turn in the planetary science arena to an all-star lineup of specialists, who are acutely aware of the importance of their research, and who have come to expect the attention and respect of their colleagues. I recall how the prospect of their first article in *Science* sent paroxysms of self-doubt through the team, but now they routinely churn out papers for one publication after another. At this meeting, Jim Head distributes no less than four recent articles he has co-authored for scientific journals, based on the profusion of new MOLA data. One concerns possible evidence of ancient oceans on Mars, another the topography of Martian polar caps, a third characterizes the terrain in a section of Utopia Planitia (Utopia Plain), and a fourth considers some observations of various Martian slopes. And there is more to come. Maria has done studies of ancient Martian solar tides, complete with an animation depicting their possible scouring effect on the putative Martian coastline. Even Sean Solomon hails the findings. "Write it up!" he calls out from the back of the room, again and again. Two years ago, when Jim Garvin stood before the team to explain his preliminary findings concerning craters on Mars, he was jeered and mocked. Now, millions of measurements later, when he stands before them to deliver an hour-long analysis of his most recent crater findings, not only does the team refrain from complaining about his long-windedness and his baud rate, they applaud.

The ride is not quite over for Mars Global Surveyor—not yet. Dave tells me that MOLA will continue for another year, or perhaps two, no one is sure. If the instrument is still working, and funding is available, it will continue all the way until 2002, or even longer. The tale of the Galileo mission, another planetary overachiever, holds some clues as to what's in store for Mars Global Surveyor. Galileo was designed in the seventies and finally launched in 1989 from the space shuttle. It was an expensive project, well over a billion dollars, and was soon forgotten by the public, since its trajectory to Jupiter took years to complete; in fact, it was almost canceled in mid-mission, until the Galileo spacecraft discovered ice-covered oceans across the entire surface of Jupiter's moon Europa. As a reward, the mission has been extended again and again, well past its life expectancy, into the new millennium, which, it is safe to say, no one would have predicted when it was conceived during the Jimmy Carter era.

It is unlikely that Mars Global Surveyor will abruptly run out of money and come to a sudden halt; that isn't NASA's way of doing things. To keep extended missions going, the agency releases funds in dribs and drabs, but Dave Smith will have to beg for every additional dollar. "A few hundred thousand will do it," he tells me. "It should keep the project co-investigators working." So the MGS mission will wind down slowly, and the scientists on the MOLA team will eventually drift off to join other projects. Sean Solomon has recently been chosen as the principal investigator for NASA's Mercury Messenger mission, which will explore the planet closest to the Sun in 2009. Mercury has been visited by only one American spacecraft, Mariner 10, in 1974–1975; and now that Mars, the moon, and Venus have been mapped and investigated, Mercury, of all the rocky planets, remains an unexplored enigma. It zips around the Sun in its eccentric orbit every ninety days, and although much of the planet bakes in solar heat, it has polar caps and probably polar ice. Mercury's climate must be what it's like when Hell freezes over. "It is thought there is water on Mercury," Sean tells me, and all that the presence of water suggests.

Eventually, Mars Global Surveyor and its laser altimeter will slowly die a space-induced death. In the frigid vacuum of space, its moving parts will stick and refuse to become unstuck. Its power supply will falter. Its instruments will return erratic, unusable readings, and ultimately its transmitter and computer will fail. At some point, NASA will formally declare Mars Global Surveyor dead and cut off funding to maintain the spacecraft, in effect taking it off life support. At the time of its official demise, the scientists and engineers associated with MGS will gather for a final beer blast and hoist a glass to a magnificent planetary mission. Long after the team members move on or retire, MGS will continue to circle Mars, a mute witness to the most spectacular landscapes to be found anywhere in the solar system. By that time, humans may be walking across the dusty surface of the Red Planet, and if they look up into the night sky and observe MGS passing overhead, they may think of it as a sentinel that helped to guide them to Mars.

"We've grown old doing this," Dave says, as he surveys his team. "We've grown old together." When he talks about the future, it becomes apparent that the Red Planet remains his first love. "Mars," he says, "is the gemstone." And under his breath, he adds, "But there are *so* many prima donnas on this team!"

Right after the meeting at Brown, Jim Garvin, a perpetual motion machine, returns to Washington to chair NASA's new Decadal Planning Team (DPT). Although its gruesome name makes it seem like another stultifying bureaucratic exercise—"those who can, do; those who can't, *plan,*" a scientist once told me—the DPT is supposed to consider NASA's course for the next twenty-five years. I gather this involves everything from exploring uses for new technologies to human space flight. Jim won't reveal the entire range of the committee's deliberations, which is unlike him; instead, I hear tantalizing tidbits: "The NASA brass wants a group of brainstormers to generate a much-needed vision to justify funding for space exploration in the future. It is a huge job, but it could serve to couple our progress in robotic exploration with human initiatives." To complicate matters, Jim suspects that Dan Goldin has another committee, a shadow organization, in place, and he fears that whatever his team eventually recommends may be discarded in favor of the mystery group's report.

Although Jim pursues science with fierce enthusiasm, he has little stomach for this type of office politics. Yet he is adept at political maneuvering. While he has his detractors, who poke fun at his boyish enthusiasm and loquaciousness, he has no enemies that I know of. He takes the idea of functioning as a good and loyal NASA employee extremely seriously, and although he often suffers keen frustration with the agency's bureaucratic limitations, there isn't a trace of cynicism about him. He realizes he could have turned down the appointment to chair the DPT, but the "invitation" came from the most senior of the NASA brass and from Goldin himself, and Jim felt he had no choice; it was an offer he could not refuse. "I like and respect Dan Goldin, but I'm afraid of him," Jim says.

At the first meeting of the Decadal Planning Team, Jim announced that he had once been a hockey goalie, and he asked the team members if they knew what he meant by that. Not everyone did, so he explained that a goalie wields a bigger stick than the other players, and nobody gets near him. He waited for them to comprehend his message. "Let's get our deliverables together and say what we believe," he urged. "After that, I'm the goalie, I have the big stick, and I'll take the hit. Not you. *I'll* take it."

The hit came sooner than expected. At his first encounter with the administrator, Jim suggested that the Decadal Planning Team adopt some of the approaches used by the Sally Ride committee, on which he had served years before. Ride's committee, formed right after the Challenger disaster, recommended that NASA revitalize itself in part by establishing a lunar scientific base, and there was even talk of a human mission to Mars, but the memory of the lives lost aboard Challenger cast a pall over the proposals. If another fatal accident occurred, the committee felt, the entire space exploration program would be set back for many years. In the end, only a few of the committee's proposals concerning robotic missions won approval. Many of NASA's leaders (including Goldin) considered the Ride Report a setback for NASA, and Garvin's suggestion was dismissed with a few well-chosen expletives.

Jim has spent years brainstorming about the future of space exploration, and he has his own ideas about what NASA should do after its latest human space flight project, the International Space Station, is fully operational. Of course, he'd like to see NASA adopt a serious plan to send people to Mars in the name of science. "As I said to Dan Goldin, 'What is the point of going to the moon and *stopping?*' We want to discover and explore the space equivalent of North America. What if the great mariners of the fifteenth century had reached the Canary Islands and said, 'Okay, that's enough exploration. Let's stop here.' It simply doesn't make sense to stop now. As I see it, NASA's role is to answer a variety of basic questions: Are we alone? How did life begin? Where are we going? That's what I think NASA ought to pursue."

No matter how far he projects himself into the future of planetary exploration, Garvin's thinking on the subject is influenced by the history of terrestrial exploration. "Exploration is a strange business," he says. "We have explored for many different reasons, some of which have laudably been for science and the discoveries we make serendipitously." He compares past exploration to "the Norse mariners who courageously ventured to the West in search of opportunity, and who had no idea how to sail or what they were looking for. That was how the Vikings encountered Iceland, Greenland, and eventually North America, thanks to their robust longboats, bravery, and good fortune."

Jim notes that the Vikings' longest voyages, which took place in the eleventh century, "were conveniently conducted in a unique period of recent Earth climate history known now as the Little Climatic Optimum, a warmer and wetter time linked to the couplings between the Sun and Earth. Later, in the early Renaissance, sailing became perfected as a technological art, and caravels well suited to tack across the wind and undertake long voyages were developed. The Portuguese put this technology to work with a visionary leader or two and challenged the seas, initially in a quest to reach Asia and the lure of the spice trade. The secret of their success was *intelligent ex-*

ploration, and with it they managed to visit the Azores, the Canaries, Cape Verde, and much of the west coast of Africa. They had their focus, but the implementation was flexible enough to permit many destinations." He'd like to see NASA do something similar. Rather than focusing on a single target, an ideal "intelligent mission" would visit many places in space.

After the renaissance of exploration in the 1400s, culminating in the voyages of Columbus and Cabot, the transportation systems involved improved, Jim notes. "Only a generation later, Magellan's circumnavigation of the Earth was accomplished, although it took three years." At the time, the ocean seemed as mysterious, menacing, and beguiling as space does now. "Drake repeated Magellan's feat with greater attention to discovery," Jim says, "but it was not until the development of the steamship that long ocean voyages became commonplace and directed. Today, when we undertake space exploration voyages of discovery, it's the same: we go to learn things, to ask questions, and to leave the doors open to discoveries, some of which are inherently unanticipated." He has an example at the ready. "When Voyager suggested Europa was a neat cue ball of a place, it took another fourteen years for the Galileo spacecraft to capture the details and catalyze the theoretical considerations that led to the quest for the sub-surface oceans on this primitive object. Similar efforts have occurred for Mars and will no doubt occur for asteroids."

"But *when* do you think we will get to Mars, or any other planet?" I demand.

His baud rate surging, Jim says, "We are very clever, adaptable, and focused, but we are still shackled by aspects of the technologies we use to go there, exist there, and work there—wherever 'there' may be. We can't do it whenever we want; instead, to save fuel, we play cosmic billiards and go when it costs the least in mass. This is *not* the way to explore space. We want to be able to reach many places with new transportation technology, perhaps traveling quickly enough to take humans—just as the kind of exploration the Vikings accomplished was very different from what Columbus achieved four hundred years later."

At present, there are three basic methods of transporting people to the Red Planet. The first, and most obvious, consists of an Apollo-style mission. You send the people, let them walk around Mars for a few days, and bring them home. It's very expensive, short-term, and involves minimal science. It's flags and footprints, the kind of approach that Jim (and many others at NASA) would like to avoid. A group of experienced engineers at Johnson Space Center favors a different scheme, tentatively set to begin sometime in the years 2014–2018. Their plan would use the International Space Station as a weightless staging area to assemble the components for a mission to Mars. Jim personally considers this strategy too risky for the crews involved in a three-year-long deep space mission. He prefers a third approach, a more deliberate, science-oriented, "punctuated" plan of development, in which succeeding missions exploit gains made along the way. Rather than sending people to Mars as an end in itself, he sees the job as part of a comprehensive plan for the exploration of space, which would include sending people to asteroids and even Europa, the moon of Jupiter thought to harbor life. Jim and others at NASA would also like to see space telescopes suspended in the magical realms between the Sun and Earth known as libration points. These are locations at which the competing forces of gravity exerted by celestial bodies cancel each other out. In these gravity-free zones, objects could, theoretically, remain indefinitely suspended, without requiring much firepower or fuel to maintain orbit.

All of these schemes involve conventional launch vehicles similar to those now in use. Garvin looks ahead to alternate methods, especially nuclear propulsion. With it, he says, "you could get to Mars in a few months, instead of nine or ten." Nuclear propulsion may sound more dangerous than conventional methods, but in some ways it's safer. If the astronauts' time in transit is drastically reduced, for example, they will be exposed to less of the lethal radiation of deep space. Jim muses about even faster routes to the Red Planet. "Antimatter catalyzed fusion is the stuff of dreams," he tells me. "It could make possible seven-day transits to Mars, but, of course, there are still serious concerns about uncontrolled fusion."

Some of Garvin's colleagues have been cultivating even more extreme ideas concerning NASA's exploration of Mars. It may soon be possible to transplant human heads onto body stems, they note. "We could take the head, keep it alive on external support, and send it to Mars—or wherever," Jim theorizes. The plan has the advantage of saving a great deal of payload weight.

I decide to take an informal poll concerning the prospect (admittedly remote) of NASA's sending human heads to the Red Planet. "That's the stupidest thing I ever heard," says my pragmatic son, when I describe the plan to him. "What would be the point?" My wife, when she hears about it, is horrified. "Unbelievable! Talk about bioethics!" she says. She fears the idea may be misused to threaten children: "Behave, or I'll send your head to Mars!" And I have a few questions of my own. Whose heads would NASA send to Mars? What would they do when they got there—and after they returned to Earth? Would they become the ultimate talking heads? Heads on Mars: I will have to get back to Jim on this one.

In a spacious office at the Goddard Space Flight Center, Jerry Soffen continues to muse about astrobiology. His speculation blooms like paper flowers expanding in water. I finally pose a question I should have asked long ago: does he think there are other worlds with life as advanced, or even more advanced, than ours? "Probably, *more* advanced," he says. "I see the probability of life elsewhere in the universe as a bell curve, and most likely we're somewhere in the middle of that curve. There are probably just as many civilizations ahead of us as there are behind us. We have no way of knowing, but that is what probability suggests. It's not completely conclusive that biogenesis started here on Earth. It might have started elsewhere." And he adds with a grin, "I'm a believer in the Rotten Tomato Theory of evolution."

"Which is?"

"In the Rotten Tomato Theory, a bunch of astronauts came here, to Earth, eons ago, from somewhere else, dumped their garbage, and we're the result. Can a tomato turn into a person?"

"What do you mean by that?"

"Evolution is so capricious, I can't imagine life on two different planets evolving the same way. That's like expecting two people to compose music the same way. Conditions are so varied they can't be repeated. It may turn out that biology is far more varied than what we know of it on Earth."

Jerry appears serene, even playful. He has accomplished a lot in his brief association with the Astrobiology Institute. He has set the program in motion, overseen the selection of the first round of proposals; in fact, he has done everything but find someone to replace him and actually run the Institute. As the search for a new head of the Astrobiology Institute continued, it became the proverbial honor that NASA could not give away, primarily because the position required residency at the Ames Research Center in California.

"What about you? Did you reconsider taking the job yourself?"

"I wouldn't go for anything."

Next, Dr. Baruch Blumberg was approached. A Nobel laureate, now in his mid-seventies, Baruch—Barry, as he is generally known—received his award in 1976 for his work on hepatitis B, one of the most common causes of liver cancer. After that, his career took him in many directions, and at the time he was approached by NASA about leading the Astrobiology Institute, he was master of Balliol College at Oxford University—"which was so different from my previous experience it was like going to the moon. I no longer had my own research team." Blumberg considered the Astrobiology Institute a grand challenge: "the most ambitious, daring research project anybody's ever thought of. It amazes me what NASA's prepared to do, what the country's prepared to do. I also liked the fact that it was multidisciplinary, a step in the direction that contemporary science *has* to take. It has been very reductionist, to good effect; we have discovered a lot of good stuff that way, but in the future, science will need to integrate different areas into fields such as astrobiology." He took the job, which meant that Jerry Soffen's work with the Astrobiology Institute was finally complete.

Although astrobiology sounds remote from medical research, Blumberg plans to apply the principles of virology to the search for life beyond Earth. Both are on the nano scale. He has given considerable thought to the fossil-bearing meteorites from Mars, examining them as a virologist, and he suggests that they do, in fact, contain evidence of extraterrestrial biological activity, as advertised. "One of the early criticisms leveled at the nanofossils was that they are too small to be artifacts of life," he reminds me. I remember that issue very clearly. "They are not too small. They are the *exact* dimensions of the hepatitis B virus, for example." While he stops short of saying the Martian meteorites bear proof of life, he feels they contain something that cannot be explained away. More than anything else, he wants to get his Institute to work on a sample of Martian soil, although the prospect of bringing a bit of Mars to Earth has alarmed some scientists, who fear its potential harmful effects on terrestrial life. It is reassuring to know that Blumberg's experience as a virologist tells him it is unlikely that life on Mars, if transported to Earth, would be harmful to us.

Shortly after his appointment, Blumberg spoke at the Goddard Space Flight Center about his plans for the Astrobiology Institute. Jim Garvin was in the audience, and by the time the presentation ended, he realized that astrobiology might be on the verge of another revolution. The next spacecraft to reach Mars, the Polar Lander, would soon touch down at the edge of the Martian south pole. If successful, the spacecraft would resume the search for life on Mars. Planetary scientists such as Garvin have become excited about the possibilities of the Martian poles—the extreme climate, the valleys and ridges, the presence of water (in the form of ice), and the evidence of volcanic activity—as potential breeding grounds for life. As Garvin puts it, "The poles are hot!"

"Go to the Jet Propulsion Laboratory for the landing," he urged me. "It will be great." He was saying something very similar, I recall, just before Mars Climate Orbiter fell off the screen.

On December 3, 1999, a thousand journalists descend on JPL to cover the arrival of Mars Polar Lander. This plague of information locusts has come equipped with laptops and cameras and microphones and cell phones and Palm Pilots clattering against each other and bulging in their pockets in technological priapism. Everyone is drawn by a single thought: this is a chance to watch space history in the making. Although the crucial events will take place more than one hundred million miles away, we will be among the first to know, if not to understand.

From time to time, America looks to NASA for national self-definition, never more explicitly than in 1969, when men—American men!—first walked on the moon. Thirty years later, NASA's time has come again. The agency has increasingly supplanted the military's impact on the national character, and NASA's culture of exploration has infused American life and language and preoccupations. Now that NASA has latched on to Mars as a worthy goal, the Red Planet is auditioning today for a role in America's destiny as the next proving ground of national character. Mars is becoming part of the national agenda, the way the moon once was. The idea is noble; the idea is foolish; and on the threshold of a new millennium, the idea is compelling.

Mars Polar Lander was launched from Cape Canaveral on January 3,

1999, three weeks after the "perfect" launch of Mars Climate Orbiter. The two spacecraft were intended to operate in tandem, one circling the planet to measure its climate, the other landing near the south pole to look for water and possible oases of life. Meanwhile, Mars Global Surveyor would continue its detailed mapping of Mars, and together, the three spacecraft would interact to produce an extremely varied portrait of a distant planet. With the disappearance of Mars Climate Orbiter on September 23, the lander acquired greater importance; NASA decided Mars Global Surveyor could fill in for the lost spacecraft, and the robotic invasion of Mars would proceed more or less smoothly. NASA confidently predicted its Mars Polar Lander websites would host twice as much traffic as Pathfinder's sites, which had set a record for popularity on the Internet two years earlier. In all, the stage was set for a defining moment in planetary exploration.

Now, Jerry Soffen, the venerable project scientist for the Viking missions of the 1970s, stands before a full-scale model of Mars Polar Lander, preening in a red vest and red tie, a pitchman for Mars, ready to relive the headiest moments of his career, giving simultaneous interviews to eager journalists and loving every minute of it. As I launch into a then-and-now discussion with Jerry, a raven-haired reporter from a local newspaper cuts me off sharply. "This is *my* interview," she says, laying claim to Jerry. It's a shock to jostle with a thousand journalists; over the last two years, I've become accustomed to the pleasant illusion of having these Marsists more or less to myself. No one else seemed to care about them, but now that the Red Planet has come back in style, all these strangers have shown up, flashing their credentials. Although I feel displaced, initially, I eventually experience a sense of relief. At last, the reinforcements have arrived, and the euphoria proves contagious.

A few feet away, at an endless press conference, NASA engineers testify to their determination to get it right this time; they proclaim they are "raring to go." The lander project manager, Richard Cook, solemn, cherubic, and casual, describes the conditions the spacecraft will shortly encounter on Mars. "At the current time, the most recent forecast calls for gusts up to twenty-eight miles per hour. We can operate with gusts up to fifty-five miles per hour. . . . We're in good shape." He describes the MPL's crucial entry-descent-and-landing sequence: "The vehicle will accelerate up to about twelve g's, and the heat shield will heat up to nearly three thousand degrees

Fahrenheit. . . . A parachute will be deployed. . . . The spacecraft is going just under a thousand miles an hour. . . . At about six miles above the surface, the lander deploys its legs. . . . The radar comes on, and then the most dramatic part of the entry occurs. . . . The lander is separated from the spacecraft and starts its descent engines. . . . It will descend very rapidly for a landing at about five miles per hour on the Martian surface, oriented in such a way that when the solar panels are deployed, they'll produce most of the electric power."

There's a question about the surface on which the spacecraft will land, and Richard Zurek, the mission's project scientist, offers a disconcerting reply: "The area we're headed to is a low and rolling terrain. Our target point is in the sweet spot of the landing ellipse. Now, we are a little further west than we wanted to be, but that terrain is within the capabilities of landing. We have a bet among the team about how many rocks we might see. If we did see a rock, that would be interesting, because we don't expect them on this kind of surface. There are some dune fields. We're looking for much harder ground, where you expect to find water ice, and that covers ninety percent of the sweet spot of the ellipse we're headed for. We really *don't* know what we're going to see on the surface. That's why we're setting the lander down."

A little later, Zurek tells me more about the purpose of the mission: "The ultimate question is where would you find habitats to look for life? If we don't see water on the surface, it if literally looks like it's dry, that will hurt the chances of where to look for water elsewhere. If we find water ice close to the surface, then we can better predict how deep it might be at other latitudes. And if we find minerals that were formed in standing bodies of water, that suggests there was a wet period on Mars, lakes and ponds, maybe a glacier, and those findings would certainly increase the chance that life could develop on the planet." He hooks his fingers into his belt and scans the auditorium, his thoughts millions of miles away. "You know, most of the uncertainty with Mars has been that maybe the wet period didn't last very long, and maybe it wasn't all that wet. Maybe the planet has witnessed only brief episodes of water bursting out onto the surface, for a relatively short time in geologic history, and an even shorter time in the biological history of the planet." If that's the case, if liquid water might have been a transient phenomenon on Mars, the best place to look for evidence of biological activity would be the polar caps.

To get a better fix on the potentially habitable environments of Mars, Zurek and his colleagues expect to reconstruct the planet's weather history, which means the record of its seasons. "Right now, Mars has almost the same inclination—the tilt of its axis to its orbit—as the Earth does, so it has the same kind of seasonality. Summer is different from winter. But the inclination of Mars could change as much as ten degrees, and that's huge." The Earth, in comparison, can change by only a degree or so. Zurek believes that Mars' changing inclination may be responsible for some of the ice age variations he's seen on that planet. To reconstruct a more detailed climate history, MPL will dig down into Martian soil, a layered terrain composed of alternating tiers of dust and ice. Much can be deduced from the thickness of the layers, in a manner analogous to studying the thickness of tree rings to understand terrestrial climate history. Even the particles of Martian soil trapped in the ice will have a story to tell about ancient conditions on Mars.

I should mention that Mars Polar Lander is a very different spacecraft from Pathfinder; it's lighter and, dare I say, flimsier. It doesn't use air bags, and it's not intended to bounce around the surface of Mars like a beach ball. Zurek explains that Mars Polar Lander isn't equipped to handle the rocky terrain Pathfinder explored. "We're expecting to land on a surface that has very few rocks. The air bags used by Pathfinder, while they protect you from some things, do have their own problems. They're very massive. Our lander is about half the weight of Pathfinder. Also, if the spacecraft sits on air bags, it's difficult to dig through them. You could get stuck out there and not be able to reach over them." That situation almost happened with Pathfinder, as I recall. "The lander gives you a better platform," he claims.

The mission's investigators take to the stage at JPL to describe the goodies with which they have stocked their spacecraft. There are the two phallic penetrators that will detach themselves from the mother ship during descent and smash into the surface of Mars to look for subsurface signs of water. There is a miniature oven to heat a soil sample and examine it with a tiny laser for signs of water. MPL also carries a microphone to capture the sound of the Martian wind; a scoop to prospect Martian soil; a camera; a weather mast to measure the polar climate and detect water in the atmosphere; radar;

sensors; and a stereo imager. Planetary science doesn't get much happier than this gleaming, fully loaded spacecraft.

The good vibes stay with me as I walk over to a large tent pitched in the JPL courtyard, where they're selling Mars Polar Lander T-shirts, sweatshirts, coasters, models, and coffee mugs. The crowd is so dense I can't get close enough to purchase souvenirs of the great event, the return to Mars at the dawn of the new millennium. Nearby, there's something more promising, a capacious white pavilion with flashing lights inside. I duck in, expecting to see a late-model Cadillac or Lincoln on a turntable. There is, instead, a full-size replica of Mars Polar Lander on display in a simulated Martian sandbox. Surrounded by curious humans, the spacecraft looks like a visitor from another planet, an artifact of a civilization far more advanced and elegant than ours. It appears to be half-machine, half-insect, about three feet high and twelve feet long, with three spindly legs tapering down to pods the size of a dinner plate. A robotic arm protrudes like a proboscis from what might be its abdomen, which contains various instruments loosely wrapped in a wrinkled gold foil skin. No one dares touch it; the creature looks intelligent and otherworldly. It's easy to believe this magic machine came from Mars, not Earth. If this gleaming and vaguely menacing replica of Mars Polar Lander began walking about the grounds of JPL and giving orders to everyone, I would be only mildly surprised.

Turning from the throng, I encounter Arden Albee, who looks thoughtful and subdued. "I hope it works," he growls. "I was on the review panel for the mission a few weeks ago, and it didn't look very hopeful. Then I was on another panel, and things looked better." He shrugs imperceptibly and does not appear convinced. Arden holds a minority opinion, I assume, until I encounter a familiar NASA engineer, who tells me that once he looked carefully into what Mars Polar Lander was supposed to do, he realized it would be a miracle if it worked. His devastating opinion, so casually expressed, reminds me of the vague reservations about Mars Polar Lander that Maria Zuber and Matt Golombek expressed a few weeks earlier, at the MOLA meeting. ("Honestly, I just don't know about MPL," said Matt. "No one knows.") At the time, I thought their comments were just gossip, but what if they were reliable scouting reports?

The packed auditorium, when I return to it, is alive with clicking shutters, whirring electronic equipment, and harsh television lights, and I realize no one here has a clue about the precariousness of the mission.

The Deep Space Network keeps in touch with Mars Polar Lander; everything is nominal, we hear; the spacecraft is right on its trajectory until it enters the critical entry-descent-and-landing sequence. Since it takes fourteen minutes for a radio signal to reach the spacecraft, real-time tracking takes on a new, elastic meaning; an instant lasts almost half an hour. The auditorium, crowded with journalists and NASA scientists and engineers, falls quiet, waiting to see the results of four years of work. Again, it all comes down to a single question: *"Do we have a mission?"* The journalists trance out in the silence, contemplating the vastness of space. At this moment, an aura hovers over JPL.

At 12:39 P.M., the first opportunity to acquire a signal from the spacecraft comes . . . and goes.

The milestone passes so quickly and quietly, it is almost unnoticed, just another unremarkable moment in time. It takes a few minutes for the realization to settle in that *something has gone wrong with the mission.* Still, there's no sense of imminent disaster, merely bewilderment and boredom in the face of a technical glitch, and it will take a while for folks to realize that the glitch isn't just a glitch—it's the failure of a quest.

In the half-empty auditorium, Richard Cook appears alone on the rostrum, resembling a stage manager appearing in front a curtain to announce that the performance has been . . . delayed. He insists it's much too early to give up hope, they may have already found the spacecraft's signal on a different frequency—if only the explanation were that simple—but as he talks he reveals that the putative signal is just a bit higher than the background noise level. Faint hope, indeed. "The 12:39 interval," NASA drones with relentless bureaucratic impersonality and optimism, "is only the first of several communications opportunities over the weekend"—the *weekend!* It hasn't occurred to the press that it could take the entire *weekend* to locate Mars Polar Lander— "where we might possibly hear from the lander for the first time. Any of several factors could delay first contact without preventing the lander from establishing communications and carrying out a full mission." He drones on about their plans to continue listening for the spacecraft during various "windows" as if this state of affairs—a missing spacecraft!—were strictly routine.

I wander out of the auditorium, past the tent, now deserted. Earlier, I couldn't fight my way to the souvenirs; now they can't give them away. Alighting in the JPL cafeteria, I bump into Bruce Banerdt, who has been following Mars Polar Lander, and he tells me he's baffled, it could go either way, but, he cautions, after the first six or twelve hours of searching for a signal, the likelihood of making contact drops off dramatically. I check my watch; over an hour has passed with no sign from the spacecraft. As I slowly fill a paper cup with luscious swirls of frozen yogurt, an imposing woman who works in upper middle management at JPL leans over and whispers, "I've never prayed for a spacecraft before, but I'm praying for this one now." I believe she would like me to join her in prayer. All around me, the talk slowly turns from the signal not heard to DVDs, HDTV, CD-ROMs and other reassuring acronyms, and then to the upcoming weekend, and finally to family concerns. It's just another day at JPL.

Back in the auditorium, the NASA engineers on the podium maintain their game faces. I huddle with Matt Golombek, who experienced the uncertainties of the Pathfinder landing. For now, red is the color of anxiety. He slouches darkly in his chair next to mine, glancing sideways at Cook, muttering, "I don't envy him." Cook, meanwhile, insists his team feels confident they will hear from Mars Polar Lander eventually; he even insists they expected this contingency. "We are a long way from being concerned," he says, with apparent sincerity, and I am a long way from believing him.

Matt shakes his head slowly, and says, "When Pathfinder landed, we picked it up immediately. We had it when it was still on the parachute, we had it on the bounce, we had it when it came to rest. It happened so fast I couldn't believe it; I thought it would take longer."

He realizes what they're up against in this mission. They're not just trying to land on Mars, but on one of the planet's most difficult landscapes. "The landing site is so different from any place we've ever been, it's hard to say what it's like, exactly. There's no Earth analogue for the Mars south pole." The more he talks about the problems involved in trying to land on the Martian poles, the more hopeless he sounds.

I recall the string of accomplishments over the past two years— Pathfinder, the orbital insertion of Mars Global Surveyor, the incredibly detailed mapping of the Red Planet, the discovery of Martian magnetic fields and plate tectonics, the observation of ghost craters—and realize once again how dangerous (and inevitable) it is to start taking success for granted. I flash back further, to childhood memories of NASA rockets—Vanguard, Mercury Redstone, Atlas Agena—collapsing in flames on their launch pads or veering out of control until they exploded, flaming fiascos televised in black-and-white. And now Mars Polar Lander has become another public failure. There aren't any German rocket aces on hand to rescue NASA these days, only their heirs, the aging Trekkies from Caltech and MIT, running down their decision trees, calmly assuring the press they expect the lost spacecraft to turn up at any moment. I hate to see the mission end this way. I know perfectly well that the success of previous missions matters more, much more, than the failures of Mars Polar Lander and Mars Climate Orbiter, but for now our hunger for Mars will feed on frustration.

It's growing dark, and I decide to suspend my vigil and leave JPL. If, by some unlikely chance, something turns up on Mars, I will return, but after my conversations with Matt and Bruce, I have come to suspect that nothing will. As I drive off into the December gloom and merge with ordinary rush-hour traffic, a sense of melancholy comes over me so palpably that I can taste it on my tongue, for I know that NASA won't fire another bullet at Mars for two years, that it will be almost three years until another American spacecraft attempts to land on the Red Planet.

That night, every news program leads with the loss of Mars Polar Lander, as they do for the next several days, holding out hope that those ingenious JPL engineers will recover their spacecraft. Despite these assurances, Mars seems farther away than ever. Going in, I knew that half of all Mars missions ended in failure. The four I have followed—Mars Pathfinder, Mars Global Surveyor, Mars Climate Orbiter, and now, Mars Polar Lander—have conformed to the statistical norm, but the perilous history of Mars exploration won't mitigate the public humiliation the agency has already begun to suffer. There is ignominy in failure of this magnitude, but there is also nobility, not that it is much in evidence or appreciated in the rush to chastise NASA, to turn wizards into incompetents. Of course, the people of NASA are neither; they are engineers and scientists and civil servants trying to solve problems, and they only look like wizards when they succeed and incompetents when they don't. Reviewing two years of intense planetary exploration, I have learned, perhaps belatedly, how humiliating space can be. There are the mundane humiliations of budget cuts and slipped deadlines, and then there are the spectacular humiliations of lost missions. I recall Jim Garvin's heartbreaking description of his reaction to the loss of Mars Observer in 1993: he cried openly. Now I know how he felt.

As the extent of the Mars Polar Lander disaster becomes apparent over the next few days, the mission's failure is mocked by Jay Leno on television ("Have you heard NASA has a new book released today? *Men Are from Mars, Women Are from Venus, Where the Hell Is the Polar Lander?*"). Outside JPL, an entrepreneur sells models of Mars Polar Lander smashed to bits on the surface of the Red Planet. "They should put one of these on the desks of everyone at JPL!" a scientist tells me. Meanwhile, newspapers offer a varied menu of speculation concerning the reasons for the loss. Potential culprits include Goldin's "faster-better-cheaper" regimen, Lockheed Martin (which built the spacecraft), the Office of Management and Budget, Congress, the cost of the International Space Station, JPL engineers, unexpectedly rough terrain on Mars, and high winds on Mars. There have been problems over the years with the separation of spacecraft from the cruise stage just before entry, or so an astronaut tells me. It is possible that MPL never recovered from a faulty separation and went tumbling into space. Meanwhile, voices at JPL continue to insist they will hear from Mars Polar Lander eventually; it just has to be there, somewhere, hobbled by technical glitches, but the voices of hope grow fainter day by day.

Mars Global Surveyor flies over the landing site, its cameras looking for evidence of MPL—a long shadow or a parachute—but sees nothing. Perhaps the spacecraft never even made it to the surface. Dan Goldin appoints a large panel, this time under the stewardship of Thomas Young, the former chief of Martin Marietta, to evaluate Mars Polar Lander. Maria Zuber, one of three scientists serving on the Young panel, discovers that meetings begin as early as 7 A.M. and often continue until midnight. On one occasion, she finds herself tied up in conference calls for ten hours, "and at the end of the day," she says, "I had to wonder what I'd learned." On March 28, 2000, the committee issues its report. It contains no "smoking gun," as Maria puts it, only a list of probable reasons for the mission's failure, the most likely of which was the premature shutdown of the spacecraft's braking engines as it was descending to the surface of Mars. Instead of lightly touching down on the surface, it might have plummeted the last few hundred yards and shattered on a distant world. Or it might have tumbled down a steep slope to destruction. The list of possible reasons for the failure of Mars Polar Lander is depressingly long.

The whole exhaustive review, Maria says, "raises really troubling questions about how something like this could have happened." Now that she has a clear idea of the risks involved in the Mars Polar Lander mission, she sounds indignant that NASA went ahead with it. Jim Head has a different perspective about the chances of failure; if a robotic isn't risky, it may not be worth doing. "If you're going to an area of a planet that's as smooth as a billiard ball, what's the point?" he asks. "Is there any science in that?" Dan Goldin and other proponents of "faster-better-cheaper" make a similar argument, saying that launching more (cheap) spacecraft makes more (cheap) losses inevitable and even acceptable. But the public outcry over the failure of Mars Polar Lander suggests that when it comes to Mars missions, especially landers, the number of acceptable losses is zero.

In this failure are sown the seeds of future Mars missions. After the Young Report, NASA tries to regroup and recover its Mars program and public support. As a start, Jim Garvin receives an appointment as the new Mars Exploration Program Scientist. This comes in addition to all his burdens as a research scientist and the Chair of the Decadal Planning Team. He abruptly moves from Goddard, the home of scientific research, to Headquarters, the home of management, planning, budgets, and other matters foreign to many scientists. He won't have much time to study Martian craters now, or to pursue the other scientific research he loves so deeply. In giving up his passionate career as a Mars scientist, he hopes he can help to direct Mars exploration for NASA, and for the ultimate end users, the American public.

"It's hard to look back now and ponder the naïveté we all had about Mars," Jim reflects in the aftermath of the loss of Mars Polar Lander. "My expectations were colored by Viking, a dramatically successful mission. I thought I knew what to anticipate, but I never suspected Mars would throw so many curves at us. It all started with Mars Pathfinder, during which I spent a lot of time with the media, commenting on the success of this mission. Then we had Mars Global Surveyor come to life in orbit. We were finally in business after twelve years of anticipation, and once MOLA and the other instruments started revealing previously unknown aspects of Mars for us, we all changed further; in my case, it energized my longstanding passion for exploring the craters of the solar system, when many of my colleagues at Goddard had strongly suggested abandoning this line of work in '93 when we lost Mars Observer." I can see them telling Garvin, "You're crazy to even *want* to do this," while he stares at them blankly. That late night in early September 1997, when we played back our first swath across Mars, changed me forever. Our first twelve thousand measurements in a weird orbit show us that MOLA worked! Those MOLA moments are very special to me. These passes showed me the world of Martian landscapes in three dimensions and captured what impact landforms really look like. Before this, no one really knew what craters were like on Mars in three dimensions. Now that I was able to understand new spatial relationships, I was stunned to learn that most polar features have been buried in ice. In January 1998, as we sifted through the small trove of data we had at the time, I realized Mars was throwing still more curves at us, and by June, with dozens more passes over North polar craters, I saw a view of Mars with huge swings in climate, which filled craters the size of Manhattan and as deep as the Grand Canyon in one short epoch, and then emptied them. What a planet! Just give us another six to twelve months, and the revolution will continue. We have only scratched the surface of Mars."

Garvin's prediction proves to be absolutely accurate. Within a few months, he sends tantalizing hints concerning a big discovery made by Mars Global Surveyor. "Mars lives and bleeds," he says, but won't reveal more for fear of jeopardizing his new job at headquarters. Speculation about the discovery begins to appear everywhere—on television, in the newspapers, and all over the web—and NASA schedules a press conference. I ride the Metroliner to Washington, D.C., and arrive at NASA headquarters on June 22, 2000, just in time to attend. Garvin, the new Mars Exploration Program Scientist, occupies the stage along with some other Mars scientists, but today's featured player is Mike Malin, the camera genius. It's been a particularly rough stretch for Malin since the loss of Mars Polar Lander in December. I've heard that three of his best people quit his small company, Malin Space Science Systems in San Diego, saying there was no future in NASA's exploration of Mars. But now, here he is: *Malin redux!* He has quite a story to tell, and it will appear on the cover of the June 30, 2000, issue of *Science;* having passed the journal's referees, it has been certified as serious stuff, not mere speculation.

Over the course of the previous six months, Malin and his assistant, Ken Edgett, studied some of the 65,000 images the Mars Orbiter Camera has acquired, and a few showed rare landforms—enigmatic, sometimes snakelike gullies or ravines cut into Martian slopes. They appeared to flow downhill from a collapsed area at the higher end, sometimes referred to as an alcove, and to fan out downhill into a formation called an apron, consisting of redistributed debris. The most dramatic picture, a composite of several images acquired on April 26 and May 22, 2000, focused on Gorgonum Crater, located on Mars' rugged Southern hemisphere, and showed textbook examples of gullies, alcoves, and aprons. Another image, of a meteor impact crater located in the Noachis Terra region of Mars, showed apparent water erosion and debris flow. When geologists see these formations on Earth, they know from experience that they are looking at evidence of flash floods. To demonstrate just how Earth-like these landforms really are, Garvin, when his turn comes, displays a photograph of the flanks of a volcano on Surtsey; it was taken during our visit there two years ago. The shapes—the gullies, the alcoves, the fans—are so similar to Malin's Martian images that Garvin must explain that he is showing landforms *on Earth*.

In sum, Malin has presented compelling evidence that liquid water recently flowed across the surface of Mars. Scientists generally agree on the presence of substantial quantities of frozen water at the poles, but this is very different: this is *liquid*. Not only that, but the apparent lack of dust accumulation on the gullies suggests that they are new—so new that Malin and Edgett raise the possibility that liquid water may exist within the upper surface of Mars even now: "They could be a few million years old, but we cannot rule out that some of them are so recent as to have formed yesterday." The idea of liquid water flooding the Martian desert is painful for them to accept, because they have always resisted the idea. Edgett says, "I was dragged kicking and screaming to this conclusion," and Malin admits, "My confidence, based on thirty years of studying Mars, was severely shaken by this discovery." Since the finding comes from confirmed skeptics rather than true believers, it is all the more striking.

Malin and Edgett say they think the liquid water resides not far below the surface of Mars, at a depth of about 300 to 1,300 feet, in aquifers similar to subterranean reservoirs. The interesting question is what's keeping it in a liquid state, given how cold the surface of Mars gets at high latitudes. But Martian water may behave quite differently from water on Earth. For example, it may be salty, and salty Martian water would freeze at lower temperatures than we are accustomed to seeing on Earth. Instead of icing over at 0° C, it may not freeze until the temperature reaches -20° C or even -60° C. And if it's moving, even very slowly, the freezing point could be lower, since the motion of the water would retard formation of ice crystals. All of this helps to explain why water could exist in liquid form in such a cold environment.

As it courses beneath the surface of Mars, the liquid water apparently emerges from cracks in cliff faces and evaporates almost instantly. In so doing, it cools the ground (already bitterly cold), and eventually freezes into a dam of ice. The water pressure builds until the dam breaks and a flood bursts forth. Since Mars' atmospheric pressure is one hundred times less than Earth's, the liquid water that has emerged begins to boil explosively within minutes, then quickly sublimates into the atmosphere, and ultimately, into space. These Martian flash floods are likely brief, violent episodes; you can picture them as speeded up, time-lapse photography of terrestrial torrents.

Despite its unusual behavior, the presence of liquid water makes the prospect of life on the Red Planet more plausible than ever. "On Earth," says Bruce Jakosky, the astrobiologist, "where you find liquid water below the boiling temperature, you find life." He won't insist that the evidence of liquid water offers proof of life on Mars, "but it is the 'smoking gun' that tells us Mars has all of the elements for life." More specifically, the discovery helps to explain the presence of carbonates and nanofossils in those three Martian meteorites of widely varying ages. Since Mars has had liquid water throughout its long history, it makes sense that evidence of water would appear in the meteorites, and since water is generally associated with life, at least on Earth, it also makes sense that nanofossils would appear in the carbonates locked inside the meteorites.

All in all, this really is a major discovery.

As far as I'm concerned, finally coming across the elusive Mike Malin is almost as good as discovering liquid water on Mars; maybe there's some synchronicity at work here. He's a serious-looking guy with a neat beard and spectacles and a dense, complicated expression. I approach, take a seat, shake his hand, reintroduce myself (he apparently has no memory of our fleeting encounter at the launch of Mars Climate Orbiter, but of course everyone at NASA would rather forget that), and right away we start to discuss the momentous question of life on Mars. "I do not believe there is life on Mars. I do not believe there ever was life on Mars," he says, but then, he once thought liquid water on Mars was impossible. His voice is youthful and reedy, like that of a kid who has just won first prize at the science fair. He suddenly evokes José Jiménez, a character popularized by a comedian named Bill Dana on *The Ed Sullivan Show* in the 1960s. José Jiménez was a bewildered, often terrified Mexican astronaut who spoke with a heavy accent. Malin quotes his favorite José Jiménez line, which concerns the likelihood of life on Mars: *"Maybe on a Saturday night!"* Malin breaks into laughter at the memory of the routine. Who knows? Everything about Mars is changing so quickly, and José Jiménez's guess is as good as anyone's.

When I ask Malin why he suddenly decided to appear today, he tells me, gallantly enough, "I finally have something to say to the American public, which has paid a lot of money to study Mars." Still, Malin is almost alone in thinking about the cost of the Mars Global Surveyor mission at the moment. One thing I've noticed: when there's a successful Mars mission, no one worries about the cost. I can't recall anyone complaining about how expensive those two Viking missions were in the 1970s because they were *successful,* but when there's a failure—Mars Polar Lander, which cost a fraction of Viking— all anyone can think about is how much money was "wasted" in the attempt. Then Malin practically apologizes for his pictures because they aren't as beautiful as those celestial visions from the Hubbell telescope. "I've suffered, because I just show people rocks and dirt." He says the media hasn't paid much attention to him, but then, he hasn't paid much attention to the media. "Everybody loves *Star Wars* and *Star Trek,* but when you show the public the International Space Station or the Space Shuttle, they don't seem as exciting as aliens." As a kid, Malin says, he was "a space cadet." Later, when pursuing his graduate studies at Caltech, "I came to the realization that you really do need to understand the technology of your experiment if you're going to be able to interpret things. I was a geologist by training, but I like technology, so I learned technology along the way." As a result, he has forever changed our impressions of Mars and its potential for life.

In the weeks following the announcement, I await the inevitable chorus of scientists criticizing the results, the methodology, NASA's Mars program, and anything else challenging the persistent image of Mars as a cold, dead planet. Mike Malin expects it, too; he almost seems to welcome it. Instead, prominent voices throughout the Mars community endorse the findings. There are suggestions that these enigmatic gullies might have been caused not by water but by some other kind of fluid partly consisting of carbon dioxide, but few people are betting on that idea; water remains far and away the most likely explanation. Now that Mars Global Surveyor has made the cover of *Science* three times, the mission is finally getting its due. MGS has rewritten the books on Mars, just as Garvin once promised, and it will affect the course of human exploration, as well.

The presence of liquid water near the surface of Mars makes sending humans to the Red Planet seem a lot more attainable. If accessible, the water will lower the mass a human mission would have to carry to Mars, and that means the overall cost of the mission will decrease. On Mars, astronauts will find 1,001 uses for the local water. Oxygen extracted from the water will enable humans to breathe. The water can also be used to shield humans from the danger of lethal radiation on the surface of Mars, whose thin atmosphere affords scant protection; water surrounding a human shelter with adequate thickness could serve as a safe haven from the radiation horrors of deep space weather. Once purified, water could also be used for irrigation in space agriculture to enable astronauts to grow their own food during extended stays on Mars. Finally, the hydrogen in the water can be used to manufacture rocket fuel to bring the astronauts home.

Garvin, along with many others at headquarters, will now try to mesh NASA's robotic and human missions to Mars, but he won't have the final word on the subject of sending people to the Red Planet. For three decades, Johnson Space Center has been the focal point for human space flight, and those who work there assume that whatever they say will become NASA policy. They won't necessarily welcome the imposition of another approach from above, not even from a new cadre of Mars program leaders at NASA headquarters. Any human expedition to Mars inevitably begins in Houston.

Chapter Twelve

MARS OR BUST

"I opened my eyes upon a strange and weird landscape. I knew that I was on Mars; not once did I question either my sanity or my wakeful-ness. I was not asleep, no need for pinching here; my inner consciousness told me as plainly that I was upon Mars as your conscious mind tells you that you are upon Earth. You do not question the fact; neither did I."

—EDGAR RICE BURROUGHS, *A PRINCESS OF MARS*

There's a rock-and-roll show going on here, somewhere. I hear the thumping bass, the shimmering hi-hat, but I can't see any musicians. I'm at a good old-fashioned Texas-style barbecue, with acres of tables, hosted by NASA for the aerospace industry and pretty much anyone else who shows up at Building 9 at the Johnson Space Center outside of Houston—way outside, a $60 cab ride outside—in the Clear Lake area, which is another way of saying "next to nowhere." If I keep going south, I'll eventually get to Mexico, but at the moment, I'm standing in a building the size of a stadium. The exterior is completely featureless, a giant cement box. If you drove past it, you might guess it holds natural gas, quarantined livestock, or something else you wouldn't make a beeline for. When you walk inside, however, there's a welcome blast of air-conditioning, the reassuring thump-thump of the band (they must be around here), and a life-size replica of the space shuttle, as well as glittering components of the new International Space Station, which may yet become one of NASA's most important projects. A few people are milling around; they are the nervous, eager-to-please NASA hosts identified by name tags, obviously expecting guests, who are in short supply at the moment, despite the barbecue, the hidden band, and—is that John Glenn over there? The astronaut-turned-senator-turned-astronaut is in glad-to-greet-you mode, but he's mostly overlooked, outdone by the scale of the surroundings, lost in the crowd. He stands beside a cavernous booth reminiscent of a carnival attraction, trying to get the hang of virtual reality. He studies a device you put over your head in the manner of a helmet. It covers your eyes and displays a computer-generated animation of space. Wearing it is like having a nightmare from which you can't awake. What could this simulation possibly mean to Glenn, who's just a couple of weeks away from the space shuttle launch that will return him to the real thing? Although his training has been covered by national television, no one here looks at him twice.

The overwhelming spacecraft serve merely as a backdrop for the Johnson Space Center's annual "Inspection Days," a giant trade show, science fair, and community-liaison event. This is a time of year when Johnson throws a party, and the festivities sprawl for miles across the Center's length and breadth. A surprising amount of technology developed by NASA is for sale here. Or it can be licensed. NASA will even give it away under the right circumstances. There's the rock band now, around the corner, dwarfed by the space shuttle looming over them; they look as if they will enter the mother ship after their next song and take off. The musicians look a bit mature, and the guests probably don't realize that the drummer, James Wetherbee, has completed four flights on the space shuttle and serves as the deputy director of Johnson and that the rest of the band consists of astronauts who are better at rocket science than chord changes. At least, I hope they are.

A somber-looking scientist strides through the crowd. He ignores everyone; his mind is obviously elsewhere, millions of miles away. On his lapel, he wears a small white button emblazoned with red letters.

MARS
OR
BUST

The phrase means something around here—something that NASA is not exactly comfortable advertising. "Mars or Bust" is the rallying cry of an active, if scantily funded, group of scientists, administrators, and engineers dedicated to sending people to Mars. This is, in fact, a skunk works. In one form or another, this group has been knocking around NASA for thirty years, back to the halcyon days of the Apollo program and even before, when the agency had all the money and goodwill it wanted, when anything seemed possible—putting a man on the moon, sending people to Mars, exploring the farthest reaches of the solar system, mapping the universe. That bold era now seems a distant memory, and the "Mars or Bust" members function as a stealth group these days, endowed with a quiet conviction about the importance of putting people on Mars. They have grown old in the service of this idea; they never imagined it would take so long to get to Mars, but they know the journey is inevitable.

"There are four things we have to do in order to send humans to Mars," Dan Goldin told me not long before my visit to Johnson, and he enumerated them in rapid-fire fashion. "First, we have to figure out how people can live and work safely and efficiently with ever-increasing productivity in space; that's why we're building the International Space Station. Next, we have to narrow the search for where to look on Mars. Where is there good geology? Where is the possibility for water? How do you generate resources to live off the land? We need to figure out how to do all this from a very remote location. Finally, do we work and play well with others so that Americans don't have to support the whole mission? When we answer those four questions we'll be ready to go to Mars." But Congress may not. The American taxpayer may not. The Russians may not. For the time being, the Mars group works in a semi-official capacity, shielded by bureaucratic euphemisms. One is BEO, Beyond Earth Orbit. That refers to distant territory, beyond the moon; it includes other planets and asteroids. HEDS—the Human Exploration and Development of Space—is another euphemism, and it refers to human space flight. (No one says "manned" anymore.) When the time comes to send people to Mars, and it's coming much sooner than the American taxpayer realizes, the guardians and elders of that program will be found right here, at Johnson, wearing their white and red "Mars or Bust" buttons.

Humboldt Mandell wears the button. He is a slight, fair, polite man, dressed in black and white, wearing gold-rimmed spectacles. He could be a preacher—except for the button on his tie. "We've been ready to start a Mars program at NASA since the early sixties," he says. "When I first came here in sixty-two, one of my jobs was working on sending humans to Mars. We just presumed that you went to the moon, and then you went on to Mars. At the time, we thought we could reach Mars eight or nine years after the moon. We had this idea to build a space station that would enable us to explore space around Earth and develop cheap transportation to get to and from the space station. It turned out to be more expensive than we thought. The next step, 'only' twenty-five years later than we expected, is going to Mars. Every time we have done this, we've started with a new crew of people, and so we've had to reinvent a lot of things, rerun the trajectories, find out how long it takes, when we can go, how much energy it takes, how much rocket fuel, the overall parameters of the mission from an engineering standpoint. We did this during the sixties, again during the seventies, the eighties, and now we're doing it once more in the nineties. The same group of people has come back together. We live together, we work together, we occupy this whole floor here. This time we think we've got a fair shot."

I wonder. There's no budget for the project. Humboldt tells me they rely on "advance mission funds" disbursed by the Office of Space Flight at NASA Headquarters, and the amount, compared to the cost of an actual human Mars mission, is tiny, just a few hundred thousand dollars a year. They frequently participate in budget exercises, requesting allocations in the tens of millions, or in the hundreds of millions, so they can begin to develop the mission in earnest, but the money, so far, does not seem to be forthcoming. "We're at a probationary period in NASA," he explains. "The International Space Station is our final exam. If we flunk, there are people who will say that NASA isn't the right outfit to take us to Mars."

Unlike the Apollo era, when a national mandate drove the space program and the size of the budget was incidental, the folks at Johnson Space Center have learned to live within budget constraints, or so Humboldt says. "We understand we're not in an Apollo world anymore," he declares with the zeal of the reformed sinner who has taken the pledge never again to squander precious resources. In practice, this means drastically changing the way NASA does business with the contractors who build the hardware to send people into space. "We found that the most economical way to produce high-quality stuff is for us to be a very smart customer, to write down exactly what we want in terms of the performance of the hardware, then have a competition to pick the right guys to do the job, then get the heck out of it," he explains, reasonably enough. Some of his colleagues discovered that companies were producing the hardware they would need at *one-sixth* the price they were thinking of paying. The cost became so inflated because

NASA has a number of special requirements. "You have to trace the materials all the way back to the mine and give us a mine ID. You have to know every person who ever touched it in any way. You have to be able to trace all these steps through the manufacturing process, yet all you really care about is if it works when you get it."

Developing hardware for space, while challenging, is fairly straightforward compared to preparing people for space. As Humboldt puts it, "The greatest technology challenge is the human being. We don't know what happens to humans when they're out in space and they can see the Earth only as a dim, blue speck in a field of stars. We don't understand human psychology." For one thing, Johnson engineers fret about the limitations of communicating with astronauts on their way to Mars. The time lag, over twenty minutes for a round-trip signal, severely diminishes the ability of Mission Control to intervene in the event of a problem. "The whole paradigm of the Mission Control Center is to do things in real time," Humboldt remarks. "That will have to change. We'll have to give the astronauts more software to enable them to troubleshoot their systems, more ability to make repairs on board. Today, worst case scenario, if something blows up on the Space Station, you de-orbit, and you're only ninety minutes away from home, but with Mars, you can be six or seven hundred days from home. So our philosophy is: make Mars the *second* safest place in the universe for people to live."

The first people to set foot on Mars, he assures me, will likely be American. The notion of an international crew, akin to those aboard the space shuttle and the International Space Station, may be politically correct or expedient, but it's not practical. "We've introduced international cultures into known environments, such as the space shuttle, and we've had problems with that. The question is: on the first mission to Mars, do you want to risk the additional complexity?" However, I gather the first planetary travelers will consist of both men and women. "I think that gender mixing introduces some uncertainties, but the positive influences outweigh the negatives. People are used to living in a mixed-gender environment. If you suddenly make it all male, you might be introducing a greater constraint." Humboldt stares at me in his matter-of-fact, we've-got-it-all-figured-out way. There is a solemnity about him reminiscent of a Grant Wood portrait, but he is not unfriendly, merely focused. Our future is his present; it's all been carefully thought out, rehearsed, and it's ready to ship. One of the last things he mentions is a fellow named Mike Duke, "probably the preeminent planetary scientist in the world. Now he's come back to be our chief scientist; he's our bridge to the scientific world. He was just here a few minutes ago."

I expect Mike Duke, when I catch up with him, to talk about science; instead, he talks about money. He talks about maneuvering. "Dan Goldin," Mike grumbles, "has this amazing capability of saying, 'I'd really like to do it for this much.' And when you give him a plan saying you will do it for that amount, he says, 'I really meant half that much.' As long as people bring him a plan for doing *anything* at a given amount of money, he knows that he hasn't pushed hard enough. It's a game he plays, and the name of the game is, 'What is the least amount of money you could imagine that it would take to get to Mars?' "

The robotic missions, Mike reminds me, will continue at two-year intervals and will prepare the way for people. They will experiment with making propellant in situ on Mars, extracting water from the soil, and measuring radiation; in short, they will lay the groundwork for the first human colony on Mars. The key is water. "Water means propellant. If you have water, you can make it into propellant. If you don't have to bring it from Earth, you save three-quarters of the cost of launching."

One last thing: Mike wants to drill deeply into the surface of Mars to look for evidence of life. "I know it sounds a little far-fetched when you consider we only got three meters into the lunar surface with Apollo, but say you drilled two or three kilometers into Mars—and it looks like that may be within the realm of possibility for a human mission—you may, at those substantial depths, find evidence of life."

Doug Cooke wears the button. To the extent that any one person embodies the embryonic human mission to Mars, he does. Doug grew up in Houston, and the year he watched lunar landings on television, his teacher was named Mrs. Moon. "That's what got me going. I mean, I signed up back then. John Glenn was one of my heroes." Doug is a young, bearded, energetic engineer, and he's convinced we're actually going. He spends his days working out how the most ambitious program of exploration in human history will be accomplished. Once you talk to him for more than a few minutes, you begin to say right along with him, "We're actually going."

"A mission to Mars will be very long compared to what we do right now, on the order of nine hundred to a thousand days long," Doug says. "That's because it takes about six months to get there and six months to get back. To make it an efficient trip, you have to leave at the right time, and you have to wait until it's the right time to come home. So that leaves you on the surface for about four hundred and fifty days. Because of that length of time, it's important to have highly reliable, autonomous systems, to a higher degree than we currently fly. To understand the reliability, we're going to want to do a fair amount of testing, not just at places like the moon, but at the International Space Station. We'll use the Space Station to test out technologies, to work out the bugs." And don't forget the moon; it will still have its uses. Astronauts can go there to assess equipment and techniques for longer missions, and get home within a matter of days. It will become a staging area, the celestial equivalent of White Sands, New Mexico. Doug also tells me there's still a lot to be learned about the moon. "The moon is a four-billion-year history book," he says, "and it has not been altered by running water or wind. Its features were present when it was formed, modified only by impacts from meteors and other bodies. Scientists can learn a tremendous amount by examining the layering of the surface and what that history is. It tells us the history of solar wind."

Doug's enthusiasm proves contagious, so I ask him how long before we send people to Mars. "Given a budget, we could go as early as 2012," he says, adding, "We could brute force it right now." The technology to accomplish the journey already exists, in his view; only the budget is lacking. There is one other concern, he allows, and that's the effect of cosmic radiation in deep space on humans. "We don't worry about it much in Earth orbit because it's deflected to a large degree by the Earth's magnetic field, but once you leave the Earth's protection, you're exposed to it," he says. "We're also going to measure radiation on the surface of Mars so we understand the effect of the Martian atmosphere. The problem with that particular type of radiation is that it consists of heavy ions that are traveling at tremendous speeds. If you try to shield that radiation with lead, as you normally do, those particles are heavy enough to hit other particles and send off secondaries, and they are many times more harmful than the initial radiation. Actually, it turns out that you can protect the crew with water or hydrogen, but we need to understand the effects of that radiation on tissue and brain cells." The harmful effects of radiation—which are inevitable, given the crew's level of exposure—raises another serious issue: the medical care of the astronauts. "The space shuttle can come home anytime," Doug reminds me. "And on the International Space Station, we'll have a crew return vehicle that allows the crew to come home if they have a medical emergency. But when we go on a thousand-day mission, they're going to have to take the diagnostics with them."

Suddenly, he's on to a new subject: nanotechnology. "If we can make smaller, more advanced computers, then we can carry all the computers to Mars that we need in a relatively compact box. And they should be wireless so that you don't have strings all over the place and create more weight. They won't use much power because they're so small. And if you don't use much power, you don't generate a lot of heat, and so your thermal systems come down." The reduction in the equipment's size has a way of simplifying everything. Not only will it weigh less and occupy less space, but it will contain fewer moving parts and consume less energy. "Like I said, we could brute force a human mission to Mars with large vehicles and heavy systems, but the technologies that we look for are going to make it more affordable and less risky."

He points out that one of the great benefits of nanotechnology will be the addition of an "inflatable habitat" for the astronauts to use on Mars. This means they won't have to remain confined within the cramped spacecraft for the duration of their stay on the surface. They will be able to stretch their legs and occupy what is in essence a dirigible designed to shelter them. Compressed for travel, the habitat will fit closely around a core eleven feet in diameter; expanded, it will be twenty-five feet in diameter, and twenty-six feet long. The astronauts will call it home or, at least, base camp. It will be surrounded by a water-filled jacket to protect the crew from perilous doses of radiation. Doug Cooke smiles. He's got it all figured out.

Doug Ming wears the button. Slender as a beanpole, Doug wears cowboy boots, khakis, and sunglasses as we walk across the Johnson campus. It's 99°, a typical Houston afternoon; you break into a sweat the moment you leave air-conditioning. Hell, they say, is a lot like Houston in the summer, except that Houston is more humid. Yet astronauts-in-training are running across the open fields, glistening with sweat, displaying superhuman indifference to the midday heat. A large, blue-green jet zips low overhead, smearing the sky with a streak of kerosene. "The Vomit Comet," he remarks, without bothering to look. "It's the plane astronauts use for weightless training. They get forty seconds of weightlessness when it flies in a parabolic arc." He walks on, as I imagine people heaving in the sky.

Doug is a life-support specialist, engaged in testing systems for astronauts to use in space, especially on the moon and Mars. That description hardly begins to convey his level of commitment to sending humans to distant planets, for he also participates in tests of the equipment, as if he were an astronaut himself. Not long before we met, he was part of a crew that volunteered to spend an entire month in an experimental chamber at Johnson, testing the life-support system on which he has worked. He was, in a manner of speaking, his own lab rat.

He tells me about the moment the chamber test began: "We walk inside the chamber . . . the morning is of course pretty hectic . . . making decisions to go into the chamber . . . getting all of the details worked out . . . it was actually a relief when we closed the door . . . of course, we all looked at each other, and that's when we said, *'What the hell are we doing in here?'* Once we got over the shock of the door closing and the managers locking it after us . . ."

"You were locked in?" I ask.

"We could, in theory, go out any time we wanted. We were in there of our own free will, as volunteers; it was not like a spacecraft, where, once you're launched, you could not open the door and return. We immediately began to settle in and make sure all our systems were functioning."

He accompanies me to the chamber itself, located in the imaginatively named Building 7 at JSC. From the outside, the chamber looks like a huge tank, three stories high. Within, it resembles the gleaming interior of a submarine, but it feels slightly more claustrophobic. I can't imagine spending a full week there, myself, let alone the months required for a transit to Mars. On the lower level, you are confronted with the work and rest areas; there is a conference table, chairs, a microwave oven, a sink, and even a washing machine: the studio apartment of the future. Other domestic touches include a television, oddly reassuring in this context, and of course ubiquitous computer monitors. You mount a ladder to reach the second story, largely given over to life-support equipment. In the penthouse, you find painfully compact sleeping quarters, each with a bed, a desk, and a few narrow shelves.

He proudly shows me the tiny lavatory that forms the heart of the chamber's recycling system, and it is quite a marvel, in its way. The commode holds plastic bags to collect solid human waste, which eventually winds up in fourteen-ounce plastic bottles. The occupants refrigerate the bottles, until a batch is placed in an airlock. Outsiders retrieve them, mix the contents with water collected from the 22,000 wheat plants in the chamber's agricultural annex, and incinerate the whole mess. The process yields both water and carbon dioxide, which are returned to the wheat plants, and the oxygen produced by the plants is piped into the test chamber proper for the occupants to breathe.

Water is also recycled, of course—not only urine and water used in washing, but condensed sweat, all of which is purified in a dandy little subsystem known as the biological water processor, which contains microbes that feed on the pollutants. The water receives additional purification in the trickling filter bioreactor, another miniature microbial purification plant. Finally, the water enters the reverse osmosis system, which removes inorganic pollutants. The system yields thirty gallons of potable water a day. Only the faint metallic trace of iodine hints at its past, and Doug tells me that no one sickened from drinking it during his stay. "It was purified, treated with iodine, and when it was pronounced safe, consumed. I'm a research chemist by training, and when I have a detailed analysis of water telling me it is purer than anything within the city limits of Houston, or any municipality in the United States for that matter, it does not bother me." My expression reveals that I would find drinking water derived from urine, sweat, and soapsuds distasteful, to put it mildly. "I suppose *some* people have psychological problems when they think about that, but our crew did not," he says. "Whenever people ask me about this, I tell them the water they're drinking was probably from somebody's urine upstream, and who knows how many times the water has been recycled? That's what the Earth does."

It is entirely possible that interplanetary space travel will be more disgusting than we ever anticipated.

"Privacy was at a minimum," Doug says of his stay in the chamber. "We did have some privacy in our little cubbyholes for sleeping, and we had a computer, so we were able to extricate ourselves from the rest of the crew to some degree. But during the entire day, there was no real privacy. We were constantly monitored from the outside by television and control personnel, so if you're up and doing something, or reading, or in the little meeting area, somebody was watching you."

During the thirty-day-long experiment, each crew member—two men and one woman—was supposed to follow a rigid, carefully planned schedule, but unexpected problems often played havoc with their timetables. "There were nights when all of us were up working, making sure the systems were running. That was during the first two weeks; during the last two weeks, we rarely had to get up during the night, so we kept normal hours, eight to five. In the evenings, if we got thirty minutes to an hour of downtime—well, that was just about all we got. We tried to take Saturdays and Sundays off, but that wasn't too successful. We were really quite busy inside the chamber."

Exercise developed into a major preoccupation. The crew members trained for at least an hour a day, as if they were astronauts in transit to Mars, fortifying themselves against the depredations of zero gravity. Once they were finished with the exercise bike, the free weights, and the treadmill, they had to make sure the system was handling the CO_2 they produced. The technology involved is the venerable Sabatier reactor, which converts carbon dioxide into oxygen. The Sabatier reactor has been around, in one form or another, for more than a hundred years, and when astronauts go to Mars, their lives will depend on it.

The chamber test demonstrated one thing quite clearly: the more you think about going to Mars in detail, the more you realize how poorly adapted humans are to space travel. It's not just the lack of breathable atmosphere, potable water, and food in space. It's the lack of gravity, the deadly radiation, and the psychological effects of separation from Earth. Whenever the Sabatier reactor broke down, a frequent occurrence during the test, the crew was up half the night fixing it. That was stressful and unpleasant, but not life-threatening during a test. Yet, if the same malfunction occurred on Mars, or in transit, the same situation would become a matter of life and death; if astronauts did not succeed in repairing their reactor, they could suffocate. There are about fifty additional systems necessary to sustain humans on Mars. Even with redundancy, one system or another is likely to fail, so the astronauts will be busy making sure they will survive. During their days in space, they will be repairing, cooking, and gardening. In fact, they will engage in space agriculture, which will be necessary to produce both food and breathable air.

Doug tells me that NASA has been developing space agriculture at a variety of centers—at Johnson, Kennedy, and Ames, to name a few. As people move off the planet to the moon, Mars, or even longer journeys beyond Earth orbit, they will have to take plants with them; agriculture will form an essential element of space travel. Plants have been adapted to various conditions on Earth; now they are being adapted to sustain human life in space. Wheat has already been engineered for space growth. Tuskegee University in Alabama is developing a sweet potato for space consumption. There are special varieties of potatoes under development at the University of Wisconsin and the Kennedy Space Center. "We have food systems here at Johnson based on the plants we plan to take into space," Doug says. "We've done the food testing and the taste testing." The sheer variety creates problems for space farmers. "When you grow five or six different plants in the same chamber, they start having interaction problems. We have to study what happens when you put peanuts and wheat together, and no doubt there will be some variables that will be difficult to understand."

In space, astronauts will consume bread baked from the wheat, which reoxidates the air; they will nibble lunar lettuce, sample native Martian tomatoes, savor space beets, taste Martian mushrooms, and feast on solar system sweet potatoes. "You cannot believe what you can do with sweet potatoes!" Doug exults. "You can make sweet potato brownies. You can even make vanilla ice cream out of what they grow here; it's just amazing what they do with this stuff." I ask whether NASA will allow astronauts to drink alcohol millions of miles from home. "That's not part of the diet, but when you are on a mission that lasts a year, it has to be taken into consideration. There's no problem brewing beer, for example, right inside our chamber. Believe me, it's been discussed." There will be no cuisine like zero-gravity interplanetary cuisine, now being served in the advanced life-support laboratories of the Johnson Space Center. Travel beyond Earth orbit will present opportunities for brand-new, genetically engineered Gardens of Eden.

After encountering various members of the Mars or Bust society, I now know that a number of NASA veterans have concrete, well-thought-out, unfunded plans for sending people to Mars, and it is equally apparent that few people outside of NASA realize the extent of the plans. Nor does everyone inside the agency endorse their ideas. "A lot of us think they must be smoking stuff from the seventies," Jim Garvin tells me. Now, that is a planetary scientist talking, protecting his robotic turf. Like any self-respecting planetary scientist, Jim knows that the moment a human mission to Mars is announced, the science component will become subservient to the human component, which is precisely what happened with Apollo. Flags and footprints! "My own point of view is that a cheap-and-dirty approach to sending people to Mars would be a waste, merely a technology demonstration," he says. "If there were an outstanding astrobiological reason to send people to Mars, I'd say, 'Let's do it!' Still, I would rather have a robotic sample return. What if there were a fungus on Mars, and humans brought it back with them to Earth, and we suddenly had a *War of the Worlds* situation, where it was harmful to life on Earth?"

Despite these hazards, astronauts have frequently advocated Mars as NASA's next big goal after reaching the moon. Shortly after returning to Earth in 1969, Neil Armstrong, the first person to set foot on the moon, confidently portrayed travel to Mars as the next logical step in space exploration. His crewmate Michael Collins published *Mission to Mars* (1990), a detailed dramatization of the rigors of travel to the Red Planet. This matter-of-fact account emphasizes the tasks astronauts must perform in space. Collins advocates yet another way to reach the Red Planet; in his scenario, the spacecraft, after leaving Earth, heads toward Venus, whose gravitational field propels it toward Mars, and a stay of about forty days. According to Collins, a Venus-Mars trajectory would take about fifteen months, and it would require less fuel than a more direct trajectory. Collins' story is not without interest, but he portrays the journey to Mars as essentially another Apollo mission, only much longer. Others perceive far greater implications—historical, philosophical, and cultural—in a human mission to Mars.

Carl Sagan considered the journey to be humanity's destiny and salvation. "Our ancestors walked from East Africa to Novaya Zemlya and Ayers

Rock and Patagonia, hunted elephants with stone spearpoints, traversed the polar seas in open boats 7,000 years ago, circumnavigated the Earth propelled by nothing but wind, walked [on] the Moon a decade after entering space," he wrote in 1994, near the end of his life, *"and we're daunted by a voyage to Mars?"* Quick to seek analogies where none may exist, he compared a journey to Mars to human migration from one continent to the next; it would be necessary if the human species is to survive its own destructive impulses. Even allowing for the expense and sacrifice involved in the journey to the Red Planet, he wondered aloud, "Isn't it possible to make a better life for everyone on Earth *and* to reach for the planets and stars?" He thought he saw humanity's great chance for interplanetary expansion about to be realized in the form of a "coherent justification for human missions to Mars. I imagined the United States and the Soviet Union, the two Cold War rivals that had put our global civilization at risk, joining together in a far-seeing, high-technology endeavor that would give hope to people everywhere. I pictured a kind of Apollo program in reverse, in which cooperation, not competition, was the driving force, in which the two leading space-faring nations would together lay the groundwork for a major advance in human history—the eventual settlement of another planet."

After Sagan's death, the role of cheerleader-in-chief for sending people to Mars fell to a former aerospace engineer named Robert Zubrin, who has strenuously devoted his energies to making human expeditions to Mars appear plausible, straightforward, and safe. He has written *The Case for Mars: The Plan to Settle the Red Planet and Why We Must,* a beguiling manifesto that updates and vastly expands upon von Braun's *The Mars Project* of half a century before. In the space travel advocacy group he has founded, the Mars Society, Zubrin has energetically lobbied on behalf of a plan he calls Mars Direct, which means, simply, sending humans straight to Mars, without stopovers at the International Space Station to assemble equipment or at a lunar staging area. Zubrin has NASA's ear, if not its imprimatur; the agency dispatches a contingent of administrators and scientists (such as Everett Gibson) to the Society's annual convention, which attracts a combination of fringe activists, restless space enthusiasts, and mainstream aerospace engineers and planners. Zubrin emphasizes how cheap his Mars Direct plan will be. "A rough cost estimate for Mars Direct would be about $20 billion to develop all the required hardware, with each individual Mars mission costing about $2 billion once the ships and equipment were in production. While certainly a great sum, spent over a period of ten years it would only represent about 7 percent of the existing combined military and civilian space budgets." It should be noted that NASA's "faster-better-cheaper" regimen approach does not, at present, extend to human spaceflight; the space shuttle and International Space Station consume billions each year in the course of simply maintaining crews in low Earth orbit, a far more modest goal than sending people to Mars. Even the zealous members of the "Mars or Bust" skunk works tend to look askance at the Mars Direct framework; it seems too brief and too expensive, given the effort and risk involved. They prefer to integrate the journey with other NASA projects, especially the International Space Station, and with overarching themes, such as the search for the origins of the universe.

At a symposium at the Space Policy Institute in Washington on November 22, 1996, Dan Goldin emphasized that he has been a Marsist all along. At that moment, NASA was basking in the afterglow of the discovery of possible nanofossils in ALH 84001, and he wanted to capitalize on renewed interest in the Red Planet. "We had a vision for the moon and we accomplished it. I think the vision for Mars has to be broader than the vision for the moon, and it has to be something that deals with the need to have a permanent presence on a planet that is not in Earth orbit; it cannot be a feel-good mission to get our adrenaline flowing and then drop back. We need to have a sustained presence."

NASA has gathered all the elements for a human mission to Mars—the rockets, the life-support systems, the Mars Rover, the habitat, the in situ propellant plant—into an elaborate Reference Mission. Although the Reference Mission still has many spaces marked "To Be Determined," it is NASA's blueprint for the future of space exploration. After realizing how serious Johnson's plans to send people to Mars have become, I return to the Kennedy Space Center in Florida to find out more about the rockets and launch activities described in the Reference Mission. There, Cristina Guidi eagerly describes how she has been preparing to convert KSC to a Mars launch center—pending funding, of course. She's got the charts, she's got the numbers (which she won't show me), and, most of all, she's got the conviction that human missions to Mars will soon make Kennedy busier than it has ever been.

Trained as an electrical engineer, a veteran of a decade in operations at NASA ("Operations deals with times and schedules, making sure everything's meeting their milestones"), Cris has become the advance person for Mars operations; in NASA-speak, she is called the exploration technologies project manager. She talks in rapid bursts, but her firm, unwavering gaze conveys intensity and the conviction that this is the way things will be. Her eyeballs appear to be wired to a 10^{21} gigabyte hard drive. Her manner seems to be saying that in the next ten minutes she will convince me that sending people to Mars is practical, efficient, at hand. "We do have to build a new mobile launching platform," she begins. "We will need a new one because we have to assume that the shuttle will still be flying. We will need more infrastructure." Although she hesitates to say it, the word "infrastructure," in space exploration, can be considered synonymous with "money."

"We have come up with a facility utilization plan to process the required elements to support each launch campaign. This is a very tight campaign, and it's a little confusing." I anticipate that she will clarify it for me, and she does. "In the first launch campaign, we will send all the equipment before the crew goes out. It's going to take us four launches, using Magnum heavy-lift launch vehicles." Magnum is a fitting name. They are nearly as big as the Saturn rockets used for the Apollo launch; each is 325 feet tall, as high as a seven-story building, and 99 feet in diameter. They look like the space shuttle blown out of proportion. "You're essentially launching two payloads and the Trans Mars Injection stage, which is really the propulsion stage for your two payloads," she says. "Once those four elements get in low Earth orbit, the appropriate payload assembles to its Trans Mars Injection stage to form a little vehicle. Then those two vehicles get onto a trajectory to Mars. Once the burn is complete, the propulsion stage drops off, and the payload continues on to Mars."

When the payloads reach Mars, one—the Earth Return Vehicle, which will eventually take the crew home—stays in orbit around the Red Planet. The other payload, the cargo vehicle, contains a propellant production plant, the science equipment, an inflatable habitat, and the rovers. It descends to the surface of Mars, and first thing, the propellant plant starts making fuel for the astronauts to use. Getting all of this material off planet Earth and up and running on Mars will take at least two years to complete. Finally, the crew will ride the space shuttle into low Earth orbit and rendezvous with the Trans

Mars Vehicle that will take them to Mars. The last phase sounds rather complex. Why don't the astronauts simply take the Magnum launch Vehicle into orbit around Earth? "Because the Magnum is not human-rated; it's an expendable," Cris says. "The cost associated with human-rating is astronomical, no pun intended." Once the crew occupies the Trans Mars Vehicle, they will spend about six months to reach Mars orbit. On arrival, they will descend to the Martian surface, close to the cargo vehicle, where they will stay for about 500 days.

All of this launching, transiting, and orbiting will start to happen before we know it, according to Cris, who displays a chart placing the launch of the first Magnum heavy lifters in 2011 and the launch of the crew in the latter part of 2013. Under this scheme, the crew will arrive on the Red Planet some time in 2014, probably in one of the flatter areas of the Northern lowlands. When their tour of Martian duty ends, the crew will climb into the cargo vehicle, which contains a small capsule awaiting them. By this time, if all goes according to plan, the propellant plant will have manufactured enough fuel to launch the capsule into orbit around Mars. The crew will then rendezvous with the Earth Return Vehicle that has been orbiting the Red Planet, awaiting them. They will transfer from the capsule to the Earth Return Vehicle and head home. By that time, subsequent human missions will already be under way. To hear Cris describe them, these complicated maneuvers sound simple, logical, and inevitable.

NASA is skittish about naming a price for sending humans to Mars, but the most frequently heard estimate comes in at around $55 billion. The agency wants to find partners to share the financial burden, but they may not be the obvious choices. Russia's space program is in disarray, and the Japanese are not inclined to be cooperative. The Italian, French, and English space agencies may all make contributions, but it could turn out that the most important partners will be international corporations rather than governments. NASA might decide to sell or license television rights to the mission to a commercial network, but that would merely be a beginning. "We're assuming our program will be ongoing," Cris tells me. "It's not going to be just a few quick steps and flags on Mars, and then we're done. It will be a colony." It is possible that Mars will ultimately be settled by companies rather than countries, and politics will trail behind commerce.

I remark that some elements of the elaborate scheme Cris Guidi has outlined sound speculative, and another NASA official curtly remarks, " 'Speculation' is probably a bad word."

"And why is that?" I ask him.

"Because, it's not just speculative."

If you wish to be selected by NASA as an astronaut, you can apply for the job. The pay is low, the work is repetitive, the hours are long, the hazards are great, and there is every likelihood you will never get off the ground. But there is a chance you will get to go to Mars in about 2014. Out of every 20,000 applications it receives, the agency invites about 2,000 to fill out a lengthy and detailed application form. Of those who do, about 200 receive a phone call inviting them to Houston for an interview. But it's not just an interview; you spend a week in Houston, and much of the time is taken up with medical tests. The actual interview lasts only forty-five minutes, and you face a board of astronauts and senior NASA managers, eight of them, and one of you. They want to see how you deal with the situation.

And then, nothing happens.

You go back to your daily life. The months pass, three, four, five months, and you try to forget all about it; you think of the long odds, knowing that only thirty people will eventually be selected, thirty out of the original 20,000 applicants. You tell yourself you could do just as well, maybe better, buying lottery tickets. And then one day a man from the FBI comes around, walking up and down your street, asking your neighbors questions about you, making them very suspicious about what you've been doing to attract such scrutiny.

Finally, you get the phone call.

They will call you at home, and if you are not at home, they will call you at work. Or they will find you if you are on the road. You hear your name being paged at the airport, as you are changing planes, sipping a cup of Starbucks coffee, and you call airport information, and they tell you to call a number. As soon as you hear the area code, your stomach tightens, because it's Houston. Suddenly, you recall hearing that if a key board member calls you, you're in, and if it's someone less important, you're out. Or you may get a call from the Chief Astronaut. If the Chief Astronaut phones, you can be pretty damn sure you made it. So you call the number in Houston, and it's not the Chief Astronaut, it's an assistant, and he puts you through, and before you recognize the voice you hear the words, "Are you still interested in coming to work in Houston?"

A day later, you are there, in training, and training will last several years. Completing the basic shuttle course, for example, takes twenty months, a combination of academics and simulations. You fly the vomit comet. You learn how all the systems work; you hear about them till you dream systems. *The power systems, the valves, the emergencies, the alarms. How do I rewire this? How do I reroute the problem without damaging other systems?* The technicians pile on the events until you are sorting out multiple failures, because in the real world, one thing rarely goes wrong all by itself, it's usually a combination of things, and the process of problem solving proceeds from a science to an art. They send you to Russia so you can familiarize yourself with their operations, as you watch the once mighty Russian space program collapse. Then they send you to the Kennedy Space Center, to inspect the International Space Station, where everything is gleaming, new, efficient, functioning. Meanwhile, you await your flight assignment with forty other astronauts in training, each one hoping to go first. You might be a specialist, or an engineer, but no matter what your background, you try to be adaptable. You count the launches—forty-three required for the International Space Station alone—and you wonder if NASA will ever get around to sending people to Mars. You ask yourself: would you volunteer for this rigorous, life-threatening, three-year-long mission?

You bet!

If you are lucky enough to get this far, you will have to be willing to accept the risks involved in a planetary journey. The most important part of it will occur right at the start, the eight and a half minutes from liftoff to Earth orbit. You will spend at least a year preparing for those eight and a half minutes. You will concentrate harder during those minutes than you ever have. And you will have to get used to the risk. As an astronaut, you will bet your life. Perhaps you thought about the risk when you were chosen to be an astronaut, but you didn't dwell on it because you were too excited. You think about it again when the training starts, but not all that much. You think about it most of the night before you go up, but by then it is too late to change your mind. You tell yourself you could die in the morning, and you ask if it's worth it, and you worry because you love your wife, or your husband, and you love your children dearly, but, you tell yourself, this is all you have ever wanted to do since you were a kid. And if you die doing this, if the rocket blows up, you know that you will have spent the last minutes of your life trying to save the rest of the crew and the vehicle, and that you died trying to do the thing you most wanted to do.

When you wake on the morning of the launch, you aren't particularly elated because you are finally going into space; there's too much to think about. You think about simple things, about taking a shower, and not slipping. You think about putting on your liquid-cooled underwear and placing your flashlight, knife, and radio in the correct pockets. After you suit up, the camera team comes in, and you give them your fake astronaut smile, but you don't feel like smiling, because there's too much to think about. You walk with the rest of the crew to the Mars shuttle, everyone in close order, you have been told, so the pictures of the smiling, happy, confident crew will be in focus, and then, as you strap yourself into your seat, you begin to think about those eight and a half minutes, and what you will do if the liquid rocket fails, if the computer fails, if the solid rocket fails. If the solid rocket fails, you can't do anything, you will probably die. You play a mental trick to avoid dwelling on the possibility. You think about specifics, and you keep on thinking about specifics all the way into orbit. And then, in zero gravity, you start playing a fresh set of mental tricks on yourself to deal with a new situation: Space Adaptation Syndrome.

"It's not like motion sickness. Your digestive tract shuts down," says Mary Cleave. A scientist turned astronaut, she's flown two shuttle missions; on her second, she helped to deploy a planetary spacecraft, Magellan. "When you get sick, it's like you've been poisoned. You don't know it's coming, and all of a sudden, *wham!*, you have to put a bag over your face. It's got a wash-and-dry so you can clean yourself off afterward. I got sick on my first flight, some people are sick quite a while, but with me, it went real fast. Someone got sick before me, and I am not real good with the smell."

She had to face other problems on that flight. "I didn't want to use my diaper. Your toilet training is incredibly strong, and it's not pleasant to wet your pants. So, the week before the flight, the crew dehydrated itself so you don't have to use a bag. If you're a male, and you're in a pressurization suit, you have a urine collection device like airplane pilots have, but it can be an extremely painful experience. The second time, I didn't get dehydrated, and I didn't get sick, either."

Mary discovered that living for an extended period in zero gravity paradoxically requires *more* rather than *less* effort. "When you're weightless, you have to make sure you don't get into the middle room and get stalled. You have to be able to push off from somewhere; if not, you get stuck. You feel like a fool. You call, 'Help, I'm stuck!' and someone will come and get you. You also have to learn not to push off from a wall too hard. My first couple of days in weightlessness, I had all these bumps on my head because I was just having a great time. I'd push off, slam into a wall. Push! Slam! You have to watch that you don't flip around too much; you work up to it gradually, or else you can get motion sickness. It's the ultimate freedom. I'm five-foot-two, and for once in my life, I didn't need a stool, I didn't have to worry about being too short for anything."

Throughout the trip, Mary warns, astronauts will have to concern themselves with the dangers of radiation. Without the protection of the Earth's atmosphere and magnetic field, they will be exposed to the full force of the Sun's radiation, which may become the single most dangerous hazard of the journey. "It's a nasty place out there," she says. "Cancer rates are higher for astronauts. You wear a dosimeter, which you're supposed to keep on your body at all times, to track the amount of radiation you receive. You can get permanently retired based on your radiation dosage. I've heard about a crew member who was concerned about his total dosage, so he took off his dosimeter and stuck it in a compartment; when that crew came back, there was one person whose dosimeter reading was a lot higher than everyone else's, and NASA wanted to know why. His dosimeter had picked up a lot of extra radiation in that compartment."

You are going to have to be able to get along with your crewmates each day, all day long, in close quarters, for three years. "Your interactions with the rest of the crew get to be what I consider familial kinds of relationships rather than your typical professional relationship, because you're winter camping, in essence. Your environment is very hostile, and you have to stay all bundled up. So you tend to be much more forthright in that situation. You get your pecking order set up. You have your little areas that you take care of. Before you go up, you practice tooth extractions, for example. You take medication for prophylactic reasons, to make sure there's no adverse reaction if you take it in space." There are immediate physical changes to contend with. In zero gravity, the fluids in the human body tend to redistribute upward. "You look at yourself in the mirror when you're in space, you look puffy in the face, and everybody has little skinny chicken legs. It does wonders for your chest in space. All the guys love it, because they look so fit.

"One other thing that's strange when you get into weightlessness is that you realize we have a very definite bias about how we approach things. For example, looking at Earth, we have a north bias. When you get up in orbit, you float toward the instrument panel as though it were up, but you're actually upside down relative to Earth. For the first couple of days, you get disoriented, and then you lose that bias. You're comfortable approaching Earth—or Mars—from any angle. You don't have the bias of north, and up." Despite the discomforts, the tedium, and the danger of the mission, Mary would go to Mars if she got the chance, like all astronauts. "If I went to Mars, I would like to try skiing across the polar ice caps. It sure beats driving a golf ball like Al Shepard did on the moon."

Ron Parise's longest journey in space lasted seventeen days. Zero gravity made each one difficult. "Not only do the muscles that you use to walk and move around atrophy because they're doing very little work in a weightless environment, but also your heart muscle, which does not really have to do much work, becomes weak," he says. "You actually lose muscle mass in your heart. Here on Earth, your heart works very hard pumping that blood that's down in your feet back up into your upper body. You get in space, all of a sudden the heart is just sitting there pumping away and it has no work to do, so your whole muscular system starts to suffer as you become weightless." That poses a serious problem for a space shuttle mission, which lasts only a couple of weeks, and when the crew returns to Earth, they have the benefit of immediate, massive medical attention aimed at rehabilitation. A mission to Mars will be much more difficult for astronauts to withstand. It will take six to nine months to reach the Red Planet, and once they get there, they won't be greeted by doctors and nurses, there won't be an opportunity for rehabilitation; they will be expected to work. No matter what precautions the crew takes in space, no matter how much they exercise and prepare, their arrival on Mars will be an ordeal.

"Your skeletal system is another problem," Ron says. "As soon as you become weightless, your body decides that it doesn't need strong bones anymore. It's not holding up anything. Calcium starts to leach out of your bones. Taking calcium supplements doesn't help because your body has already decided it doesn't need the calcium. There has to be some way of counteracting that, but right now we don't know how. We have a treadmill that we used to fly in the shuttle; it had a harness with bungee cords that would pull you down with about one hundred and twenty pounds of force.

"Then there's the psychological problem of separation from the Earth for a long, long time," he adds. We really don't have any experience with that situation; we don't know how astronauts will react so far from home. "Even if you put somebody in the Space Station for a year, they're still in real-time video and voice communication with people on the ground," Ron tells me. "At least once a day, you can see your wife on television or talk to your kids, but when you get out toward Mars, round-trip communication time becomes immense, so you can't have a voice conversation." You can imagine the stilted conversations between the spacecraft and home this situation would create. An astronaut says, "Hi, honey, how are you?" and twenty minutes or more will pass before he hears the reply: "Doing fine." Or, "Not so good." They will probably wind up recording a longish report, or an e-mail, which may only reinforce the sense of separation and isolation.

In space, simple comforts won't be simple anymore. In a zero-gravity environment, just falling asleep turns out to be a complex physical and psychological chore. "On my shuttle flight, I had a lot of difficulty sleeping, at first," Ron says. "The problem is that even though you may be mentally exhausted from working a twelve-hour shift, your body has been doing essentially nothing, just floating there, and you aren't physically exhausted, so it's difficult to sleep. When I started spending an hour on the exercise bike a day, working as hard as I possibly could, I started sleeping much better. Plus, it's an unusual feeling to be in a sleeping bag in a little compartment and completely weightless. It's sensory deprivation. When you're in your bed on Earth, although it may be pitch black, you have a sense of lying on the bed; you know where you are. When you're weightless and inside a sleeping bag, you are floating. You have no visual sensory information and you also have no sense of feel of where you are, of up and down. You become completely disoriented. You don't know whether you've rolled over. You have no idea until you open the door or turn on the light for visual reference."

Sheer tedium presents another hazard for astronauts. "When you're busy, you don't dwell on where you are and the problems you might have, or the fact that you're stuck in this little can with these same people all the time. I recall, on my last mission, that we had to delay our landing because of weather. We had to shut down all the payload operations, and we had absolutely nothing to do. We had no film left to take pictures out the window. We had nothing. That was a very slow day and a half. You float there looking out the window, and when you have nothing to do but think about where you are, you start thinking about the fact that there's the Earth out there and I'm two hundred and fifteen miles away from it, and there's a vacuum right outside the window. On a trip to Mars, you're going to have a lot of time to think once you get into the cruise phase. Sure, you will have a few medical experiments to do, and there's going to be daily ship maintenance, but you're not going to be real busy. And it will take nine months to get there." Despite the hardships, the expense, the time, and the risk to life involved, Ron believes in the necessity of going to Mars. In fact, it's an article of faith with him. "Let me put it this way," he advises. "I think it's an awesome undertaking and it would be an incredible adventure, and if I were young enough and single enough, I'd do it in a minute."

There will be another level of problems entirely, problems that few of the public—and very few people at NASA—will ever know about: the intimate lives and emotional needs of the astronauts. The crew will include women and men, as I've heard. Some will be single, some married; some may even be married to each other. NASA will have to face the prospect of sex in space.

If you talk long enough with astronauts, you hear the rumors about sex in space, aboard the shuttle, or even in the vomit comet during its brief weightless interludes—imagine! On a three-year mission to Mars, NASA and its astronauts will inevitably confront sex. Although sex can help to reduce stress on the crew, as psychologists are quick to point out, it raises concerns such as pairing and jealousy. If you are a married crew member, and your spouse is about 132 million miles away, and the last video conversation the two of you had was sour and unproductive, if "issues" are surfacing and there's nothing you can do until you get home two years hence, and meanwhile two other astronauts on the mission have found each other and are having relations in very close quarters, you may become a sullen astronaut, or a resentful one, as the days of your enforced celibacy slowly unwind. On the other hand, there is some evidence that prolonged weightlessness may lower humans' sex drive; perhaps no one will want to have sex in space, perhaps they will tell each other, "Not on this transit, not this year, dear. Wait till we get *home.*"

Sex in space, especially in zero gravity, will pose special problems. It will be like having sex in a crowded submarine, only more complex. If two astronauts are having sex, if they are making noise, if they are banging up against the bulkhead in the small weightless environment, all the other crew members are going to know about it. Simply docking will prove difficult, because as one body pushes against another, the two of them are liable to go flying off in any direction and land against a rigid wall. That could hurt. It will be impossible to remain motionless without some sort of restraint, and while the thought of being strapped down in order to have sexual relations may excite some, it would deter others who'd rather flail freely. If two astronauts do manage to surmount the mechanical problems posed by zero-G sex, NASA will inevitably want to study their reactions; they will want samples of fluids, they will want sperm counts—it's enough to strip the prospect of sex in space, or on Mars, of its romantic appeal. Nonetheless, it is bound to occur.

And sex in space, for all its complications, may prove helpful for astronauts. In her study of the problem of "enforced intimacy" in long space journeys, Lara Battles, a West Coast psychologist, writes, "Skin hunger and sexual deprivation are among the primary complaints of pair bonded astronauts who fly without their wives, and this very lack is one of the primary causes of loneliness and depression. Emotional intimacy is enhanced through sexual contact. Affectional needs are met, and the biochemical side effect of the orgasmic experience is a diminishment of depression and of obsessing, and a general release of 'feel good' and psychological bonding chemicals, thus enhancing couple *and* crew cohesion. It reduces stress and tension. It provides a basis for a good deal of humor. Sex by oneself is simply not the same, nor nearly as effective in combating the negative psychological side effects of space flight, as it is with one's partner."

After your nine-month transit, Mars will loom reassuringly close, only a few hundred miles away. The atmosphere is too thin to obscure anything on the surface of Mars. At the edge of the Martian night, as it creeps across the globe, faint clouds are visible, and giant columnar dust devils, many miles high, spring up without warning and obscure the surface, but in general, the view is the clearest you have ever seen, clearer than anything on Earth; every feature on the surface of Mars is diamond-hard. When the Sun coasts into the proper alignment, light glints off the polar caps. Passing silently below you, they have a peculiar splendor. They await their first bard.

As a crew member, you will check your velocity and note that aerobraking has begun. Each orbit brings you closer to the landing site in the northern hemisphere, the equivalent of the Bonneville Salt Flats. Only redder. It's another day on Mars. You are suddenly gripped by the urge to be on land; your head is playing a strange game, reactivating latent instincts to orient yourself by fixed points. Up and down—what a concept! You can suddenly see yourself walking on a solid surface, the Martian crust, as solid as the Earth's. How reassuring it would be to jump and kick without making your environment shake. How fine it would feel to pick up a rock and throw it as hard you could, and watch it disappear in the distance.

3,000 miles per hour . . . 2,000 . . . 1,000 . . . now you are traveling slower than the speed of sound . . . you might as well be walking . . . but there is little time to appreciate the sensations of entry, for these will be the busiest minutes of the flight since liftoff. You are in your pressured suit, strapped in your console, wholly engaged in a long checklist of rehearsed, memorized tasks. They follow events as closely as they can, back on Earth, but they are fourteen minutes behind Martian time.

You feel the slight kick of the guidance rockets as the lander orients itself, you look out the window, hoping to catch a glimpse of the propellant plant; without it, there will be no real mission, only an emergency return to Earth, but you can't see the propellant tower on this pass, maybe on the next. The automatic sequences proceed rapidly, and you feel the jolt of the parachute as it deploys, and the lander suddenly swings beneath it, and you feel sick to your stomach, about to vomit, and you realize, *gravity*. It's been sneaking up on you, and now here it is, all at once, and you feel exhausted, you can barely lift your arm, it feels so heavy, even breathing requires effort. You become light-headed and feel you may black out at any second, but you aren't exhausted, you are exhilarated; it's only the return of gravity that you must overcome. Fortunately, it's only three-eighths as strong as the Earth's.

There it is now, visible outside the window, the propellant plant, your ticket home. A thought flashes through your mind: we are landing in a *desert*. There are rocks, dried-up riverbeds, caverns, and boulders; it's all so Earth-like, an undiscovered world about to yield its secrets. You will employ your geological training to analyze descent images, comparing local Martian vistas with those you have analyzed on Earth. You recall training missions to the island of Surtsey and Meteor Crater, in Arizona, and realize you have seen all this before. If NASA intended to instill a sense of déjà vu in its Mars-bound astronauts, the training has succeeded. The strangeness and mystery of landing on Mars for the first time have been minimized, downplayed, routinized. There is no time now for philosophical speculation; that will come later.

The rest of the landing sequence is automatic, even the emergency procedures are second nature by now. You've rehearsed the event a thousand times on Earth and again in space, you've accomplished the feat in your dreams, settling softly onto the surface of Mars. The red rocks of Chryse Planitia rise to greet you. The spacecraft settles on the surface with a satisfying thump.

Confirmation of landing, as well as streams of audio and video, are transmitted to Earth, and eventually, Houston replies with congratulations and instructions. How strange to think that humans now inhabit two planets, yet there's no time to contemplate the enormity of the achievement. The spacecraft whirrs, clicks, and hums as it adapts itself to the Martian environments, and you check and recheck the systems necessary for survival on an alien world.

Martian gravity, though less than half of Earth's, has a persistent and debilitating effect on the crew. Those who diligently exercised during transit cope with the demands of gravity relatively well; the others pay dearly for their lack of self-discipline. Tasks that might have taken an hour to complete in the weightlessness of space require two full days of painful effort to accomplish on Mars. Stowing equipment, checking systems, making minor repairs, preparing the inflatable habitation module all require uncanny effort and willpower. The crew's emotions swing wildly between the exhilaration of landing safely on Mars and the anguish of adapting to gravity. Who would have thought that brushing your teeth would cause muscle spasms, that a wrench could weigh so much, that one astronaut would have to help another get dressed. Everyone aches, smiles, and cries in apparently random sequence; even terrestrial emotions need to be reprogrammed on Mars, at the outpost now known as New America.

By the third sol on Mars, the crew has begun to adapt to gravity, and Mission Control back in Houston okays the first extravehicular activity. With great care, you put on your space suit, pausing to rest every fifteen minutes; it takes nearly four hours to get the job done, and you are exhausted by the time you finish. But the vast Martian terrain, pitiless and pristine, beckons you. It is time to walk into a new epoch in history. The hatch swings open, and the bright Martian day surrounds you. You glance up at the distant, shrunken Sun; you glance down at the desiccated, lava-like surface of Mars. You descend the ladder, step by step, six steps in all, recording impressions for posterity, trying to keep your words—the ones they will play and replay back on Earth—purely analytical, although your mind races in a million directions. You reach the last rung on the ladder, and step down on the surface, feeling the Martian soil crunch underfoot . . .

And the paradigm shifts.

STARGAZING

We are standing on a hilltop on the island of Nantucket, thirty miles off the coast of Massachusetts, on a clear August night. The date is arbitrary, a moment in infinity, but our position is exact: 41° 17′ North, 70° 6′ West.

I am with my daughter, two old friends, their children, and two dozen other families. We walk across the grounds of a small observatory named for Maria Mitchell, America's first woman astronomer and a Nantucket native, born in 1818. Nantucket was a whaling port of worldwide influence during the nineteenth century. Navigation mattered to Nantucketers, and that meant astronomy mattered. Maria Mitchell's father encouraged his daughter's ambition to study the sky and took pride in her accomplishments. On October 1, 1847, William Mitchell wrote in his diary: "This evening at half past ten Maria discovered a telescopic comet five degrees above Polaris. Persuaded that no nebulae could occupy that position unnoticed it scarcely needed the evidence of motion to give it the character of a comet." A telescopic comet can be seen only with the aid of a telescope. Maria Mitchell was the first to discover a comet in this manner.

Looming in the dark, their windows faintly glowing, are the solid gray shingled homes of Nantucket, including the one in which Maria spent her childhood. At the bottom of the hill, a country road circles the cemetery where she is buried.

We are all looking up.

For the moment, the vastness of the sky and the multiplicity of stars silence the urgent noise of the day. The stars are scattered like shards of brilliant glass on black velvet. You cannot help but wonder what is out there. Here and there, recognizable constellations leap out at us. The Milky Way's immense smudge extends across the sky, softening the harsher contrasts. Spinning end over end, a satellite crosses overhead, reflecting light from the sun, revolving through the darkness like a celestial acrobat.

We are near the peak of the Perseids, the late-summer meteor shower, and the sky is filled with fireworks. My daughter catches sight of the faint blue streak of a shooting star. "You have to keep your head up," she says, mostly to herself. Another shooting star appears, much brighter this time, quivering and flaring. "They're like sparklers," says one of my friends. "There are so many I can't keep up with all the wishes."

I step up to the eyepiece of the telescope. It is stubby and small, a Schmidt-Cassegrain design that folds and reflects incoming light to form an image at the eyepiece. And it is computerized; you enter data indicating place, time, and the celestial object you seek, and the computer locks the telescope onto the target. I am about to look at an object ninety million miles away.

It is so bright I can see it with the naked eye. There it is now, low in the Western sky, to the left of the moon, drifting over a telephone pole. That is what I am after. Once I isolate it, the object appears to be a different hue than the blue-white stars surrounding it. The light from the stars, distant point sources, scintillates, but this object casts a topaz beam, a steady, remote search-light. Through the telescope, the object appears fuzzy at first glance, and then it gradually coalesces into a well-defined disc. It seems to float in the glass, a pumpkin-colored dot: Mars.

Several yards away, at another telescope, the moon attracts a crowd of curious families, but I stay with the mysterious little pumpkin floating in the eyepiece of the telescope. It seems to dance, to tantalize, and to provoke. If I keep a promise to myself to look for Mars on August 27, 2003, when it will be only 34.6 million miles away, it will seem as large as it ever does from Earth.

My friend Jane looks through the telescope. When she was growing up, she wanted to be an astronaut; she fully expects to reach the moon in her lifetime and to see people walk across the surface of Mars. "I see a guy on Mars waving," she announces. "He's saying, 'Loved the rover!' " My friend Ken, her husband, peers through the telescope with evident interest but without comment.

Now it's my daughter's turn to look.

"See any canals?"

"No, Dad."

"Little green men?"

"No."

"Did you see Mars?" everyone in our party asks as we leave the observatory and stumble through the darkness to our cars. "You saw Mars, right?"

Yes, we saw Mars.

"They were wonderful men, the early astronomers," Maria Mitchell wrote in her diary on the day after Christmas, 1854. "Sometimes we are ready to think they had a wider field for speculation, that the truth being all unknown it was easier to take the first steps in its path, but is the region of truth limited? Is it not infinite?"

ACKNOWLEDGMENTS

I logged about 75,000 miles in pursuit of NASA's campaign to explore Mars, visiting the agency's centers in places as diverse as Houston, Iceland, Washington, and Pasadena. Along the way, scores of people at NASA offered extraordinary cooperation and access to closed scientific meetings and conversations in which they freely discussed their speculations and motivations, their ambitions and frustrations. I am keenly appreciative of their hospitality and generosity of spirit. Despite the candor of the many individuals at NASA whom I interviewed, no one attempted to alter my opinions or conclusions in any way.

The following scientists, managers, and astronauts took the time to talk with me, generally at considerable length:

At NASA's Goddard Space Flight Center, Greenbelt, Maryland: Jim Abshire, Mario Acuna, Rob Afzal, Mary Cleave, Al Diaz, James Garvin, James Hansen, Mark Hess, Greg Neumann, Ron Parise, Claire Parkinson, Jan Ruff, James Sahli, Susan Sakimoto, Dave Smith, Maria Zuber, and Joe Zwally.

At the Johnson Space Center, Houston, Texas: Carl Allen, Jacklyn Allen, Paul Buchanan, Mark Cintala, Doug Cooke, Mike Duke, Everett Gibson, Steve Hoffman, Kent Joosten, Marilyn Lindstrom, Humboldt Mandell, David McKay, Doug Ming, Larry Nyquist, Hugh Ronalds, Piers Sellers, Kathie Thomas-Keprta, and James Wetherbee.

At the Kennedy Space Center in Florida: Omar Baez, David Breedlove, Bruce Buckingham, Mike Dunkle, Cris Guidi, William Larson, Mica Parenti, and David Taylor. Special thanks to Mike O'Neal for allowing me to visit the launch pad and control room during the launch of Mars Climate Orbiter on December 11, 1998, from Cape Canaveral Air Station, and to Patricia Marx, who brought her unique perspective all the way to Florida.

At Lockheed Martin Astronautics in Denver, Colorado: Ed Euler, Cindy Faulconer, Noel Hinners, Roman Matherne, and Joe Vellinga for their hospitality during my visit.

At the Lunar and Planetary Institute, Houston, Texas, Steve Clifford welcomed me to the Mars Polar Conference he co-hosted near Houston in October 1998.

At NASA Headquarters, Washington, D.C.: Dan Goldin, NASA's administrator, and, in addition, Baruch Blumberg, Joseph Boyce, Wes Huntress, Michael Meyer, Carl Pilcher, Jerry Soffen, and Ed Weiler for discussing aspects of their work with me.

At the Jet Propulsion Laboratory, Pasadena, California: Diane Ainsworth, Arden Albee, Richard Cook, Matt Golombek, Jennifer Harris, Norm Haynes, Doug Isbell, Bridget Landry, Rob Manning, Dan McCleese, Kenneth Nealson, Franklin O'Donnell, Donna Shirley, Suzanne Smrekar, Tom Thorpe, Sam Thurman, and Richard Zurek.

In Iceland: Asrun Elmarsdottir, Sturla Friðricksson, Bill Krabill, Jon Sonntag, Thorsteinn Tómasson, and the Keflavik Naval Air Station for hospitality during my stay in Iceland.

I also wish to thank other distinguished members of the MOLA team: Jim Head, Duane Muhlemen, Roger Phillips, and Sean Solomon.

To keep the information and concepts in this book as current and accurate as possible, a number of NASA scientists, generally experts in their fields, reviewed the scientific content. I am grateful for their help and their patience. They have done their best, and any errors in fact or judgment that have crept through the review process are my responsibility alone.

Many other scientists discussed aspects of their work on Mars and space exploration with me: Lara Battles, Frazer Fanale, Kate Fishbaugh, David Fisher, Rejean Grard, Anton Ivanov, Gil Levin, Madeline Mutch, Bill Nye (the Science Guy), John Nye, Hector Perez-de-Tejada, Joe Rayo, Jim Rice, Mark Skidmore, Pat Smith, and Becky Williams. Others who generously furthered the cause include Darrell Fennell, Curt Greer, C.M. Hartman, Meredith Palmer, C. L. Reeve, Joseph Thanhauser. Daniel Dolgin belongs in a class of guardian angels all by himself. I also wish to thank the Maria Mitchell Association of Nantucket, whose observatory and library proved both useful and inspiring. Heidi Hammel, of the Space Science Institute, made a number of helpful suggestions. Special thanks to my friends Linda and Phil Lader, who set the stage for this book. My wife, Betsy—my first reader as always—has contributed much in ways both large and small.

At Riverhead Books, Celina Spiegel, my editor, won my gratitude for the confidence she showed in this idea and contributed much to the project from start to finish. Susan Petersen Kennedy, my publisher, made it all happen in the best possible way. Sharp-eyed Chris Knutsen patiently assisted me in honing the manuscript. Marilyn Ducksworth was another welcome source of encouragement, and Erin Bush proved helpful at every turn. Finally, I am grateful to Phyllis Grann for her support.

At HarperCollins in England, Val Hudson once again contributed editorial support, friendship, and good cheer. It is a privilege to work under her good auspices. Thanks also to Leeza Morley and Monica Chakraverty for their assistance.

At ICM, I owe a big debt of thanks to my spirited and perceptive literary agent, Suzanne Gluck, as well as to her colleagues Karen Gerwin, Sloan Harris, and Caroline Sparrow. This book has benefited greatly from their advice. In London, Gillian Coleridge has my appreciation.

The Graduate Writing Division of Columbia University's School of the Arts generously provided a capable researcher, Victoria Gomelsky, under the Hertog Research Assistanceship program, and for this, Richard Locke and Patricia O'Toole have my appreciation. I also wish to extend thanks to Mayo Gray and Betsy Cummings for their meticulous transcriptions of my recorded interviews.

Finally, I must reiterate my gratitude to two remarkable NASA scientists: Dr. James Garvin and Dr. Claire Parkinson. Without the countless hours they devoted to discussions with me about Mars and NASA, and their exemplary dedication to their calling, this book would not exist.

"The scientist does not study Nature because it is useful: he studies it because he delights in it, and he delights in it because it is beautiful," Henri Poincaré, the mathematician, wrote. "If Nature were not beautiful, it would not be worth knowing. And if Nature were not worth knowing, life would not be worth living."

SELECT BIBLIOGRAPHY

Bilstein, Roger E. *Orders of Magnitude: A History of the NACA and NASA, 1915–1990.* Washington, DC: NASA, 1989.

Bradbury, Ray. *The Martian Chronicles.* New York: Bantam, 1979 (originally published 1950).

Bradbury, Ray, et al. *Mars and the Mind of Man.* New York: Harper and Row, 1973.

Burrough, Bryan. *Dragonfly: NASA and the Crisis Aboard Mir.* New York: HarperCollins, 1998.

Carr, Michael H. *The Surface of Mars.* New Haven, CT: Yale University Press, 1981.

A Clementine Collection. Washington, DC: Naval Research Laboratory, 1994.

Collins, Michael. *Mission to Mars: An Astronaut's Vision of Our Future in Outer Space.* New York: Grove, Weidenfeld, 1990.

Cooper, Henry S. F., Jr. *The Evening Star: Venus Observed.* New York: Farrar Straus & Giroux, 1993.

———. *The Search for Life on Mars: Evolution of an Idea.* New York: Holt, Rinehart and Winston, 1980.

Davies, Paul. *The Fifth Miracle: The Search for the Origin and Meaning of Life.* New York: Simon & Schuster, 1999.

Dick, S. J. *Plurality of Worlds: Origins of the Extraterrestrial Life Debate from Democritus to Kant.* London: Cambridge University Press, 1982.

Doody, David, and George Stephan. *Basics of Space Flight.* Pasadena, CA: JPL, 1995.

Ezell, Edward, and Linda Ezell. *On Mars: Exploration of the Red Planet, 1958–1978.* Washington, DC: NASA, 1984.

Ferguson, Kitty. *Measuring the Universe.* New York: Walker and Company, 1999.

Fisher, David E. *The Third Experiment: Is There Life on Mars?* New York: Atheneum, 1985.

Flammarion, Camille. *La Planète Mars et ses Conditions d'Habitabilitié.* Paris: Gauthier-Villars et Fils, 1892–1909.

Gell-Mann, Murray. *The Quark and the Jaguar.* New York, W. H. Freeman, 1994.

Glasstone, Samuel. *The Book of Mars.* Washington, DC: NASA, 1968.

Gould, Stephen Jay. *The Panda's Thumb: More Reflections in Natural History.* New York: W. W. Norton, 1980.

Hancock, Graham. *The Mars Mystery: The Secret Connection Between Earth and the Red Planet.* New York: Crown, 1998.

Horowitz, Norman H. *To Utopia and Back: The Search for Life in the Solar System*. New York: W. H. Freeman, 1986.

Hoyt, William Graves. *Lowell and Mars*. Tucson: University of Arizona Press, 1996.

Jakosky, Bruce. *The Search for Life on Other Planets*. Cambridge: Cambridge University Press, 1998.

Journal of Geophysical Research 97, no. E5 (May 1992).

Kieffer, Hugh H. et al., eds. *Mars*. Tucson: University of Arizona Press, 1992.

Kuhn, Thomas S. *The Structure of Scientific Revolutions*. Chicago: The University of Chicago Press, 1996.

Lemonick, Michael D. *Other Worlds: The Search for Life in the Universe*. New York: Simon & Schuster, 1998.

Lovelock, James. *The Ages of Gaia: A Biography of Our Living Earth*. New York: W. W. Norton, 1988.

Lovelock, J. E. *Gaia: A New Look at Life on Earth*. New York: Oxford University Press, 1979.

Lowell, Percival: *Mars*. Boston: Houghton Mifflin, 1895.

———. *Mars and Its Canals*. New York: Macmillan, 1906.

———. *Mars as the Abode of Life*. New York: Macmillan, 1908.

Ming, D. W., and D. L. Henninger, eds. *Lunar Base Agriculture: Soils for Plant Growth*. Madison, WI: American Society of Agronomy, 1989.

Mutch, Thomas A., et al. *The Geology of Mars*. Princeton, NJ: Princeton University Press, 1976.

The NASA Astrobiology Institute: A Compendium of the Science of Its Initial Set of Members. Washington, DC: NASA, 1998.

Neufeld, Michael J. *The Rocket and the Reich: Peenemünde and the Coming of the Ballistic Missile Era*. New York: The Free Press, 1995.

Parkinson, Claire L. *Breakthroughs: A Chronology of Great Achievements in Science and Mathematics*. Boston: G. K. Hall, 1985.

———. *Earth From Above: Using Color-Coded Satellite Images to Examine the Global Environment*. Sausalito, CA: University Science Books, 1997.

Parkinson, Claire L., et al. *Arctic Sea Ice, 1973–1976: Satellite Passive-Microwave Observations*. Washington, DC: NASA, 1987.

Pritchett, Price, and Brian Muirhead. *The Mars Pathfinder Approach to "Faster-Better-Cheaper."* Dallas: Pritchett & Associates, 1998.

Raeburn, Paul. *Uncovering the Secrets of the Red Planet: Mars*. Washington, DC: National Geographic Society, 1998.

Robinson, Kim Stanley. *Red Mars*. New York: Bantam, 1993.

Sagan, Carl. *Billions and Billions: Thoughts on Life and Death at the Brink of the Millennium*. New York: Random House, 1997.

———. *Cosmos*. New York: Random House, 1980.

————. *Pale Blue Dot: A Vision of the Human Future in Space.* New York: Random House, 1994.

Sagan, Carl, and John Norton Leonard. *Planets.* New York: Time Inc., 1966.

Sheehan, William. *The Planet Mars: A History of Observation and Discovery.* Tucson: University of Arizona Press, 1996.

Shirley, Donna. *Managing Martians.* New York: Broadway Books, 1998.

Shklovskii, I. S., and Carl Sagan. *Intelligent Life in the Universe.* New York: Dell, 1967.

Taylor, Michael Ray. *Dark Life.* New York: Scribner, 1999.

Von Braun, Wernher. *The Mars Project.* Urbana: University of Illinois Press, 1991 (originally published in England in 1953).

Wallace, Alfred Russell. *Is Mars Inhabited?* London: Macmillan, 1907.

Wells, H. G. *The War of the Worlds.* New York: Signet, 1986 (originally published 1898).

Wilford, John Noble. *Mars Beckons.* New York: Alfred A. Knopf, 1990.

Zubrin, Robert. *The Case for Mars: The Plan to Settle the Red Planet and Why We Must.* New York: Touchstone, 1996.

INDEX

Abshire, Jim, 223
Acuna, Mario, 123–24, 125, 127, 129, 141, 168–72, 238, 250–51, 256
Aerobraking, 105–6, 113, 124, 135, 138, 171, 222, 239, 334
Aerogel, 205
Afzal, Rob, 163–68, 245–46, 248
Agassiz, Louis, 9
Airy-O Crater, 156
Albee, Arden, 129, 130, 134–35, 236–37, 238, 242, 247, 262, 268, 298
ALH 84001 (meteorite), 44–49, 265, 271
  criticism of findings from, 53–55, 272–75
  discovery of, 44–45
  evidence for life in, 47–48, 52, 54, 56–57, 181, 185–86, 203, 204, 211, 265–66, 272–74, 323
  Gibson's defense of findings from, 265–68
  media and, 265–67
  NASA and report on, 51–52
  Sagan on significance of, 53–54
  Science paper on, 48–52
American Geophysical Union, 259, 260–61
Ames Research Center, 201, 293
Announcements of Opportunity (A/Os), 236–37
Antimatter catalyzed fusion, 291
Apollo 11, 14, 81
Apollo program, 12, 62, 67, 246, 321, 324
Ares Vallis, 75, 96
Armstrong, Neil, 62, 321
Astrobiology, 24, 129, 181, 221, 243, 268, 292
  complexity theory and, 200–201
  as discipline, 183–84
  paradigm shift in, 181–83
Astrobiology Institute, 183–84, 195, 197, 201–2, 216, 275, 292–93, 294
Astronauts
  crew interactions and, 329–30
  emotional needs of, 332–33
  exercise by, 319
  inflatable habitat for, 316
  life-support chamber test for, 317–19
  loss of calcium by, 330–31
  medical care of, 316
  privacy and, 319

  psychological problems of, 331–32
  in return to gravity, 334–36
  risks and, 327–28
  selecting of, 326–27
  sensory deprivation and, 331
  sex in space and, 332–33
  sleep problems of, 331
  Space Adaptation Syndrome and, 328–29
  weightlessness and, 329–33
  zero gravity and, 329–31, 333
ATP molecule, 207

Babbitt, Bruce, 253
Bacteria, 21, 24, 40, 175
  hyperthermophile, 208
  nanofossils of, in ALH 84001, 47–48, 55–57, 185–86, 203, 211, 272–74, 323
Baez, Omar, 229, 230, 231–33
Baltimore Sun, 90
Banerdt, Bruce, 222, 223, 261, 281–82, 300, 301
Basilevsky, Sasha, 128
Battles, Lara, 333
Beer, Wilhelm, 156
Beyond Earth Orbit (BEO), 311
Biological water processor, 318
"Biomarkers in ALH 84001???" (Treiman), 274–75
Biomathematics, 209
Biospherics Incorporated, 193
Black smokers, 133
Blumberg, Baruch, 293–94
Bogard, Donald, 45
"Book of Life," 210–11
Bradbury, Ray, 36–37, 179, 180
Breakthroughs: A Chronology of Great Achievements in Science and Mathematics (Parkinson), 152
Buckyballs, 209
Burroughs, Edgar Rice, 178–79, 180, 309
Burton, LeVar, 92–93
Bush, George, 62–63, 64, 65, 72, 213

California Institute of Technology, 106–7, 265
Carbon, 188–89, 209
"Carl Sagan Memorial," 89
Carr, Michael, 103, 282

Case for Mars, The: The Plan to Settle the Red Planet and Why We Must (Zubrin), 321
Cassini, Giovanni, 155
Cassini spacecraft, 143
CBS Evening News, 51
Center for Dynamic Exoecology (CDEX), 197
Central Intelligence Agency (CIA), 114
Ceres, 144
Challenger disaster, 29, 62, 169–70, 212, 288–89
Channeled Scabland, 75
Chryse Plain (Chryse Planitia), 75, 282, 335
Cintala, Mark, 43
Cleave, Mary, 328–29
Clementine spacecraft, 126–27, 143
Clifford, Steve, 41–43
Clinton, Bill, 50, 51
CNN, 267
Cold War, 52, 125–26, 169, 237, 245–46
Collins, Michael, 321
Complexity theory, 199–201
Congress, U.S., 6–7, 51, 126, 143, 213, 281, 302, 311
Cook, Richard, 295, 299–300
Cooke, Doug, 314–16
Cooper, Henry S., Jr., 101, 186–87
Copernicus, Nicolaus, 130–31
Cosmic dust, 204–6
Craters, cratering, 102–3
    on Earth, 120, 147
    Garvin on, 119–21, 122, 145–46, 176–77, 225, 243–44, 259, 269–70, 286
    ghost, 263–64
    life and, 264
    water under, 145–46
Curie point, 172

Darwin, Charles, 25, 130, 131, 161, 162, 181
Decadal Planning Team (DPT), 287–88, 303
Deep Space Network (DSN), 78, 80, 83, 85, 94–95, 214, 299
Defense Department, U.S., 125–26, 214
Deimos, 5, 39, 223–24
Delta-v burns, 113
Delta II rocket, 108, 113, 168, 228–29
    of Mars Climate Orbiter, 230, 233–34
Diode pumped laser, 165
Discovery Channel, 253–54
Discovery program, 65
DNA, 207, 215, 271
Doppler shift, 84, 262
Duke, Mike, 314

Earth, 5, 145, 208
    ancient exploration of, 289–90
    Copernican view of, 130–31
    craters on, 120, 147
    emergence of life on, 24–25, 41, 174, 183
    geoid of, 261
    magnetic field of, 124
    Mars's resemblance to, 4–5, 96, 275–76
    ozone layer of, 191
    panspermia theory and, 203–4
    plate tectonics and, 251–52
    space dust and, 205
    water on, 41–42
Earth Return Vehicle, 324–25
Edgett, Ken, 305–6
Edison, Thomas, 279
EETA 79001 (meteorite), 55
Einstein, Albert, 181
Engineering & Science, 271
Europa, 221, 260, 271, 286, 290, 291
    water on, 186, 194, 207–8
Evolution, 131, 188, 210
    Rotten Tomato Theory of, 292
"Exobiological Strategy for Mars Exploration, An" (NASA), 202–3
Exobiology. *See* astrobiology
Extraterrestrial life
    ALH 84001 and, 47–48, 54, 56–57, 181, 185–86, 204, 265–66
    architecture of, 209–11
    ATP molecule and, 207
    Best Guess Scenario for, 38–43
    cosmic dust and, 203–5
    and emptiness of space, 205–6
    genesis question and, 22–23, 38–39, 55, 208
    hyperthermophile bacteria and, 208
    on Mars, 24, 39–43, 129–30, 132–33, 177–78, 243–44, 250, 251, 264, 265–66, 294
    MGS workshop discussion on, 129–30
    nanofossils as evidence of. *See* ALH 84001
    and need for redefining life, 188–90
    panspermia theory of, 23, 203–5, 276
    proposed plan for discovery of, 197–98
    search for, 270–71
    tensegrity principle and, 209–10
    Viking mission experiments and, 186–87, 189–95, 196, 197, 202
    virology principles and, 293
    water and, 16, 208, 221

"Face on Mars," 142, 226–28, 242
False color, 164
Feynman, Richard, 107
Fifth International Conference on Mars, 264–65, 268–70, 272–76
    ALH 84001 discussions at, 265–68, 272–76
Flammarion, Camille, 156, 158, 160

Fontana, Francisco, 155
Ford, Peter, 117–18
Fossilization, 38, 57
Fourth International Conference on Mars, 268
Friðricksson, Sturla, 25
Fuller, Buckminster, 209

Gagarin, Yuri, 128
Galileo Galilei, 131
    astronomical observations by, 154–55
Galileo planetary mission, 94, 143, 286, 290
Garvin, Cindy, 35, 90, 103, 260, 269
Garvin, Danica, 103, 269–70
Garvin, Jim, 3, 5, 6, 8, 9, 25–34, 38, 49–50, 72,
        101, 103, 129, 134, 135–36, 137, 138–39,
        141, 162–63, 168, 222, 228, 235, 256,
        257–58, 268, 285, 291, 294, 302, 321
    Astrobiological Institute proposal submitted
        by, 197–99, 216
    author's first meeting with, 26–27
    on craters and cratering, 119–21, 122,
        145–46, 176–77, 225, 243–44, 259,
        269–70, 286
    ghost crater announcement of, 263–64
    in Iceland survey, 29–36
    on Levin, 196–97
    on life on Mars, 177–78, 243–44
    on loss of MPL, 303–4
    Martian polar cap data reviewed by, 172–75
    MGS data reviewed by, 144–50, 172–75,
        248–50
    named to DPT, 287
    in Navassa Island expedition, 252–55
    Pathfinder mission assessed by, 97–98
    Pathfinder talks by, 89–90
    on quality of MOC images, 259–60
    Solomon's exchanges with, 115–17,
        121–22, 242–43, 244
    on space exploration, 289–90
    in Surtsey expedition, 11–24, 25
Garvin, Zachary, 35, 137, 228
Gas chromatograph-mass spectrometer (GCMS)
        experiment, 190, 191, 193–94, 202
Gas-exchange experiment, 190–91
Gell-Mann, Murray, 107, 199, 200–201
Genesis Question, 22–23, 38–39, 55, 208
Genetics, 188
Geoid, 261–62
Geology of Mars, The (Mutch), 27
Ghost craters, 263–64, 300
Gibson, Everett, 46, 50, 53–54, 264, 322
    ALH 84001 findings defended by, 265–68
Glamour, 97
Glenn, John, 10, 229, 309–10, 314
Goddard, Robert Hutchings, 139–40

Goddard Space Flight Center, 26, 71–72, 89,
        135, 139, 201, 216
    described, 141–42
    testing spacecraft at, 163–64
Goldin, Dan, 7, 9, 50, 51, 71–72, 78, 85, 183,
        194, 195, 201, 232, 238, 281, 288, 289,
        302, 303, 311, 314, 323
    and reform of NASA, 211–16
Golombek, Matt, 72–75, 84–86, 94, 96, 222,
        226, 244–45, 268, 282, 298, 300, 301
Gorbachev, Mikhail, 237
Gore, Al, 126
Gorgonum Crater, 305
Graphite epoxy motors (GEMs), 229
Gravity, 120, 262
    adapting to return to, 334–36
    libration points and, 291
    zero, 329–31, 333
Great Galactic Ghoul, 102, 103, 222, 278, 280,
        285
Ground truth, 87
Guggenheim, Harry, 140
Guidi, Cristina, 323–26
Gulliver's Travels (Swift), 224

Hall, Asaph, 224
Hancock, Graham, 227
Harris, Jennifer, 59–62, 85, 87–88, 91, 93–97
Head, James William, III, 27, 115, 137, 219–22,
        223, 225, 228, 243, 264, 269, 283, 284,
        285, 286, 303
Hellas Planitia (Hellas Crater), 256, 257–58
Herschel, William, 155–56
Hinners, Noel, 232
Hohmann, Walter, 101–2
Hohmann transfer orbits, 101–2
Horowitz, Norman, 188–89, 190, 192
Houston Chronicle, 48
Human Exploration and Development of Space
        (HEDS), 311
Human heads plan, 291–92
Human space travel
    costs of, 312–13, 325
    gender mixing in, 313
    international agencies and, 325
    length of time of, 315
    Mars Direct plan for, 323–24
    nanotechnology and, 316
    NASA's Reference Mission for, 323–26
    radiation problem in, 315–16, 329
    science component vs. human component
        in, 321
    Space Adaptation Syndrome and, 328–29
    space agriculture and, 320
    tedium of, 332

water and, 314, 318
  weightlessness and, 329–33
  zero gravity and, 329–31, 333
  See also astronauts
Huygens, Christiaan, 155
Hyperthermophile bacteria, 208

Iceland, 3
  geology of, 5–6
  jökulhlaup disaster in, 15–16
  1986 Reagan-Gorbachev summit in, 237
  volcanic eruptions in, 11, 20–21, 31
Ingber, Donald, 209–10, 211
International Space Station, 215, 289, 291, 302, 307, 309, 311, 312, 313, 315, 316, 322, 323, 327
Internet, 5, 36, 51, 201, 227, 238, 240, 295
  Pathfinder phenomenon on, 66, 77–78, 89, 90–91
Is Mars Inhabitable? (Wallace), 161
Ivanov, Boris, 127

Jakosky, Bruce, 129, 130–32, 276, 306
Jet Propulsion Laboratory (JPL), 58–59, 60, 61, 65, 67, 71, 74, 79, 80, 81, 88, 97, 104, 107, 123, 176, 213, 236, 239, 258, 281–82, 294
  and loss of Mars Climate Orbiter, 277–78
Johnson, Pratt, 45
Johnson Space Center, 43, 44, 54, 56, 59, 74, 272, 291, 308–9, 312
  Inspection Days at, 310
  Mars or Bust group at, 310–11
Jökulhlaup disaster, 15–16
Journal of Geophysics Review, 138, 192
Jupiter, 84, 143, 144, 186, 194, 207–8, 221, 286, 291
  Galileo's observations of, 154–55

Kallemeyn, Pieter, 85
Kármán, Theodore von, 58
Kennedy, John F., 63
Kennedy Space Center, 59, 76, 320, 327
"Kevorkian Syndrome," 95
Kirschvink, Joseph, 54
Korolev, Sergei Pavlovich, 127–28
Korolev Crater, 177
Kuhn, Thomas, 181

Labeled release experiment, 190–91, 193, 195
Laboratory for Terrestrial Physics, 175
Landry, Bridget, 79, 81–82, 85–86, 87, 91–96
Langjökull glacier, 32

Laser altimeter, 26, 107–8, 223, 228
  Smith's work on, 175–76
  See also Mars Orbiter Laser Altimeter
Lasers, technology of, 164–68
Launchcam, 234
Lava, 15
Lederberg, Joshua, 182, 184
Lee, Wayne, 104
Leibovitz, Annie, 97
Leno, Jay, 302
Letterman, David, 109
Levin, Gil, 191, 193–96
Lewis and Clark expedition, 238
Lewis Center, 212
Libration points, 291
Life
  architecture of, 209–11
  ATP molecule and, 207
  from common ancestor, 206–7
  complexity theory and, 200
  Copernican view of, 130–31
  Darwinian revolution and, 131–32
  emergence of, 24–25, 41, 174, 183
  extraterrestrial. *See* extraterrestrial life
  extremophile, 133, 174–75, 177, 197–98, 221, 268
  genetics and, 188
  handedness and, 207
  hyperthermophile bacteria and, 208
  Miller-Urey experiment and, 22
  paradigm shift in view of, 181–82
  Ptolemaic view of, 130–31
  redefining of, 188–90
  requirements for, 55–56
  as system, 188
  tenacity of, 21–22
  tensegrity principle and, 209–10
  water and, 16, 208, 221, 260
Lindbergh, Charles, 140
Lindstrom, Marilyn, 53
Little Climate Optimum, 289
Lockheed Martin, 113–14, 141, 205, 246, 280, 302
Lowell, James Russell, 157
Lowell, Percival, 157–62, 180, 187, 192, 195
  authors influenced by, 178–79
Lunar and Planetary Institute, 41
Lunar and Planetary Science Conference, 148, 244, 285
  of 1995, 47
  of 1999, 57

McAuliffe, Christa, 62
McCleese, Dan, 230–31

McDonnell Douglas, 165, 166, 167
McKay, David, 46–47, 50, 52, 53–54, 56–57, 181,
    196, 216, 265, 267, 272–74, 275
Mädler, Johann von, 156
"Mad Martian" scenario, 71
Magellan mission, 105–6, 221, 222–23, 239
Magma, 149
Magnum heavy-lift launch vehicle, 324–25
Malin, Mike, 114, 148, 225–26, 230, 235, 240,
    259, 277
    and liquid water evidence on Mars, 304–7,
        308
Malina, Frank, 58
"Malin Rule," 283
Malin Space Science Systems, 304
Mandell, Humboldt, 311–14
Manning, Rob, 65–66, 70, 83–85, 87, 93
Mariner missions, 10, 36, 190
Mariner 4, 102
Mariner 9, 103, 160
Mariner 7, 191
Mariner 10, 287
Mars
    ancient shorelines of, 19–20
    asteroid bombardment of, 40, 42–43
    atmosphere of, 191, 197, 230–31, 333–34
    canals of, 156–60, 162
    changing inclination of, 297
    climate of, 231, 304
    colonization of, 206
    craters on. See craters, cratering
    day (sol) of, 5
    difficulty of launching rockets to, 101–2
    early astronomical observations of,
        154–57
    Earth's resemblance to, 4–5, 96, 275–76
    experiences in landing on, 333–36
    "Face" on, 142, 226–28, 242
    first drawing of, 155
    flash floods on, 306
    geoid of, 261–62
    geology of, 6–7, 27–28
    gravity of, 262–63, 335–36
    human heads plan and, 291–92
    Iceland's resemblance to, 4–5
    life on, 24, 39–43, 129–30, 132–33,
        177–78, 243–44, 250, 251, 264, 265–66,
        294
    liquid water on, 304–7, 308
    Lowell and popularization of, 157–62
    magnetic anomalies of, 123–24, 169,
        171–72, 238, 250–51, 252, 256–57
    measuring gravity of, 262–63
    missions to. See specific missions
    moons of, 5, 223–24

NASA's Reference Mission to, 323–25
panspermia theory and, 204
plans for human transportation to, 290–92,
    308
plate tectonics and, 251–52
polar caps of, 108, 155–56, 172–74, 192,
    243, 248, 250, 257
Sagan on life on, 40
science fiction inspired by, 178–79
seasonality of, 297
slope of southern hemisphere of, 257–58
spallation from, 43–44
surface of, 185, 195, 197
temperatures on, 86
ultraviolet radiation of, 191–92
volcanic activity on, 6–7, 103, 145, 150–51,
    243, 251, 294
water on, 16–17, 19–20, 41–42, 96–97, 108,
    133, 135, 145–47, 150, 159, 174, 177,
    189, 231, 250, 251, 257, 294, 295–96,
    304–7, 308
weather on, 5, 86
winds on, 21, 39–40, 137, 192
winter on, 231, 297
Mars (Lowell), 159
Mars as the Abode of Life (Lowell), 161–62
Mars Climate Orbiter, 7, 223, 229–35, 294–95,
    301
    Delta II rocket of, 230, 233–34
    equipment on, 230
    launch of, 231–35
    loss of, 276–81
    pounds-newtons miscalculation and,
        280–81
Mars color imager (MARCI), 230, 277
Mars Direct, 322–23
Mars Environmental Survey (MESUR), 65
Mars Express, 237
Mars Global Surveyor (MGS), 26, 30, 49, 123,
    142–43, 221–26, 230, 235, 258, 269, 277,
    279, 295, 300, 301, 302, 303
    aerobraking maneuver of, 105–6, 113, 124,
        135, 138, 171, 222, 239, 334
    antenna debate and, 222–23, 228, 239
    budget question and, 109–10
    cost of, 261, 307
    on cover of *Science,* 308
    data release issue and, 117–18, 225–26
    data returned by, 240, 247–49
    "Face" controversy and, 227–28
    images returned by, 114–15, 146–47,
        172–74
    launch of, 101
    liquid water flow discovered by, 304–7
    Magellan mode crisis and, 222–23, 228

mapping by, 236, 239–41
Mars Orbiter Camera and, 114–15
in orbit, 104–5
Phobos mission of, 223–24, 228
slow demise of, 286–87
transmission error and, 241
See also MOLA science team
Mars Global Surveyor Geoscience Workshop,
    127, 132–33
    extraterrestrial life discussion in, 129–30
    magnetic field discussion in, 123–24
Mars Mystery, The (Hancock), 227
Mars Observer mission, 29, 71, 72, 124, 127,
    165, 175–76, 211, 222, 247, 268, 278,
    302
    Announcement of Opportunity for, 236–37
    conspiracy theories on, 141–42
    failure of, 109, 237–38
Mars Orbital Insertion (MOI), 277–78
Mars Orbiter Camera (MOC), 16, 114–15, 148,
    240, 259, 279, 283, 305
Mars Orbiter Laser Altimeter (MOLA), 101,
    256, 303–4
    Garvin's description of, 107–8
    See also MOLA science team
Mars or Bust society, 310–11, 321
Mars Polar Lander, 7, 223, 235, 241, 277, 279,
    307
    entry-descent-landing sequence of, 295–96,
    299
    equipment carried by, 297–98
    Garvin on loss of, 303–4
    landing site debate and, 244–45, 258,
    282–83
    launch of, 294–95
    loss of, 299–303
    purpose of, 296–97
    replica of, 298
Mars Polar Orbiter, 235
Mars Project, The (Das Marsprojekt) (von Braun),
    63–64, 322
Mars Rover, 323
Mars Rover Sample Return, 72
Mars Society, 322
Mars Underground, 28–29
Martian Chronicles, The (Bradbury), 36–37, 179
Martin, Jim, 68–69
Meagher, L. D., 256
Melosh, Jay, 43
Mercury, 6, 287
Mercury Messenger mission, 287
Mercury missions, 229
Mercury Redstone rocket, 301
Meteor Crater, 335
Meteorites, 29

ALH 84001. *See* ALH 84001
    fossilized bacteria in. *See* nanofossils
    Genesis Question and, 38–39
    SNC, 45
    spallation model and, 43–44
Method of Reaching Extreme Altitudes, A
    (Goddard), 139
Meyer, Michael, 203–6, 216
microspacecraft, 216
Milky Way, 337
Miller, Stanley L., 22
Miller-Urey experiment, 22
Ming, Doug, 316–20
Mission Control Center, 313
Mission to Mars (Collins), 321
Mitchell, Maria, 38, 337, 339
Mitchell, William, 337
Mittlefehldt, David, 45–46
Mittman, David, 83
MOLA science team, 105–8, 110–22, 123, 138,
    220, 279
    cratering discussion in, 119–22
    data debates in, 106, 117–18, 224–26,
    240–44, 269, 281–84, 298
    "Face on Mars" controversy and, 226–28,
    242
    financial problems of, 261, 284
    Magellan mode debate in, 222–23, 239
    mapping discussion in, 238–40
    MGS's aerobraking maneuver and, 105–6,
    239
    MGS's slow demise and, 286–87
    MPL landing site debate in, 244–45,
    282–85
    press conference of, 256–57
    public recognition and, 258
    Science report and, 134–35
    sense of family in, 285
    See also specific members
Moon and Mars Day, 14
Moore, Hank, 75
Morris, Dick, 50–51
Ms., 97
Muhleman, Duane "Dewey," 127, 128
Murchison meteor, 53
Murray, Bruce, 102
Mutch, Madeline, 28
Mutch, Tim, 27–28, 115, 220, 228, 260

Nakhla meteorite, 56–57, 267, 272–73, 274
Nanofossils, 47–48, 52, 55–57, 185–86, 203, 211,
    272–74, 323
Nanotechnology, 316
National Aeronautics and Space Administration
    (NASA), 3, 24, 27, 93, 107, 168

ALH 84001 report and, 51–52
Announcements of Opportunity used in, 236–37
anti-science bias of, 202
astronaut selection process of, 326–27
"Book of Life" guidebook of, 210–11
bureaucracy of, 126, 170, 246
Challenger disaster and, 29, 169, 212, 288–89
Cold War space race and, 125–26, 169, 237, 245–46
Decadal Planning Team (DPT) of, 287–89
"Face" controversy and, 227
funding of, 7–9, 109, 126
Goldin's reform policies and, 211–16
and humiliation of lost missions, 301–2
Levin's criticism of, 193–94, 196
Mars Underground in, 28–29
materials requirements of, 312–13
meteorite collection of, 44
nuclear material used by, 143
Pathfinder and new open-ended approach of, 61–62
politicization of, 245–46
pounds-newtons miscalculation of, 280–81
private sector and, 170
and Reference Mission to Mars, 323–25
Sagan's visionary influence on, 9–10
space exploration role of, 289
Viking report of, 202–3
Young Report and, 303
National Air and Space Museum, 146
National Institutes of Health, 185
National Public Radio, 97
Natural selection, 25, 131
Nature, 36, 46, 117, 119, 138
Navassa, 252–55
Nealson, Ken, 270–71, 272, 275
Neptune, 144, 162
Neumann, Greg, 109, 117, 125–27, 130, 141, 142–43, 192, 241, 261, 282, 283–84, 285
Newton, Isaac, 120
New York Times, 139, 258
Nix Olympica. See Olympus Mons
Noachis Terra, 305

Occult Japan (Lowell), 157
Office of Space Flight, 312
Olympus Mons, 6–7, 31, 103, 256
    measurement of, 250
On the Revolutions of the Celestial Spheres (Copernicus), 131
Oscar (pilot), 4, 7, 11
Oyama, Vance, 190
ozone layer, 191

panspermia theory, 23, 203, 276
    cosmic dust and, 204–5
Parade, 227
Parise, Ron, 330–32
Parkinson, Claire, ix, 149–54, 168, 197
Pathfinder mission, 5, 26, 29, 49, 60–98, 121, 195, 239, 244, 245, 247, 258, 268, 282, 300, 301, 303
    assessment of, 97–98
    demise of, 93–96
    emotional and physical stress of, 81–83
    engineering problems of, 65–66
    final trajectory maneuver in, 84–85
    images returned by, 88–89
    as Internet phenomenon, 66, 77–78, 89, 90–91
    in landing on Mars, 86
    landing plan of, 65–66
    landing tests of, 69–70
    launch of, 79–80
    media and, 78, 89–90, 97–98
    open-ended approach to, 61–62
    origin of, 62–64, 65
    preparation and testing of, 76–77
    rover air bag crisis in, 86–87
    rover debate and, 66–71
    science component of, 69, 73
    "Star Trek" culture and, 91–93
    target site selection for, 72–75, 86
    von Braun's vision and, 63–64
Paul III, Pope, 131
Pedro (mechanic), 33–34
Perseid meteor shower, 337–38
Perspicillum (telescope), 154
Peter (pilot), 8
Phobos, 5, 39, 187, 223–24, 227, 228
Photosynthesis, 190, 195
Pickering, Edward, 158
Pickering, William H., 158
Pilcher, Carl, 206–7
Planetary geodesy, 175
Planète Mars et ses Conditions d'Habitabilité, La (Flammarion), 156, 158
Plate tectonics, 251–52, 266, 300
Pluto, 162
Pressure modulator infrared radiometer (PMIRR), 230
Princess of Mars, A (Burroughs), 179, 309
Ptolemy, 130–31
pyrolitic release experiment, 190

Q-switch, 166
Quarantine orbit, 239
Quark and the Jaguar, The (Gell-Mann), 199

Reagan, Ronald, 237
Reference Mission, 323–25
Reis Crater, 146
Religion, science and, 153–54
Ride, Sally, 29, 288
Ride Report, 288–89
RNA, 271
Romanek, Chris, 45–46
Rotten Tomato Theory, 292
Rowlands, Sherry, 50–51
Russia, 215, 325

Sabatier reactor, 319–20
Sagan, Carl, 46, 51, 66, 112, 142, 179, 180,
    182–83, 187–88, 189, 191, 192, 202, 206,
    221, 268, 270
    "Face on Mars" debunked by, 227
    on life on Mars, 40
    NASA influenced by, 9–10
    on significance of ALH 84001, 53–54
    space exploration advocated by, 321–22
Santa Fe Institute, 201
Sasquatch Crater, 177
Saturn, 143, 271
Saturn rocket, 324
Schiaparelli, Giovanni, 156–57, 158, 159, 160
Science
    Occam's razor approach of, 196
    paradigm shift in, 181–82, 186
    religion and, 153–54
    revision process in, 276
Science, 36, 48–52, 117, 138, 192, 241–42, 261,
    285–86, 305
    Mars Global Surveyor on cover of, 308
    MOLA Report in, 134–35
    peer review process at, 118–19
Score, Roberta, 44
Search for Life on Mars, The (Cooper), 186–87
Search for Life on Other Planets, The (Jakosky),
    130
"Search for Past Life on Mars: Possible Relic of
    Biogenic Activity in Martian Meteorite
    ALH 84001," 49
Shepard, Alan, 229, 330
Shergotty meteorite, 49, 267, 274
    evidence of life in, 272–73
Shirley, Donna, 67–71, 79, 94–95, 97
Shklovskii, Iosif S., 187, 227
Shock, Everett, 53
Sidereus Nuncius (Starry Messenger) (Galileo), 155
Smith, Dave, 112, 113, 115, 117, 121, 129, 134,
    141, 220, 224, 236, 242, 243, 245, 247,
    256, 259, 285, 286, 287
    described, 110–11
    laser altimeter work of, 175–76

in MGS data debates, 118, 222–23, 238–40,
    261–62
    Solomon's exchanges with, 118, 239–40,
        261–62, 284
Smith, Pete, 88–89
Smithsonian Institution, 139
SNC meteorites, 45
Soffen, Gerald, 181, 183, 185, 189–92, 194,
    195–96, 198–99, 201, 216, 292–93, 295
Sojourner Truth (rover), 26
Solomon, Sean, 119, 127, 129, 134, 222, 223,
    225, 235, 280, 281, 286, 287
    Garvin's exchanges with, 115–17, 121–22,
        242–43, 244
    Smith's exchanges with, 118, 239–40,
        261–62, 284
    Zuber's exchanges with, 241–42, 284–85
Sonntag, Jon, 33, 34–35
Soviet Union, 127–28, 146
    Mars '96 mission of, 237
Space Adaptation Syndrome, 328–29
Space agriculture, 320
Space Exploration Day, 14
Space News, 50, 226
Space Policy Institute, 323
Space Shuttle, 307, 324
Spallation model, 43–44
Spear, Tony, 67–68
Sputnik, 128, 236
Stardust mission, 205
Starry Messenger (Sidereus Nuncius) (Galileo), 155
"Star Trek," 91–93
Stickney Crater, 224
Structure of Scientific Revolutions, The (Kuhn), 181
Sublimation, 263
Surtsey, 11–19, 25, 33, 149, 177, 305, 335
Swift, Jonathan, 224
Syrtis Major, 155

Telescope, 154
Tensegrity principle, 209–10, 211
Texas Monthly, 51
Thomas-Keprta, Kathie, 47–48, 54–55, 267–68,
    274
Thompson, D'Arcy Wentworth, 210–11
Thurman, Sam, 277–78
Time Interval Units (TIU), 148
Time Machine, The (Wells), 255
Titan, 271
Titan missile, 113–14
"Topography of the Northern Hemisphere of
    Mars from the Mars Orbiter Laser
    Altimeter," 134–35
Trans Mars Injection stage, 324–25
Treiman, Allan, 274–75

Truth, Sojourner, 89
TRW, 141, 213
Tuskegee University, 320

Ultraviolet radiation, 191–92
Universe, 6, 182–83, 210
Uranius Patera, 137, 149–50
Uranus, 155
Urey, Harold, 22, 272
Utopia Planitia (Utopia Plain), 286

Valles Marineris, 245, 256
Vanguard rocket, 301
Vanity Fair, 97
Venus, 6, 105–6, 115, 118, 127, 128, 145, 160,
    221, 287, 321
Verne, Jules, 139
Vesta, 144
Vietnam War, 169
Viking Lander, 4, 66, 68–69, 72, 73, 260
Viking 1, 14, 147, 190
Viking program, 27, 36, 45, 88, 98, 104, 111,
    121, 126, 132–33, 142, 144, 146, 181,
    263, 269, 295, 303, 307
    images returned by, 190
    life detection experiments on, 186–87,
        189–95, 196, 197, 202
    NASA's report on, 202–3
    success of, 184–85
Viking 2, 27, 190, 191
Vishniac, Wolf, 189
Von Braun, Wernher, 63–64, 93, 140, 322
Voyager mission, 290
V-2 missile, 63, 140

Wallace, Alfred Russel, 161
Wallace, Matt, 83
War of the Worlds, The (Wells), 139, 177

Water
    under craters, 145–46
    on Earth, 41–42
    on Europa, 186, 194, 207–8
    ghost craters and, 263–64
    life and, 16, 208, 221, 260
    on Mars, 16–17, 19–20, 41–42, 96–97, 108,
        133, 135, 145–47, 150, 159, 174, 177,
        189, 231, 250, 251, 294, 295–96, 304–7,
        308
    sublimation and, 263
Webb, James, 221
Weightlessness, 329–33
Wells, H. G., 139, 177, 255
Wetherbee, James, 310
Wild 2 comet, 205
Wisconsin, University of, 320
Wolf Trap experiment, 189

Young, Thomas, 302
Young Report, 303

Zero gravity, 329–31, 333
Zhamanshin Crater, 146
Zuber, Maria, 72, 114, 127, 129, 222, 223, 240,
    248, 256, 259, 269, 286, 298
    background of, 111–12
    on exploration of space, 112–13
    "Face on Mars" controversy and, 226–28
    on life on Mars, 132–33
    on measuring gravity of Mars, 262–63
    at MOLA press conference, 257–58
    MOLA team role of, 111
    in PML landing site debate, 282–85
    in review of MPL loss, 302–3
    Solomon's exchanges with, 241–42, 284–85
Zubrin, Robert, 322
Zurek, Richard, 278, 296